Materials and Chemistry of
Flame-Retardant Polyurethanes
Volume 1: A Fundamental Approach

ACS SYMPOSIUM SERIES **1399**

Materials and Chemistry of Flame-Retardant Polyurethanes Volume 1: A Fundamental Approach

Ram K. Gupta, Editor

Department of Chemistry, Kansas Polymer Research Center
Pittsburg State University
Pittsburg, Kansas, United States

American Chemical Society, Washington, DC

Library of Congress Cataloging-in-Publication Data

Names: Gupta, Ram K., editor.
Title: Materials and chemistry of flame-retardant polyurethanes / Ram K. Gupta, editor, Department of Chemistry, Kansas Polymer Research Center, Pittsburg State University, Pittsburg, Kansas, United States.
Description: Washington, DC : American Chemical Society, 2021- | Series: ACS symposium series ; 1399, 1400 | Includes bibliographical references and index. | Contents: volume 1. A fundamental approach -- volume 2. Green flame retardants.
Identifiers: LCCN 2021049619 (print) | LCCN 2021049620 (ebook) | ISBN 9780841298026 (hardcover) | ISBN 9780841298019 (ebook other)
Subjects: LCSH: Fire resistant polymers. | Fireproofing agents. | Polyurethanes.
Classification: LCC TH1074.5 . M38 2022 (print) | LCC TH1074.5 (ebook) | DDC 628.9/223--dc23/eng/20211201
LC record available at https://lccn.loc.gov/2021049619
LC ebook record available at https://lccn.loc.gov/2021049620

The paper used in this publication meets the minimum requirements of American National Standard for Information Sciences—Permanence of Paper for Printed Library Materials, ANSI Z39.48n1984.

Copyright © 2021 American Chemical Society

All Rights Reserved. Reprographic copying beyond that permitted by Sections 107 or 108 of the U.S. Copyright Act is allowed for internal use only, provided that a per-chapter fee of $40.25 plus $0.75 per page is paid to the Copyright Clearance Center, Inc., 222 Rosewood Drive, Danvers, MA 01923, USA. Republication or reproduction for sale of pages in this book is permitted only under license from ACS. Direct these and other permission requests to ACS Copyright Office, Publications Division, 1155 16th Street, N.W., Washington, DC 20036.

The citation of trade names and/or names of manufacturers in this publication is not to be construed as an endorsement or as approval by ACS of the commercial products or services referenced herein; nor should the mere reference herein to any drawing, specification, chemical process, or other data be regarded as a license or as a conveyance of any right or permission to the holder, reader, or any other person or corporation, to manufacture, reproduce, use, or sell any patented invention or copyrighted work that may in any way be related thereto. Registered names, trademarks, etc., used in this publication, even without specific indication thereof, are not to be considered unprotected by law.

Foreword

The ACS Symposium Series is an established program that publishes high-quality volumes of thematic manuscripts. For over 40 years, the ACS Symposium Series has been delivering essential research from world leading scientists, including 36 Chemistry Nobel Laureates, to audiences spanning disciplines and applications.

Books are developed from successful symposia sponsored by the ACS or other organizations. Topics span the entirety of chemistry, including applications, basic research, and interdisciplinary reviews.

Before agreeing to publish a book, prospective editors submit a proposal, including a table of contents. The proposal is reviewed for originality, coverage, and interest to the audience. Some manuscripts may be excluded to better focus the book; others may be added to aid comprehensiveness. All chapters are peer reviewed prior to final acceptance or rejection.

As a rule, only original research papers and original review papers are included in the volumes. Verbatim reproductions of previous published papers are not accepted.

ACS Books

Contents

Preface ix

1. **Materials and Chemistry of Polyurethanes** 1
   Felipe M. de Souza, Muhammad Rizwan Sulaiman, and Ram K. Gupta

2. **Green Materials for the Synthesis of Polyurethanes** 37
   Ziwei Li, Kaimin Chen, and Mingwei Wang

3. **Overview on Classification of Flame-Retardant Additives for Polymeric Matrix** 59
   Mattia Bartoli, Giulio Malucelli, and Alberto Tagliaferro

4. **Self-Extinguishing Polyurethanes** 83
   Tuhin Ghosh and Niranjan Karak

5. **Highly Flame-Retardant Polyurethane** 103
   Young Nam Kim, Hyunsung Jeong, Sooyeon Ryu, and Yong Chae Jung

6. **The Role of Polyurethane Foam Indoors in the Fate of Flame Retardants and Other Semivolatile Organic Compounds** 125
   Mesut Genisoglu, Sait C. Sofuoglu, and Aysun Sofuoglu

7. **Halogen-Based Flame Retardants in Polyurethanes** 141
   Nycolle G. S. Silva, Noelle C. Zanini, Alana G. de Souza, Rennan F. S. Barbosa, Derval S. Rosa, and Daniella R. Mulinari

8. **Mechanistic Study of Boron-Based Compounds as Effective Flame-Retardants in Polyurethanes** 173
   Saptaparni Chanda and Dilpreet S. Bajwa

9. **Two-Dimensional Nanomaterials as Smart Flame Retardants for Polyurethane** 189
   Emad S. Goda, Mahmoud H. Abu Elella, Heba Gamal, Sang Eun Hong, and Kuk Ro Yoon

10. **Flame Retardant Polyurethane Nanocomposites** 221
    Wen-Jie Yang, Chun-Xiang Wei, Hong-Dian Lu, Wei Yang, and Richard K. K. Yuen

11. **Industrial Flame Retardants for Polyurethanes** 239
    K. M. Faridul Hasan, Péter György Horváth, Seda Baş, and Tibor Alpár

12. **Recycling of Polyurethanes Containing Flame-Retardants and Polymer Waste Transformed into Flame-Retarded Polyurethanes** 265
    Marcin Włoch

Editor's Biography 285

Indexes

Author Index 289

Subject Index 291

Preface

The polyurethane (PU) industry is one of the fastest growing industries, with commercial applications in many areas, such as automotive materials, construction materials, medical items, packaging materials, furniture, thermal materials, electrical materials, and vibration insulation. The high flammability of PU foams is a concern, and it restricts many of their valuable applications. The research and progress in PUs are growing rapidly, as witnessed by a rapid increase in the number of publications in this area and wide industrial applications. However, a lack of any books covering the materials, chemistry, and mechanism of various flame retardants (FRs) for PUs under one title makes this book very unique. This book provides current state-of-the-art knowledge on the materials and chemistry of novel FRs for PUs. The synthesis, characterization, and applications of halogen, nonhalogen, and eco-friendly materials as FRs for PUs and their FR mechanisms are explored to provide in-depth knowledge in this field. The FR mechanisms based on the types of FRs in PUs, along with their eco-friendly synthesis, are explored. Environmentally friendly approaches to recycling PUs and their future aspects are discussed.

This book contains a total of 25 chapters, which are divided into two volumes. In this volume (entitled *Materials and Chemistry of Flame-Retardant Polyurethanes Volume 1: A Fundamental Approach*), 12 chapters are included. The first chapter introduces basic concepts, materials, chemistry, and various approaches for the synthesis of PUs. A wide range of applications of diverse types of PUs—including rigid, flexible, thermoplastic, ionomer, and waterborne PUs—is discussed. This is followed by a chapter highlighting recent advances in bioderived renewable materials for the synthesis of PUs. The subsequent chapters discuss various types of FRs, synthesis, and their applications in reducing the flammability of PUs. Advantages and disadvantages of halogen-based FRs, along with those of many FRs as substitutes for halogen-based FRs, are covered. The last few chapters in this volume cover nanomaterials, nanocomposites, industrial FR-PUs, and approaches used to recycle FR-PUs.

The second volume of this book (entitled *Materials and Chemistry of Flame-Retardant Polyurethanes Volume 2: Green Flame Retardants*) deals with eco-friendly FRs for PUs. The first few chapters cover natural resources and polysaccharides as green FRs, followed by some chapters covering carbon-based FRs such as carbon nanotubes and graphite. Metal oxides, metal hydroxides, and nitrogen- and phosphorus-based FRs for PUs are explored in detail. The mechanism of flame retardancy of these materials is discussed to provide better insight for the development of new and efficient FRs. The last couple of chapters cover synergism effects in FRs, as well as future perspectives and challenges of FR-PUs.

I would like to thank the authors for their valuable contributions and the reviewers for their productive comments and recommendations to improve the quality of this book. I also gratefully acknowledge the support from the ACS Books editorial team—particularly Anne Brenner, Amanda Koenig, James Antley, and Tracey Glazener—for their timely responses and expert handling of the submitted chapters.

I would like to dedicate this book to my parents, family, teachers, and students, without whom none of my success would be possible.

Ram K. Gupta, Associate Professor of Polymer Chemistry
Department of Chemistry
Kansas Polymer Research Center
Pittsburg State University
Pittsburg, Kansas 66762, United States

Chapter 1

Materials and Chemistry of Polyurethanes

Felipe M. de Souza,[1] Muhammad Rizwan Sulaiman,[1,2] and Ram K. Gupta[*,1]

[1]Department of Chemistry, Kansas Polymer Research Center, Pittsburg State University, Pittsburg, Kansas 66762, United States
[2]Department of Engineering Technology, Pittsburg State University, Pittsburg, Kansas 66762, United States
*Email: rgupta@pittstate.edu

This introductory chapter summarizes the main concepts related to polyurethane (PU) chemistry. It gives proper insight into the vast number of materials incorporated into the experimental design of this class of polymer. The chapter includes a discussion of petrochemical and biobased materials, emphasizing different chemical approaches and their characteristics. These approaches include epoxidation/ring opening, hydroformylation, ozonolysis, and thiol-ene. Furthermore, a section is dedicated to explaining the structure–property relationship. A wide range of applications of diverse types of PUs, including rigid, flexible, thermoplastic, ionomer, and waterborne, is discussed. This chapter also aims to provide new ideas for the readers with an in-depth and straightforward discussion of the possibilities that will evolve PU chemistry.

Introduction

Polyurethanes (PUs) are polymeric materials that contain a urethane linkage [NH–C(O)–O]. They are formed by the reaction between components containing hydroxyl (–OH) groups, such as components that contain polyol and isocyanate (–N=C=O). The isocyanate component should have at least two isocyanate groups to guarantee the PU polymerization process (1). The general reaction for PU synthesis is given in Figure 1.

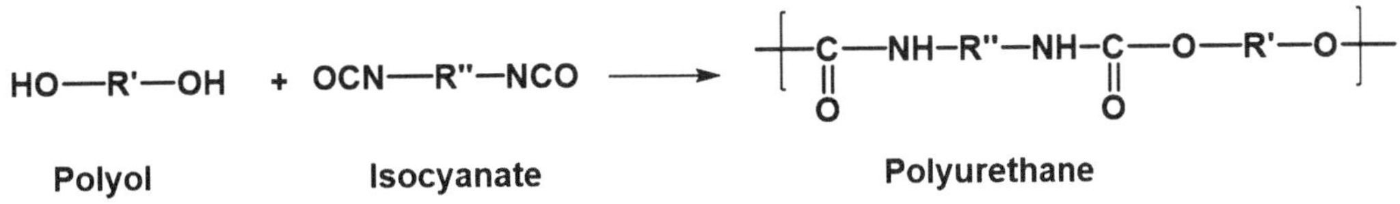

Figure 1. General addition polymerization reaction between a polyol and diisocyanate reaction to yield a PU.

© 2021 American Chemical Society

The first report on PU was published in 1947 when Dr. Otto Bayer and coworkers reported the reaction between a polyester diol and an aliphatic diisocyanate (2). Initially, PU did not find many industrial applications, but after further development by Bayer, the industrial production of flexible PU began in 1952. Subsequently, DuPont introduced the polyether polyols, such as poly(tetramethylene ether) glycol. This development allowed PU production with multiple benefits, such as low cost, easy handling, and high hydrophobicity. The polyether polyols also enhanced the PU polymer's stability against the environment and moisture (3). Later, the introduction of diisocyanates with aromatic and rigid structures, like polymethylene diphenyl diisocyanate, resulted in the development of rigid PUs. The rigidity in PUs comes from the growth of a highly cross-linked structure accompanied by rigid segments (4). The wide range of mechanical properties from flexible to rigid PUs intrinsically depends on the chemical structure of the starting materials, mainly polyols. Polyols complement many starting materials used for PU synthesis, as opposed to isocyanates, for which use is very limited. Polyols having aliphatic chains, higher functionality (i.e., around 2 or 3), and high molecular weight (i.e., a degree of polymerization higher than 300), which lead to elastomeric PUs. The creation of elastomeric PUs occurs because the chains can move past each other with ease, and aliphatic segments have innate mobility.

The typical example of this type of polyol is polyether polyols, which present weak intermolecular interactions, predominantly Van der Waals interactions. Due to this characteristic, polyols are classified as the soft domain of the PU chain. However, isocyanate displays the opposite behavior. For example, polymethylene diphenyl diisocyanate is a polyisocyanate and contains an aromatic ring, which offers low mobility and high rigidity to the polymeric structure. Additionally, when isocyanate reacts with the hydroxyl group to form the urethane linkage, an increase in hydrogen bonding between the chains is observed, which drastically increases the packing capacity of the PU and helps create hard domains. The schematics for both soft and hard domains are provided in Figure 2. Hence, highly flexible PU can be obtained by reacting polyols and isocyanates with linear structures, high molecular weight, and low functionality. However, for rigid PU structures, the monomers should possess a low molecular weight, low flexibility, and high functionality. Today, the PU materials designed by researchers and industry lie between the two extreme cases, providing an extensive range of tunable properties for different applications (1).

Other components, such as chain extenders or cross-linkers, are also added to the PU formulation to achieve the desired characteristics. The chain extenders are low-molecular-weight diols that extend the polymeric chain, usually to add more flexibility. Some examples are 1,4-butanediol, triethanolamine, ethylene, diethylene, propylene, dipropylene, and neopentyl glycols. The cross-linkers are polyols with a functionality that can go from 3 to 8, creating a network structure. Higher functionality leads to a higher degree of cross-linking. It also affects the chemical structure by, for example, decreasing mobility, elasticity, modulus, and insolubility, since cross-linker polymers only swell when in contact with proper solvents. Some examples of cross-linking agents are glycerol, sucrose, sorbitol, and xylitol (1). A crucial component for the synthesis of PU is catalysts, which speed up the rate of reaction. The improvement in reaction rate is a determining factor for industrial applications, since the faster the reaction, the faster the production. The catalysts used in PU production can be divided into two major groups: amine-based and metal complexes. The amine-based catalysts are mostly tertiary amines, the activity of which is related to their basicity and molecular architecture. Higher catalytic activity can be achieved if the catalyst has higher basicity and less steric hindrance on the N atom. These types of catalysts form a complex bond with isocyanate by donating the pair of electrons from the tertiary amine to the partially positive C in the isocyanate group (4). The mechanism for amine catalysis is given in Figure 3a. Some examples of amine-based compounds are 1,4-diazabicyclo[2.2.2]octane, triethylenediamine, dimethylethanolamine, and dimethyl cyclohexylamine (3). The low volatility of metal complex catalysts presents an advantage over amine-based catalysts. The catalysis mechanism of the metal complex involves the formation of complex bonds between the metal center and oxygen-rich species, which can take place either from the hydroxyl or isocyanate group. These bonds further lower the energy levels and promote the formation of the urethane linkage, as shown in Figure 3b. Some examples of metal complex materials are dibutyltin dilaurate, stannous octoate, and other complexes based on Bi, Zn, Hg, and Pb. Because of the high toxicity of Hg and Pb, there is a tendency to shift to less toxic catalysts (3).

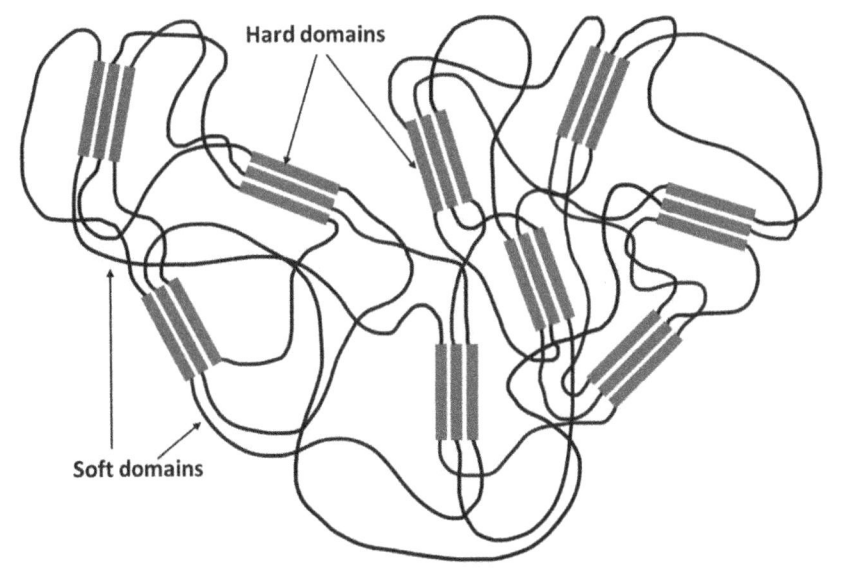

Figure 2. Schematic representation of the soft and hard domains originated from the structure of polyols and isocyanates, respectively.

One of the main ingredients for PU production is the blowing agent. Blowing agents are used to form PU foams, which define an essential aspect of the PU morphology, cell structure, and physical properties. The increased use of blowing agents produces less dense PU foams with inferior mechanical properties. The resulted polymer may become fragile or develop a low compressive strength when exposed to a load. Hence, the optimum amount needs to be defined to obtain optimal properties such as regular cellular structure, optimal density, and strong mechanical properties. The blowing agent most commonly used today is water, but others include CO_2, *n*-pentane, and chlorofluorocarbons, which are now banned (5). Despite the quick reaction between isocyanates and polyols, the reaction components usually lack proper mixability. Therefore, surfactants are used to guarantee homogeneous reaction media and are employed for both foaming and nonfoaming processes. Silicon-based copolymers or oils are usually used for this purpose, such as polyether polysiloxane.

Thus, there are many possible combinations to produce versatile PU properties. Beyond the mechanical properties, it is also possible to control PU density, closed-cell structure, thermostability, flame retardancy, chemical resistance, and biocompatibility. By controlling these PU properties, different materials can be produced to be used in a wide range of sectors. PUs have already proven useful in a vast range of contexts because of the many possible synthetic routes and the large set of properties they yield. High flexibility and low density allow the employment of PUs in furniture, bedding, car seats, dampers for shoes, and cushioning. Rigid PUs, accompanied by its inherent thermal insulation properties, are used in the construction of walls, coolers, freezers, and packing. The list of applications goes on, as PUs are also used in textiles, electronics, biomedical devices, and coatings.

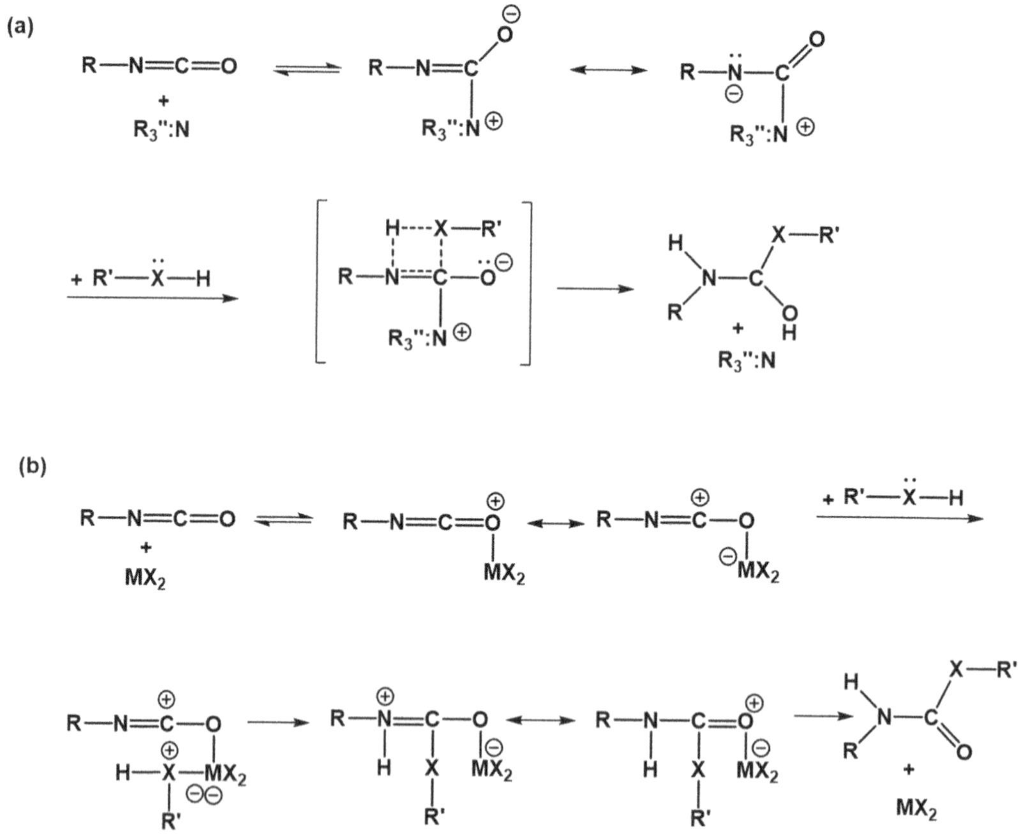

Figure 3. The mechanism for the catalyzed reaction between the isocyanate and hydroxyl groups through (a) tertiary amine and (b) organometal complexes. Adapted with permission from reference (4). Copyright Sharmin and Zafar, some rights reserved; exclusive licensee InTech. Distributed under a Creative Commons Attribution License 3.0 (CC BY), https://creativecommons.org/licenses/by/3.0.

This multitude of properties is reflected in the market for PUs, which has been on the rise since they became commercially available. In 2012, for instance, the PU market accumulated around $43 billion, presenting an annual growth of 7.4% that resulted in $66.5 billion in 2018. The market is still growing due to its relevance to the overall industry. The American Chemistry Council estimated that around 260,000 jobs were created by the PU market, which accounted for $89.3 billion in output in 2019 (6).

Materials and Chemistry of PUs

Most of the starting materials used in industrial PU production are derived from petrochemical sources. There are two major types of polyols that have been widely used: polyether and polyester. The process of synthesizing polyether polyols includes an epoxidation step of olefins (e.g., ethylene, propylene, and butylene oxides) followed by an alkoxylation reaction. The latter process can be performed by a substance containing several hydroxyl groups, such as glycerin, sucrose, or pentaerythritol. When used for this purpose, these materials are called starters, and they define the functionality of the desired polyol. For example, glycerin has a functionality of 3; therefore, it yields a polyol that has the same functionality (7). The conventional polyether polyol derived from this process can react with an aromatic isocyanate to yield a PU, as shown in Figure 4. The polyether polyol given in the example can be further functionalized by its reaction with ethylene oxide, which increases its molecular weight and leads to more flexible structures. Another factor is that the reaction with ethylene oxide provides a primary hydroxyl group in the end chain, which is more reactive to isocyanate than secondary or tertiary hydroxyls (8).

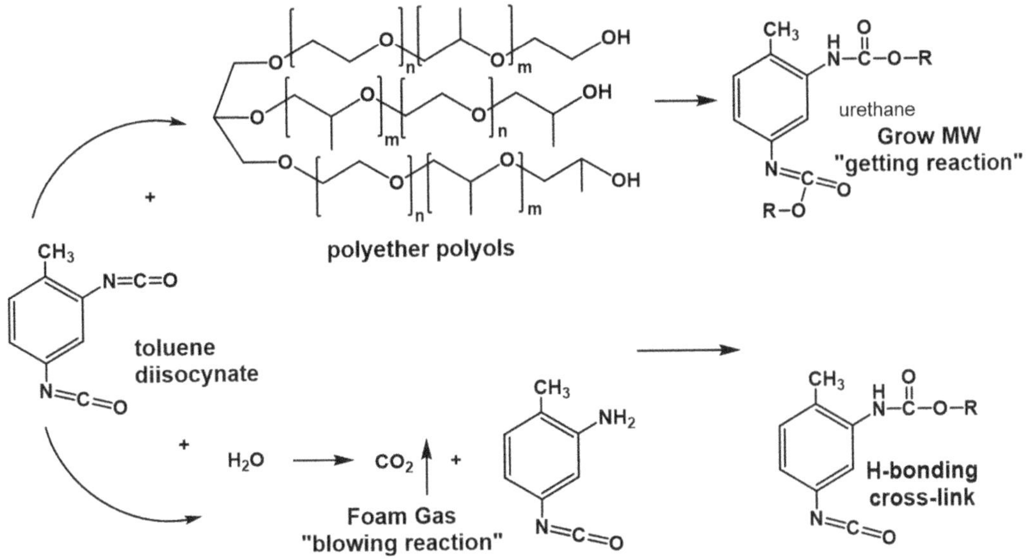

Figure 4. Schematics of the synthesis of conventional polyether polyols through epoxidation followed by an alkoxylation (ring-opening) reaction. Adapted with permission from reference (9). Copyright 2011 Springer Nature.

Another example of a classical polyether polyol is poly(tetramethylene ether) glycol, obtained through the ring opening of tetrahydrofuran catalyzed by acid, as shown in Figure 5. Because of their linear structure, polyether polyols are used to synthesize PU as thermoplastics, elastomers, and fibers (10). The overall advantage of polyether polyols lies in their high hydrophobicity, low cost, low viscosity, and increased flexibility. These factors made polyether polyols the market leader, and their popularity is still growing annually.

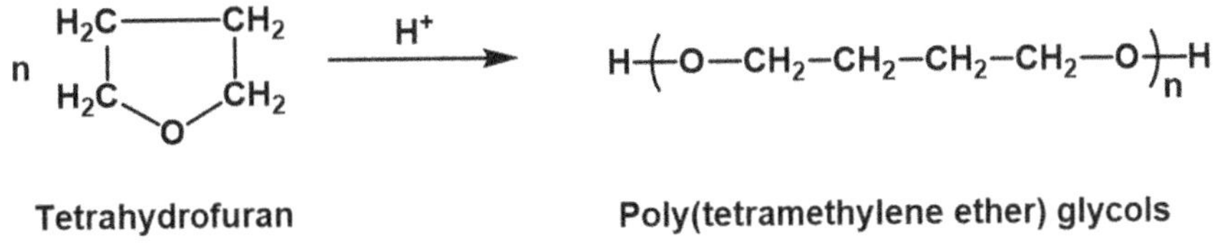

Figure 5. Ring-opening reaction of tetrahydrofuran catalyzed in acid media to yield poly(tetramethylene ether) glycol. Adapted with permission from reference (10). Copyright 2015 Springer Nature.

The second most significant class of polyols is polyester-based polyols. They are synthesized through a polycondensation reaction that follows a step-growth mechanism. The process uses starting reagents that consist of dicarboxylic acids, its derivative esters, or anhydrides. Some common examples of these materials are adipic acid, phthalic anhydride, and terephthalic acid. The latter two are widely used to produce aromatic polyester polyols. The synthetic route for this type of polyol is expressed in Figure 6. The introduction of the aromatic ring into the PU backbones provides higher rigidity and thermal stability, which are desirable properties for many applications.

On the other hand, polyester polyols are mostly waxy materials, as some of them are solid at room temperature. This characteristic is reflected in the high viscosity and demands more processing, which results in incremental increases to the final cost. Also, the presence of carboxyl groups in its polymeric chains makes polyester polyols more susceptible to chemical attacks than ether polyols. The industry uses polycaprolactone polyols, mostly ε-caprolactone, as a viable reagent for the synthesis of polyester polyols and to overcome the drawbacks of high viscosity and improved moisture resistance. The reaction is performed similarly to that of ring opening by using a starter that defines the functionality of the polyols. This example is illustrated in Figure 7.

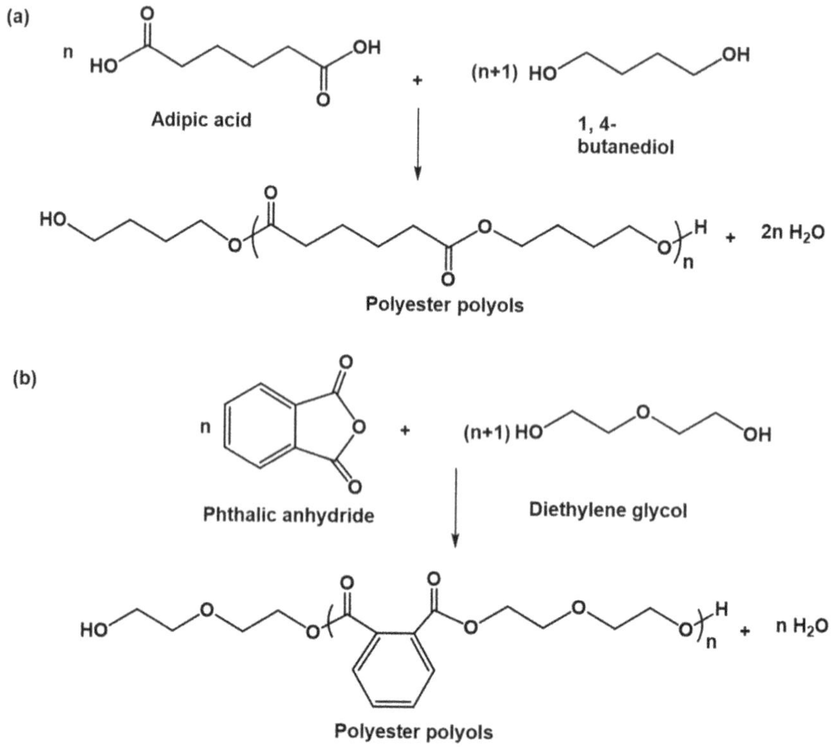

Figure 6. Polyesterification reaction for the synthesis of polyester polyols through (a) adipic and 1,4-butanediol and (b) phthalic anhydride and diethylene glycol. Adapted with permission from reference (10). Copyright 2015 Springer Nature.

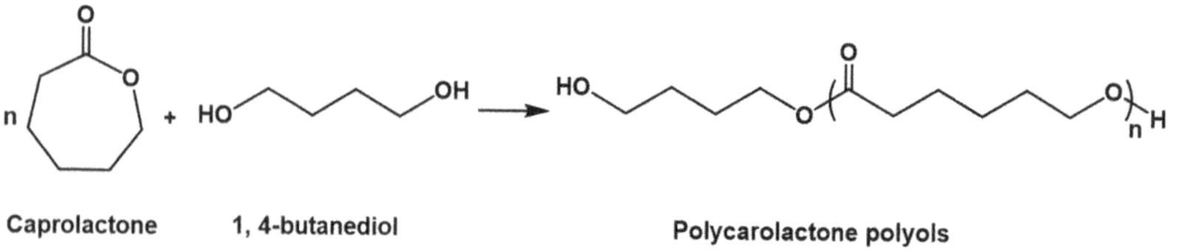

Figure 7. Ring-opening reaction in a caprolactone using 1,4-butanediol as starter yielding a polycaprolactone polyol. Adapted with permission from reference (10). Copyright 2015 Springer Nature.

The silicon-based polymers known as polysiloxanes possess the unique property of low glass-transition temperature around −123 °C. This characteristic allows polysiloxane-based materials to retain their mechanical and thermal properties under a wide range of temperatures. Hence, functionalizing their end chains with hydroxyl groups yields polysiloxane polyols, which can react with isocyanate and yield elastomeric PUs that offer properties similar to polysiloxanes. Thus, a whole new set of materials can be synthesized for high-performance applications, offering high thermal and mechanical stability over an extended range of temperatures and presenting significant resistance to corrosion and the passage of electrical current (*11*).

Another class of polyols that is designed mostly for high-performance coatings is acrylic polyols. They are produced by radical addition copolymerization using acrylic- or methacrylic-based monomers. However, the structure of the starting materials must contain hydroxyls as a side group. Despite that limitation, most of the materials used for PU synthesis in the industry are fossil derived. Soybean-based acrylic polyols are also produced. The process consists of an epoxidation reaction that converts oxirane rings into acrylate. After that, the reminiscent epoxy groups react with water leading to hydroxyl groups (*12*). The reaction process is shown in Figure 8. Due to the presence of terminal unsaturation, the polyol can undergo a UV-curing process to make high-performance coatings. Generally, there are many possible approaches to polyols due to the wide range of commercially available monomers, such as methyl methacrylate. Methyl methacrylate can infer UV stability, hardness, moisture resistance, and good adhesion to metals, which explains its application in coatings and adhesives. Styrene is another option that provides the same properties; however, it degrades when exposed to UV light for an extended period.

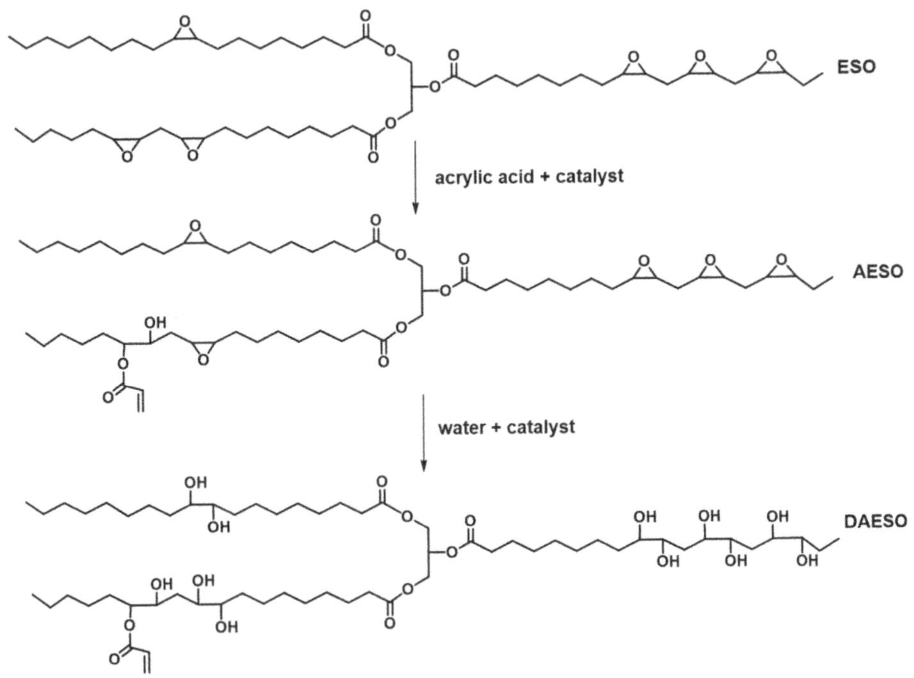

Figure 8. Synthetic route for an acrylic polyol derived from soybeans. The process involves the epoxidation that yields the epoxidized soybean oil, followed by an acrylic acid reaction to form acrylate-epoxidized soybean oil (AESO). Finally, AESO's reaction with water to open the remaining oxirane rings produces hydroxyl groups. Adapted with permission from reference (12). Copyright 2015 Royal Society of Chemistry.

Polybutadiene polyols are similar to acrylic polyols. This group of polyols is also produced by radical addition polymerization in which a monomer such as 1,4-butadiene is polymerized, and hydroxyl end groups are chemically attached in the final step. The advantage of these polyols is their flexibility even at lower temperatures. They may also present a high solvent resistance when their double bonds are hydrogenated (3). As environmental concerns grow, the development of new biobased materials that are environmentally friendly poses a challenge for scientific communities and industries. The shift to ecofriendly materials is mostly due to fossil fuel instability in terms of its price and continually decreasing availability. However, there is a great possibility that biobased materials will offer the required properties, such as several chemical structures, sustainable sources, and overall lower cost as the technologies for their synthesis become more widespread. Some biobased components that have been presented in the PU industry are glycerin, sucrose, and sorbitol, as previously mentioned. They work either as starters or cross-linkers and are added at various steps of the synthetic process.

One of the sources that has been significantly explored for the synthesis of biopolyols is seed oil triglycerides (9). Castor oil is an example that has become popular at the industry level, since its addition to the formulation of PU imparts environmental resistance, satisfactory performances, and low cost. The general structure of seed oils consists of glycerin that is esterified with a fatty acid chain. Most of these chains present a certain degree of unsaturation, which is the key element that allows several types of chemical modifications, such as the introduction of hydroxyl groups to convert the oils into polyols. The fatty acids such as oleic, linoleic, and linolenic are composed of an 18 carbon chain and one, two, and three double bonds, respectively (9). These essential oils are found in a variety of natural sources. Castor and lesquerella oils are examples that already present hydroxyl groups in their structure, reaching a functionality of 3 without any chemical treatment (13). The fact that the castor plant is adaptable, grows quickly in different environments, and presents a low quantity of protein encourages its production entirely for industrial purposes. Another advantage of using castor oil is that the oil extracts are mostly ricinoleic, decreasing the purification process cost (9). One of the disadvantages is that castor plants produce a highly toxic protein named ricin, which imposes a barrier for its use as a ration for feedstock or human consumption (14). Castor oil-based PU is still new in the market, yet promising results have been reported (15–17).

Similarly, corn and soybean oil have conquered an adequate space within the scientific community and industry as an abundant and low-cost source for PU production. Among various investigations, several works regarding vegetable oil as a source of PU have been published (18–24). Scientists have shown that biobased materials have great competitive potential, and trademarked products such as BiOH and Merginol polyols have been introduced within the market (9, 25). Other classes of natural compounds that can be used for polyol synthesis are terpenes and turpentines. They are biosynthesized by plants, which can play the role of repellent to avoid predators or attractant to induce pollination, as well as contributing to other metabolic activities. A fair share of these compounds presents double bonds that allow chemical functionalization in the same way seed oils do. Some of their chemical structures are shown in Figure 9. Another compound that can be generously found in plants is a biomacromolecule called lignin. It provides physical support and transports nutrients within the plant. Its chemical structure is a combination of several small molecules called monolignols. Some of these main molecules are *p*-coumaryl alcohol, coniferyl alcohol, and sinapyl alcohol, as shown in Figure 10 (26).

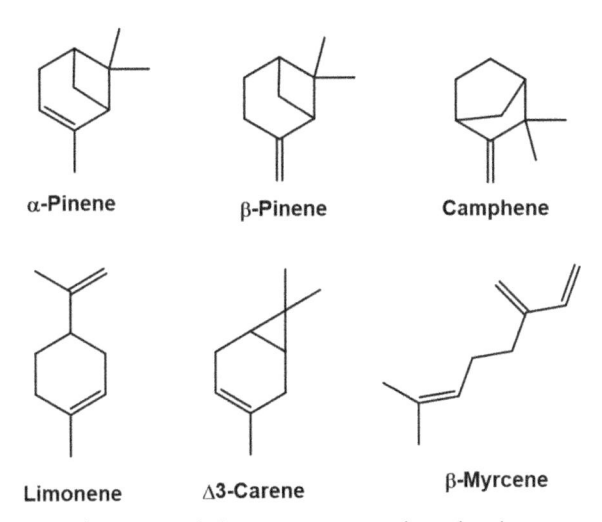

Figure 9. Chemical structures for commonly known terpenes. Adapted with permission from reference (27). Copyright 2020 American Chemical Society.

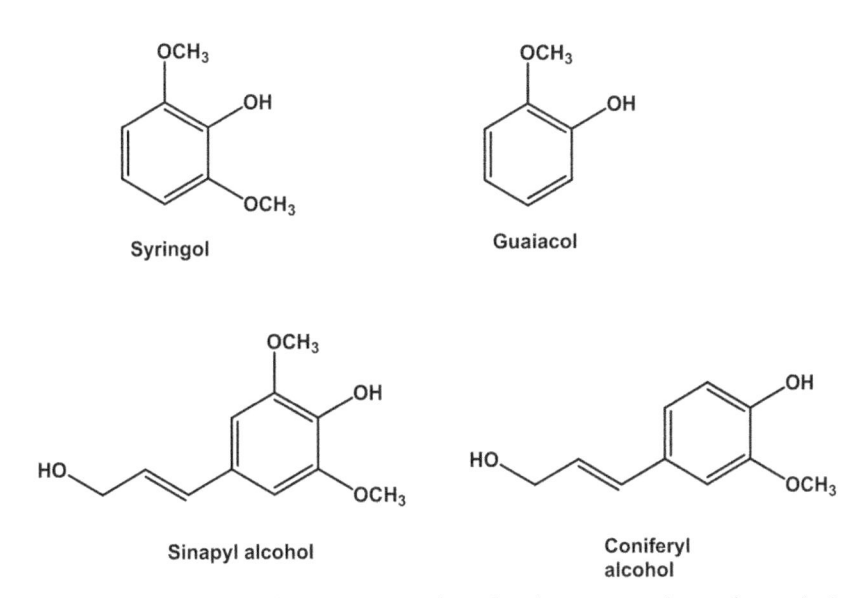

Figure 10. Phenolic derivatives from lignin sources. Adapted with permission from reference (26). Copyright 2008 Elsevier.

Lignin is produced mainly for the papermaking industry. In papermaking industries, the separation process of lignin from cellulose must be performed to make high-quality paper. This process is known as delignification and can be executed in several ways, such as via the kraft process, the sulfite process, or steam explosion. For context, kraft lignin is obtained by placing biomass in a high-temperature environment with an aqueous alkaline solution containing NaOH and Na_2S. These conditions cause the lignins to undergo hydrolysis and allows them to be soluble in aqueous media, which accounts for around 95% of its removal from wood sources (9). Sulfite lignin is obtained through the reaction between lignin and sulfite or bisulfite salts containing alkaline metals.

This process introduces sulfate groups in the lignin's structure that increase its solubility in aqueous media and facilitate its removal (28). The steam explosion process consists of applying high pressure under a wood source followed by a quick release of pressure, which causes the cellulose fibers to separate from lignin (29). Even though lignins are a byproduct of paper production, they have many applications, including as humectants, sequestrants for water cleaning, emulsion additives, and dispersants (30). Another essential lignin application is as a renewable source for phenolic compounds that can be used to synthesize phenolic foams. The inherent presence of hydroxyl groups and double bonds in the lignin allows chemical modifications. Based on that possibility, many studies have used these biomaterials as a suitable source for polyol synthesis. Several technological advancements for the chemical modification of double bonds to introduce hydroxyl groups have been developed. Some examples are hot air oxidation, epoxidation followed by ring opening, ozonolysis, hydroformylation, and thiol-ene.

The hot air process is one of the simplest methods and involves the blowing of hot air into seed oils to convert the double bonds into epoxy groups and later into hydroxyl. Despite lacking a high yield and broad polydispersity, it is a cheap and facile process that has applications in many areas within the PU market (9, 18). Epoxidation followed by ring opening is a process similar to the hot air process, but it is performed using chemical reagents. One common method for this process occurs through an in situ reaction between formic acid and hydrogen peroxide that forms performic acid (31). Performic acid is an unstable, highly reactive intermediate that reacts with an unsaturation, leading to the formation of epoxy groups or the straight introduction of hydroxyl groups depending on the conditions. An example of this reaction performed in cardanol oil can be seen in Figure 11.

Ozonolysis is another technology that has been known for almost 70 years. It is based on a complex mechanism and is an interesting tool for synthesizing polyols using olefins. It is a highly oxidative process that can cleave the double bond leading to a primary hydroxyl group (32). Therefore, reactivity is improved at the cost of decreased functionality. This scenario can be observed in the case of soybean oil. When hydroxyl groups are introduced through ozonolysis in the soybean oil structure, the oil reaches the maximum functionality of 3. However, if the same oil is chemically modified through epoxy/ring opening, a functionality of around 4.5 can be achieved. A general example of the chemical structure for a polyol derived from ozonolysis is shown in Figure 12.

Hydroformylation uses CO and H_2 along with catalysts under some pressure to convert double bonds into formaldehyde groups. Then, a reduction follow-up process is performed by using H_2 to reduce aldehyde groups into hydroxyls. It is an effective technique to introduce primary hydroxyl groups into seed oils, with the drawback being that there are specific requirements for the occurrence of the reaction, such as high pressure and the presence of catalysts (34). Figure 13 describes the hydroformylation process of a triglyceride.

Thiol-ene is a method that has gained a lot of recognition in the scientific community due to its simplicity, effectiveness, and speed in comparison to other methods. It is based on the radical addition reaction between mercaptans (S–H) with a double bond when exposed to UV light in the presence of photocatalysts. It holds tremendous potential, since it can be performed at room temperature and is not influenced by pressure. Its drawback is the inherent toxicity of mercaptan compounds and the introduction of a C–S bond at the end of the reaction, which may not be desired in some cases (35). An example from the literature demonstrates the thiol-ene coupling reaction between limonene and 1-thioglycerol given in Figure 14.

Figure 11. Epoxidation and ring-opening reaction performed in cardanol oil using performic acid. Adapted with permission from reference (31). Copyright 2013 American Chemical Society.

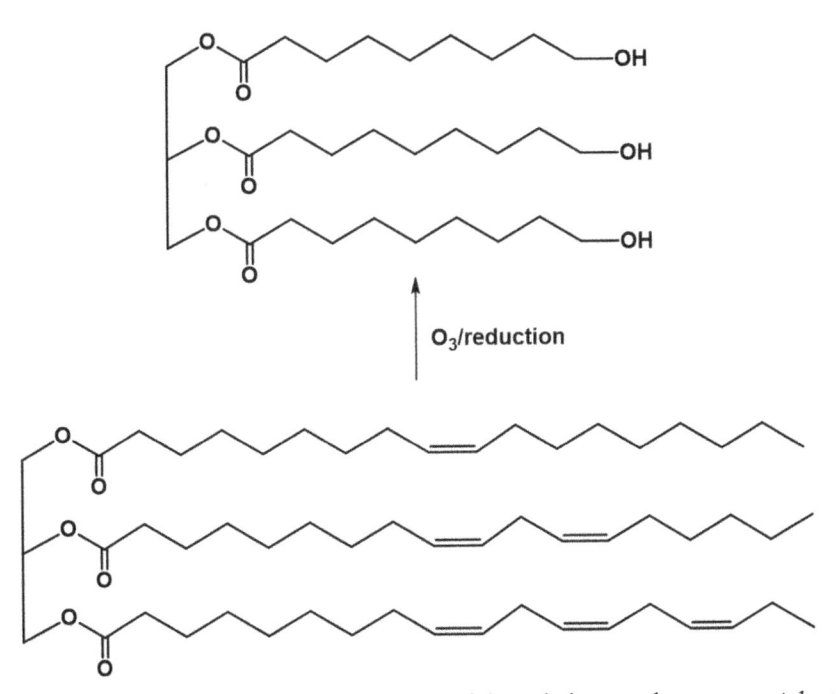

Figure 12. The general structure of a polyol is synthesized through the ozonolysis process. Adapted with permission from reference (33). Copyright 2008 Taylor & Francis.

Figure 13. Hydroformylation process of a triglyceride to introduce primary hydroxyl groups followed by esterification with formic acid. Adapted with permission from reference (33). Copyright 2008 Taylor & Francis.

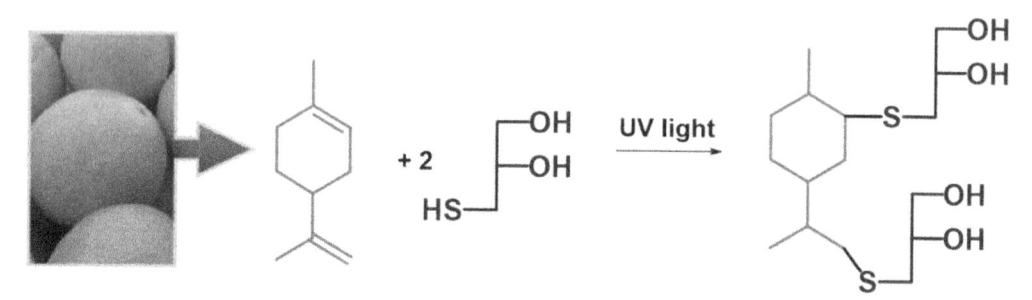

Figure 14. Example of the thiol-ene coupling reaction between limonene (derived from the orange peel) with 1-thioglycerol, yielding a limonene-based polyol. Adapted with permission from reference (36). Copyright Ranaweera et al., some rights reserved; exclusive licensee Tech Science Press. Distributed under a Creative Commons Attribution License 4.0 (CC BY), https://creativecommons.org/licenses/by/4.0.

The preceding technologies show the many possible routes adopted to design polyols, allowing scientists to explore many paths that have not been described yet. Isocyanates are the key components to produce PU. Their chemical function is composed of the group R–N=C=O. The high electronegativities of N and O toward C cause a strong withdrawing effect, which makes C partially positive. Thus, isocyanates are susceptible to nucleophilic attacks. This explains why isocyanates show a reactive behavior toward hydroxyls, amines, phenols, and so on. However, there is a difference in reactive behavior when aliphatic isocyanates are compared with aromatic ones. The latter present a lower reactivity due to their resonance structures, making them more stable and less reactive (4). The resonance structures for isocyanate are given in Figure 15.

Despite the much smaller number of commercially available isocyanates compared to polyols, there is still a large enough variety of them to achieve different properties in the final polymer. Their structure can be aromatic or aliphatic, which affects their properties. The two most common isocyanates in use in industries are methylene diisocyanate and toluene diisocyanate (TDI). These isocyanates have a relatively lower cost and exhibit higher reactivity (37). TDI possesses aromatic rings known to implement rigidity into the structure, which allows its use in rigid PU. However, thermally insulating and even flexible foams also use TDI to produce soles of shoes, cushions, mattresses, and car seats. A range of properties can be achieved by properly defining the formulation to obtain a polymeric structure with the desired soft/rigid segments ratio.

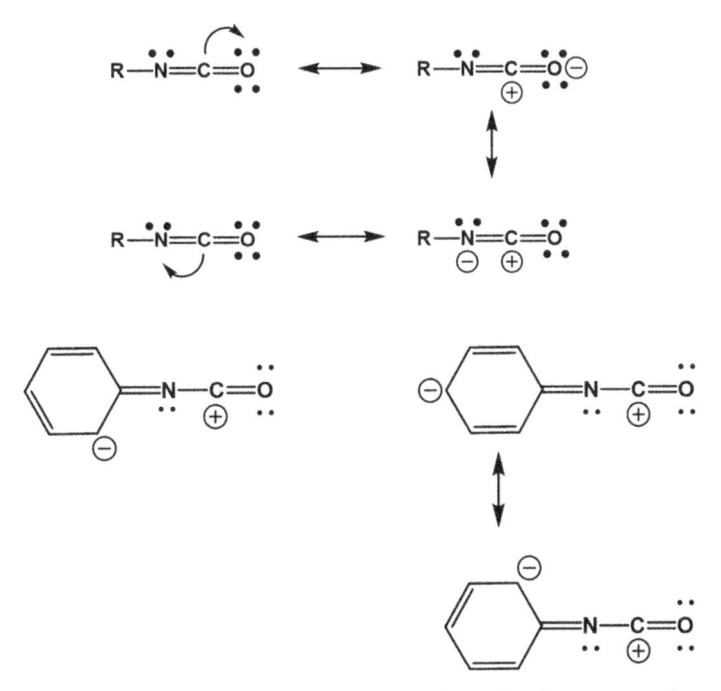

Figure 15. Resonance effect into an aromatic isocyanate. Adapted with permission from reference (4). Copyright Sharmin and Zafar, some rights reserved; exclusive licensee InTech. Distributed under a Creative Commons Attribution License 3.0 (CC BY), https://creativecommons.org/licenses/by/3.0.

Aliphatic and cycloaliphatic isocyanates such as hexamethylene diisocyanate, 1,6-hexamethylene diisocyanate, and isophorone diisocyanate are examples of commercially available isocyanates that are mostly used as reagents for coatings. The aliphatic structure permits the free movement of the polymeric chains, making these polymers more flexible. Furthermore, their inherent chemical stability, aesthetics, glossy aspect, and color retention are essential factors for their use in industries, unlike aromatic isocyanates, which may not retain color and may turn darker or yellowish when exposed to light (38). The chemical structures of leading isocyanates are described in Figure 16.

Isocyanates are very convenient reagents because of their high chemical versatility and reactivity with several organic functions. However, their toxicity and petrochemical origins are a concern to scientists, which drives the research to find isocyanate-free routes for PU synthesis. To provide an alternative for the formation of urethane linkage without the use of isocyanate, a group of researchers reported the synthesis of a limonene-based PU that consisted of a three-step process (39). First, the unsaturation of limonene was epoxidized. Second, CO_2 with the presence of catalysts at high temperature and pressure was used to react with the oxirane ring to form a carbonated group. Third, the cyclic carbonate was opened by reacting with a diamine that yielded the urethane linkage. The steps for this reaction are described in Figure 17. Another path that followed the same concept used a vegetable oil that was epoxidized and followed up with a carbonation reaction. An amine siloxane-based coupling agent was then used to open cyclic-carbonated groups, forming a urethane linkage (3). The presence of a Si–O bond derived from the siloxane group allows further chemical modification. The development of these new technologies expands the possibilities for PU, showing that alternative routes can be created to reach sustainable production.

Figure 16. Chemical structures of most common aromatic and aliphatic isocyanates.

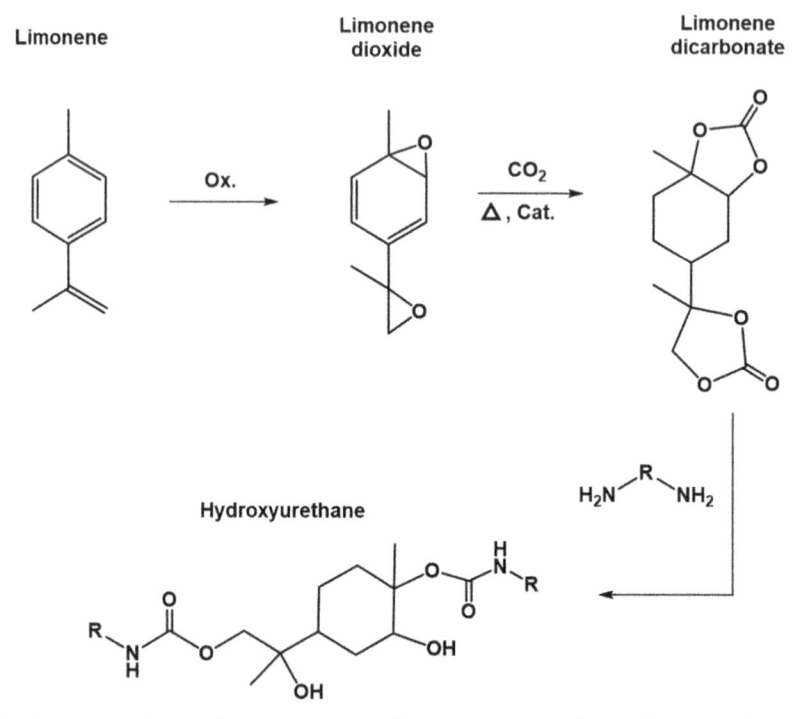

Figure 17. Isocyanate-free synthetic route using a three-step process of epoxidation, carbonation, and reaction with a diamine leading to the urethane linkage. Adapted with permission from reference (39). Copyright 2012 Royal Society of Chemistry.

Applications of PUs

As discussed up to this point, there are various ways that PU can be synthesized. The facile procedures and the significant number of reagents available for its production make this polymer suitable for many different applications. The PU can be rigid, flexible, thermoplastic, ionomer, or waterborne. Each of these classes presents widely distinct usages, which is why PUs are arguably one of the most versatile polymers in synthetic approaches and applications.

The rigid PU composes one of the most important industry segments, mostly due to its appreciable mechanical properties accompanied by low density and thermal insulation. These factors make this class of PU almost ideal for specific applications. A classic example is its use in houses and buildings. Its properties, such as low weight and strong mechanical properties, decrease a construction's overall weight with proper support. On top of that, thermal insulation aids not only in maintaining a pleasant environment but also in saving energy used by heaters and air conditioners. The introduction of thermal-insulating PU could significantly decrease the total energy usage of a house. This concept applies in the same manner for automobiles; the rigid PU's light weight promotes ease in the car movement, and the rigid structure provides a safety aspect, since it can dissipate the mechanical energy from an impact. The production of freezers and coolers also has PU as one of its core components. The high thermal insulation of PU can be further improved by decreasing the size of its cellular structure (40). The PU has these properties because of porosity, which creates a heat dissipation barrier that reduces the heat that passes through irradiation. By controlling the addition of blowing agents, optimum foam cell size, density, and thermal insulation can be attained (41).

Another widely used strategy to decrease thermal conduction involves nucleating agents that reduce the cell size to either block the infrared radiation or create a temperature diffusion barrier (42). For this reason, some fillers such as calcium carbonate, talc, or carbon black can be used. However, a higher concentration of these microparticulates may disrupt the cellular structure, harming the closed-cell content and mechanical properties of PU (7, 43). Some parameters such as innate compatibility with the polymeric matrix, homogenous dispersion, and method of incorporation are important to find the most effective filler and concentration for each polymeric matrix. The addition of filler should not cause reagglomeration of particles and should maintain a high closed-cell content (41). Based on this idea, a group of scientists has developed composites to increase the inherently high thermal insulation and mechanical properties of PU. The experiment was performed by adding cellulose nanocrystals (CNCs) through ultrasonication without any solvent. The implementation of only 0.4 wt % of CNCs provoked a reduction of 5% in the thermal conduction, which demonstrated a facile and sustainable way to improve the properties of PU (44). Several beneficial factors for using CNCs were also observed, such as their low cost, abundance, ecofriendly nature, and potential covalent cross-linking with the PU matrix that increases the mechanical properties and thermal barrier effect (44, 45). A decrease in average cell size during the foaming process was observed, which occurred due to the anisotropic alignment of the CNCs within the PU that limits its size growth without disrupting the cellular structure. This is reflected in the results, which showed that the highest closed-cell content of around 98% followed the lowest thermal conductivity, 24.5 mW/mK. The same improvement was observed for the mechanical properties, which yielded the highest compressive strength for parallel (0.21 MPa) and perpendicular directions (0.17 MPa) at 0.4 wt % of CNCs. A visual representation of this mechanism is provided in Figure 18.

The flexible PUs are usually composed of block polymers, and their flexibility relies on increasing the soft/rigid ratio segments (46). The lower degree of cross-linking accompanied by an aliphatic, linear, and weak intermolecular interaction with each polymeric chain is an essential factor that shifts a PU from rigid to more flexible material. Some well-known examples are polyethers and polyesters, which contain long alkyl chains with terminal hydroxyl groups that enable these oligomers to act as diols to synthesize flexible PU. These polymers can also be used as chain extenders in some cases. An increase in flexible segments leads to a less packed structure that would likely make the PU amorphous unless a different processing method is applied to induce crystallinity, such as annealing. Elasticity can also be increased in a PU by using a diol with high molecular weight and a linear structure, since the soft/rigid segment ratio would be higher, therefore allowing higher mobility into the PU. Using these types of polyols would increase the elasticity and affect the mechanical properties. Thus, controlling the soft/rigid ratio is crucial to defining PU's mechanical properties. Controlling the morphology through blowing agents also plays an essential role, since the higher number of micropores tends to decrease density and can increase the material's overall flexibility. The industrial sectors related to packing, cushioning, furniture, interior parts for cars, dampers, soles for shoes, and more sophisticated applications, such as in biomedicine and electronics, take advantage of properties offered by flexible PUs (47–49). Their high elongation strength and deformation through energy absorption are critical properties for those applications. These properties are related to the flexible polymeric segments that increase the mobility and free volume of the polymer, allowing it to effectively absorb high mechanical strain or compression without reaching its yield point. In other words, flexible PU can receive a greater load while still obeying Hooke's law and retaining its elastic properties. On top of that, flexible PU presents a relatively high chemical resistance. These PUs contain long aliphatic segments that are relatively inert against acid and basic environments and lack reactive sites. However, current research in this area deals with the challenge of improving mechanical properties, such as cut, tear, and abrasion resistance, which are essential for practical application. This obstacle can be bridged by developing various composites by incorporating fillers such as glass fiber, aramids, woven fabrics, or basalt. This process can be optimized by applying techniques to the polymers and fillers through different hybrid classifications, such as inter- and intraply, selectively placed, intimately blended, and superhybrid. The first two processing techniques are more balanced regarding cost and efficiency; however, that may vary depending on the application and material. Some other viable engineering processes to further enhance mechanical resistance against tearing, abrasion, and cut are laminating and stitching. Based on these concepts, Yan et. al. (47) developed a flexible PU composite that possesses enhanced properties through the addition of laminated Kevlar and glass, as demonstrated in Figure 19.

Scanning electron microscopy micrographs and photocopies of the aforementioned flexible foam composite are shown in Figure 20. The engineering process performed by Yan et al. offered an overall improvement in the mechanical properties of flexible PU composite compared to its neat sample. The obtained results showed improvement in the tensile strength (from 139 to 660 N), tear strength (from 222 to 310 N), static puncture (from 30 to 127 N), and dynamic puncture strength (from 48 to 75 N) and demonstrated effective processability. The porosity of the composite foam affects the mechanical property, as uniform and smaller pores provide better mechanical support for the foam than larger pores with different sizes. Surfactants play an important role, since they increase mixability between the polyol and isocyanate to ensure a homogenous foaming process. The blowing agent is a major component in introducing porosity in the foam due to the formation of bubbles. In that sense, it is important to control the quantity of blowing agent. The optimal amount of blowing agent should form uniform pores with equal distribution. CNCs in PU provided physical support for the foam, which compensated for the variation in cell size and therefore enhanced the mechanical properties of the foam (44). Hence, properly mixing the components to decrease phase separation and adding the optimal amount of blowing agent should lead to a foam with a regular and uniform cell structure.

Besides applying different processing techniques and methodology, another economically viable approach is the partial incorporation of biobased materials into flexible PU (50). Low-cost materials such as soybean-, rapeseed-, and palm-oil-based polyols can incrementally increase the thermal stability and mechanical properties more efficiently than their petrochemical counterparts (51–53). An interesting approach was analyzed by Zhang et al. (50); they focused on the influence of increasing additions of different cross-linker polyols up to 30% into a conventional polyether polyol formulation. The components used as cross-linkers were soybean-oil-based polyol obtained through epoxidation/ring opening, styrene acrylonitrile, and a commercial cross-linker. Impressively, the addition of 10% of the soybean polyol resulted in similar values for compression and modulus compared to other cross-linkers.

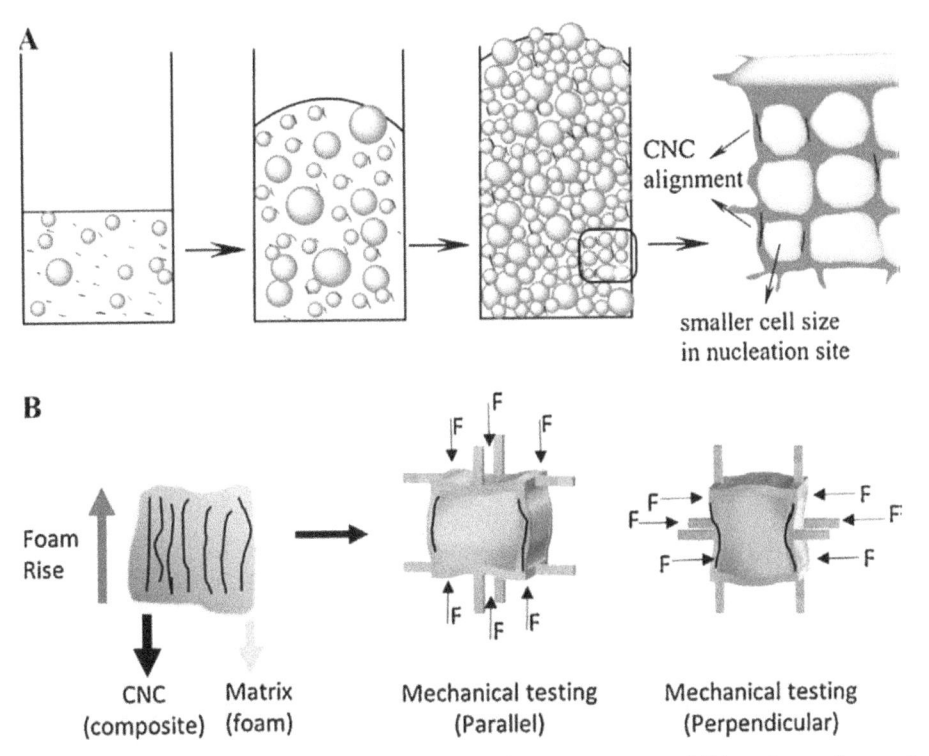

Figure 18. Scheme for (a) CNC arrangement during the foaming process and (b) mechanical strengthening of CNCs into the rigid PU under an applied load. Adapted with permission from reference (44). Copyright 2017 Elsevier.

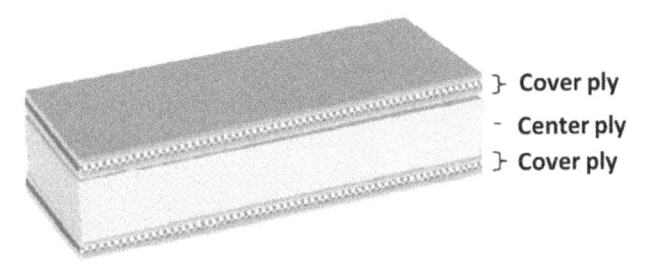

Figure 19. Schematics for the general design of a laminated hybrid composite. Adapted with permission from reference (47). Copyright 2015 Elsevier.

Similarly, the addition of 30% of the soybean polyol presented a drastic improvement in such properties and easily surpassed the petrochemical cross-linker polyols. Despite the overall increase of mechanical properties for all the fillers, it occurred through different factors. According to the authors, the cross-linker soybean polyol decreased the concentration of large hard segments (HSs) by dispersing them into smaller ones. This dispersion caused these shorter hard domains to blend within the polymeric matrix and decreased electronic density contrast. This phenomenon was correlated to a small-angle X-ray scattering characterization, which suggested a transformation of large HSs into a higher concentration of small hard domains. This effect was believed to improve the mechanical properties (50).

The strength-to-weight ratio is an important analysis for designing materials. Industries add reinforcing materials to improve the mechanical properties, which may add to the cost. Therefore, finding low-cost sources to introduce preferred mechanical properties is desired. With that perspective, a study was performed to analyze the addition of lignin, which is a low-cost and abundant biomaterial, into a PU to understand its effect on mechanical properties (54). Unmodified lignin with a molecular weight in the range of 600–3600 g/mol with a varying concentration of 5–40 wt % was added to PU. The highest mechanical strength was observed when 40 wt % of the 600 g/mol lignin was added, which provided 33 MPa of tensile strength and 1394% strain at break. The better results obtained from lignin with lower molecular weight could be attributed to the better dispersibility of the smaller layers of lignin into the foam structure to provide mechanical support as confirmed through differential scanning calorimetry and transmission electron microscopy. In addition, the presence of hydroxyl groups in the lignin structure led to higher hydrogen bonding to provide a physical cross-link. Approaches like this show a reasonable way to fabricate high-performance materials derived from biosources. Along with the mechanical properties, the thermal insulation is also a relevant aspect for the use of PU. A group of researchers incorporated a silica aerogel into a PU through a coprecursor method to decrease the thermal conductivity from 0.25 to 0.032 W/mK (55). Another similar study demonstrated that the addition of silica into a PU matrix decreases the cell size, which decreases the heat conduction through radiation (56). This could be because of the low inherent heat and electrical conductivity of silica due to the lack of free electrons in its structure.

Despite these properties, PUs carry an inborn issue, which is their susceptibility to catching fire. This takes place because of the porous structure of PU that increases the surface area and allows easy oxygen diffusion. Additionally, the highly flammable organic polymer content causes a fire incident to perpetuate and quickly becomes a safety issue. The heat releases during the fire, and the toxic fumes such as dioxanes, carbon monoxide, and cyanate can impose a life-threatening situation. Most fire incidents occur in residential areas, with a higher mortality rate among the elderly or the young. The scientific communities and industries worldwide have been investigating flame-retardant (FR) PU to tackle this problem. Flame retardancy can be achieved by adding FR additives into the PU matrix that can self-quench the fire or increase the ignition temperature of the material.

FRs can be divided into two major groups, blended or reactive FRs. The blended FRs consist of materials that are physically mixed into the PU matrix. Due to its overall efficiency and ease in processing, this approach is usually desired among industries (57). However, the efficiency of blended FRs may decrease over time and jeopardize mechanical properties, since the microparticles may disrupt the cell structure. The reactive FRs are reagents that are covalently bonded with the PU's polymeric chain. Despite requiring an extra synthetical step, this approach presents a lasting effect, and the addition of the FR may increase other properties (15). Based on these two approaches, an extensive number of FR components have been developed, which cover metal oxides (i.e., aluminum trihydroxide, magnesium dihydroxide, borates), organophosphorus (i.e., phosphate esters, phosphinates, red phosphorus), nitrogen-based materials (melamine and its derivatives), carbon materials (expandable graphite, carbon black, fullerene) and halogen-based materials, which are no longer allowed because of high toxicity (58–60). More than one FR can be combined within the same PU matrix to improve the flame retardancy, which provides a synergistic effect. This method offers a large number of possibilities and enhanced properties. Some examples include combining N and P compounds such as ammonium phosphate, melamine polyphosphate, and melamine cyanurate encapsulated with red phosphorus. Others are aluminum hypophosphite, C and P, and so on (61–64). The primary objective of developing FR PU is to provide a quick response in terms of fire quenching, preventing the perpetuation of the fire and dripping, improving thermal stability, and reaching the lowest level of heat and smoke release.

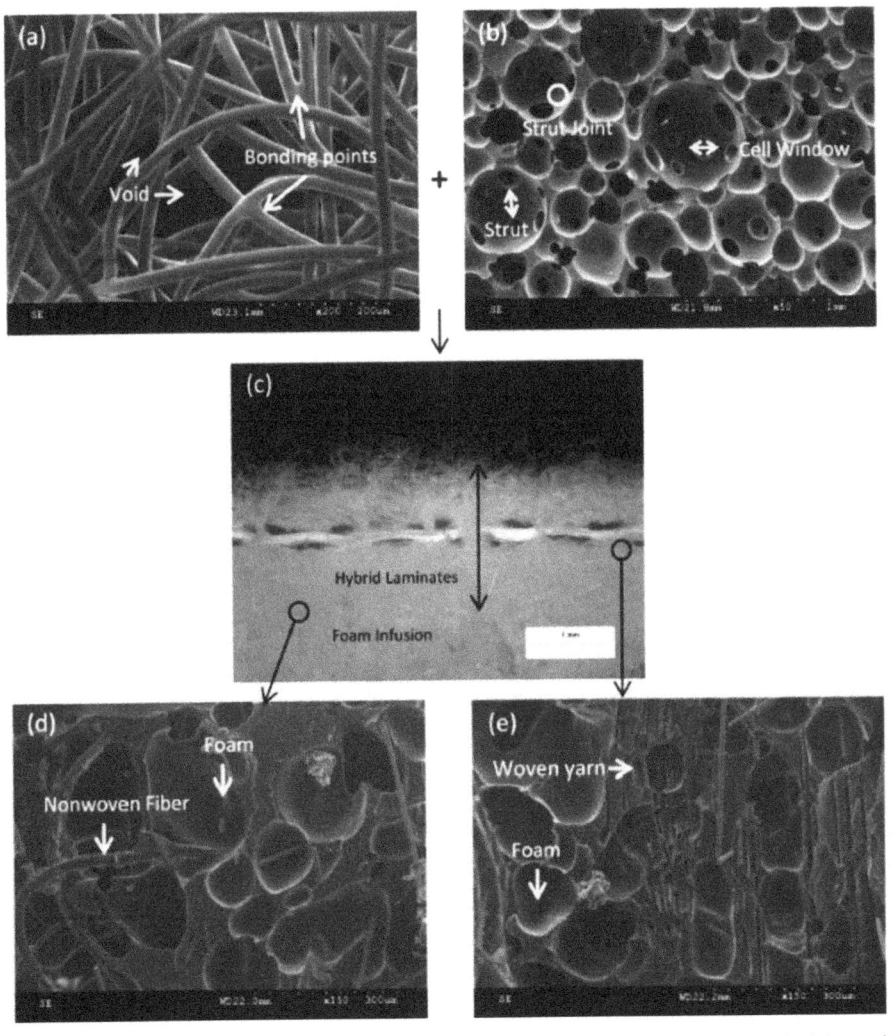

Figure 20. Micrographs and pictures of the flexible PU composite showing (a) the intraply fabric of Kevlar and glass, (b) the neat flexible foam's cellular structure, (c) the hybrid laminated fabric–foam composite, and (d–e) the laminated Kevlar and glass embedded into the foam. Adapted with permission from reference (47). Copyright 2015 Elsevier.

PU's weak fire resistance pushes scientists to develop new materials or chemical routes that improve the flame retardancy and maintain the properties that make PUs so versatile. A study by Zhi et. al. showed successful incorporation of a novel FR based on molybdenum sulfide (MoS_2) and 9,10-dihydro-9-oxo-10-phosphaphenanthrene-10-oxide (DOPO) named MoS_2–DOPO (65). The simple incorporation of 9 g of FR per 100 g of polyol provided an effective improvement in fire-resistant behavior by decreasing the peak heat release rate, total heat release, and maximum smoke density by an average value of 36.5%. Despite provoking a considerable decrease in compressive strength, the tensile strength remained constant regardless of the addition of MoS_2–DOPO. These results demonstrated an effective and facile way to improve the flame retardancy of flexible PU. The FR mechanism can occur mainly in two ways. The first is referred to as the solid-phase mechanism. This type of flame retardancy can be observed through the addition of MoS_2 nanoparticles, which forms a physical barrier over the PU's surface. Hence, it prevents reactive radical species and oxygen from diffusing into the inner layer to propagate the fire. The second way is the gas-phase mechanism, which is observed simultaneously. In this case, the source of P, such as DOPO, releases a PO• species, which is a low-energy radical able to capture highly reactive radicals, such as H• and OH•, derived from the combustion process. Because of the higher energy of the latter species, their reaction with the PU matrix is an exothermic reaction, which promotes the fire. Thus, the PO• prevents that promotion by reacting with these reactive species before reaching the PU's surface and converting them into stable substances. This work is a good example that fully describes the synergy effect between the two FRs and presents an exciting approach for blending nanomaterials within the PU without causing much deterioration of mechanical properties (65). Although PUs possess an inherent problem of flammability, there are several ways in which this issue can be countered, allowing their use as FR material for buildings, houses, and fire suits.

Thermoplastic PU (TPU) is another essential class of elastomeric polymer. Perhaps the greatest advantage of these materials is their melt-processing capabilities, which allows their preparation through many techniques, such as injection molding, blow, extrusion, and compression (66). Furthermore, TPU has an essential set of properties for industrial applications, such as flexibility, elasticity, and resistance to abrasion, chemical attacks, and UV radiation. Hence, it can be found in various products, such as caster wheels, medical equipment, shoes, sporting materials, interior panels in cars, and many others. The chemical structure of TPU is the key element that makes these polymers fall between rigid and flexible PUs, and yet it allows TPU to be manufactured in several ways. The glass-transition temperature, the temperature range at which amorphous or semicrystalline polymer becomes fluid, is a critical parameter for defining the processability and application areas. This rheological behavior occurs because the TPU's chemical structures are composed of polymeric diols such as polyethers, polyesters, and polycarbonates with 1000–3000 monomer units attached. Either aliphatic or aromatic isocyanates can also be employed in TPU. Like flexible and rigid PU, the number of soft segments (SSs) from the polyol compared to the HSs from the isocyanates significantly influences the final properties. The two main synthetic routes for the synthesis of TPU are the one-shot and two-shot methods. As the name suggests, the one-shot method is performed by mixing all the reagents into a single reactor, which is later placed in a mold. It is a facile procedure; however, it leads to random block polymers that may demonstrate lower performance because of high polydispersity, depending on the TPU application. The two-shot technique is also referred to as the prepolymer method. In the first part of this method, a prepolymer is synthesized through the reaction between a diisocyanate and an oligopolyol. The formulation is set to obtain isocyanate as the end groups, followed by chain extender polyols, which are the short-chain diols such as ethylene, diethylene, propylene glycol, or 1,4-butanediol (67). The NCO/OH ratio is extremely relevant for the processing of TPU. When this ratio is lower than 0.96, low-molecular-weight PU is obtained, which may perform poorly because of lacking properties. On the other hand, when the ratio is higher than 1.1, high molecular weight and a high cross-linking degree are obtained. This effect is owned by the higher reactivity of isocyanates in comparison to hydroxyls. However, it can often lead to side reactions and hardships in processing, so an in-depth analysis of the processing is required for optimum production (68).

Another factor that must be considered is the interaction between the HS and the SS. It is a well-established phenomenon that the HSs originate from the urethane linkages that form hydrogen bonding with each other. The SSs are derived from the weak intermolecular interactions between aliphatic and nonpolar domains. Because of this difference, HSs and SSs are insoluble in regular conditions. However, by increasing the temperature, their mixability increases, followed by the formation of a homogenous melt, which allows several processing methods. When the TPU has cooled down, the different domains come back to their previous configuration (67).

Most industries that produce TPUs use petrochemical-based sources because they demand high-molecular-weight polymers such as polyethylene or polypropylene that are often used as diols. However, with the scientific trend of finding new renewable sources for economic and environmental purposes, some research has demonstrated alternative paths for the synthesis of TPU. For example, Alagi et al. (69) used a CO_2-based oligopolymeric diol and a cycloaliphatic diisocyanate, named 4,4'-methylene bis(cyclohexyl isocyanate), for the synthesis of TPU. The product presented high corrosion resistance and thermal response, shape memory, and a broad range of workable mechanical properties that allowed this material to function as an effective hard coating. The properties were controlled by adding different amounts of poly(tetrahydrofuran) diol (PTMEG). The increase in CO_2 content led to rigid carbonate groups that increased the rigidity and glass-transition temperature. This is expected, since the C=O confers an sp^2 electronic configuration, which is planar. Hence, there is a decrease in the mobility of that session that resulted in increased rigidity. Further, the introduction of carbonate groups may form hydrogen bonding with the urethane linkage of other polymeric chains, resulting in a further decrease in mobility.

Conversely, the increase of PTMEG content decreased the glass-transition temperature because it contains large alkyl groups that are more flexible and promote elasticity of the material. Hence, this approach demonstrates a feasible tuning process to control the rigidity and glass-transition temperature of the PU coating, allowing it to cover a wide range of workable temperatures. On top of that, it consumes CO_2, which tackles a critical environmental issue and uses an abundant and low-cost substance as a reagent to obtain high-performance hard coatings to protect several types of substrates. CO_2-based TPU's chemical structure is expressed in Figure 21. Understanding the thermal properties of PU is also an important aspect of defining its applications. In the aforementioned study, thermal gravimetric analysis was performed in the CO_2-based polycarbonate TPU, and a three-step degradation process was observed. The first degradation occurred around 210–250 °C and was related to the decomposition of carbonate units. The second degradation occurred in the range of 270–340 °C, which is a signature range for the disruption of urethane linkages. Finally, the third degradation around 400–450 °C was related to the decomposition of PTMEG segments. Even though the carbonate groups degraded earlier than the urethane linkage, it was observed that there was a smaller amount of toxic fumes produced (70).

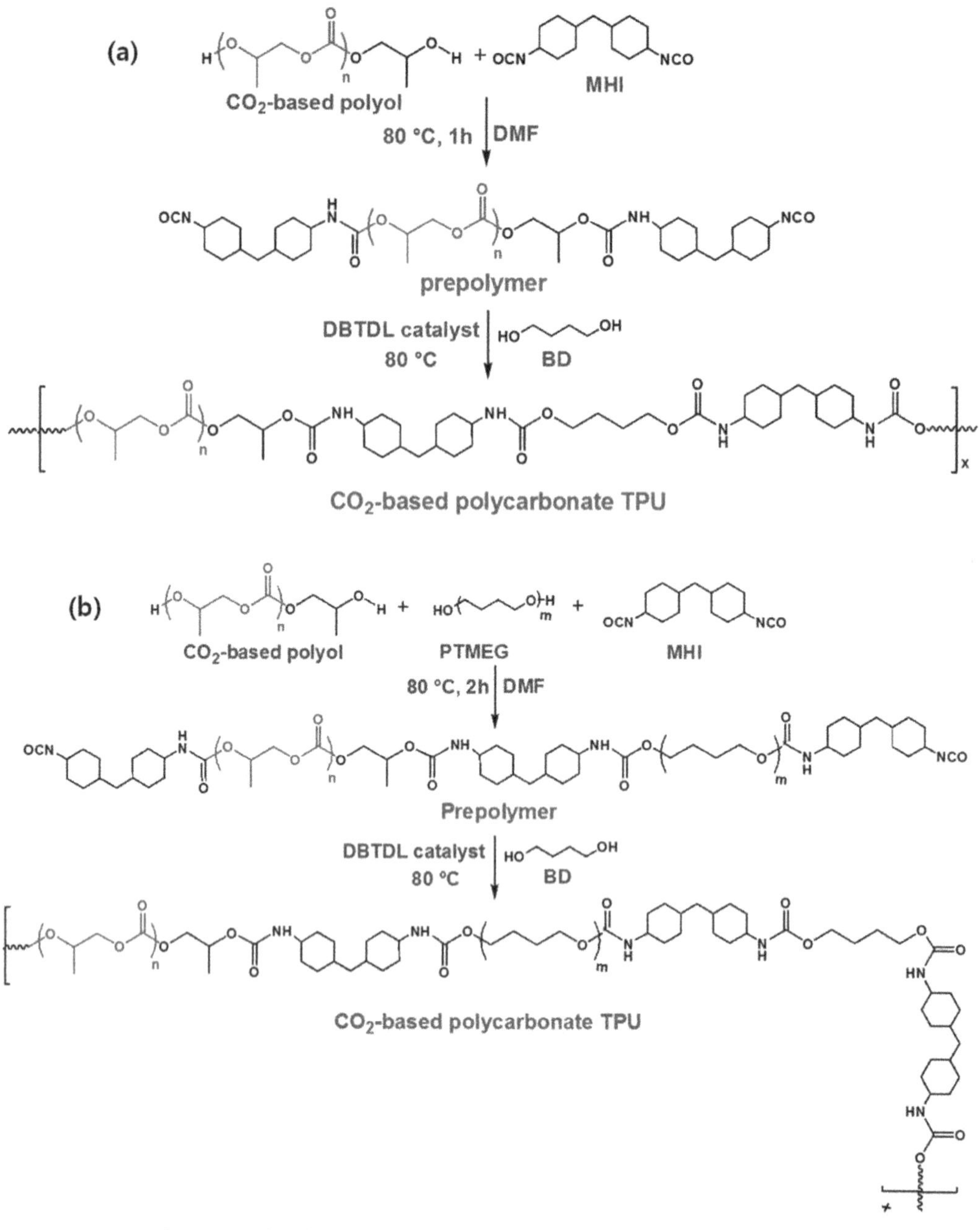

Figure 21. Synthetic procedure for (a) CO$_2$-based TPU and (b) CO$_2$–PTMEG block polymeric TPU. Adapted with permission from reference (69). Copyright 2017 American Chemical Society.

Viable PU applications in the medical field demand high performance and efficiency for their end use, which may often be challenging. One common yet essential concern is finding ways to avoid the adhesion of bacterial or fungal microorganisms onto biomedical devices because this can quickly lead to infection or a response from the body's immunologic system that causes inflammation. An effective strategy to accomplish this goal lies in using zwitterions, which are molecules that contain equivalent numbers of positively and negatively charged functions, such as amino acids. Incorporating chemical segments into these types of molecules helps create antifouling surfaces that prevent undesirable biomaterial adhesion.

As an example, catheters are important medical equipment used to transport nutrients, vitamins and medication, monitor hemodynamics, and remove toxins from the body. Hence, it is of great importance that this device remains clean to prevent the growth of microorganisms. For this reason, a TPU was developed from on a triblock polymer based on 1,4-butanediol, a commercial polyether polyol, and 3-allyloxy-1,2-propanediol. At the same time, the urethane linkage was formed using 4,4′-diisocyanato-methylenedicyclohexae (71). The double side bonds were used to chemically attach a synthesized zwitterion, which contained a quaternary amine for the positive charge and a sulfate group for the negative charge, through a thiol-ene reaction. The overall scheme for the synthesis is provided in Figure 22. Mild conditions were used to synthesize a polymer for a specific and high-performance utilization that showed a viable path for scaling up the process. It was observed that bacterial adhesion was reduced from 40 to 50%, accompanied by good biocompatibility. These results suggest that synthesized TPU has promising short-term uses for medical equipment. Further research on this topic may require the integration of antibacterial properties to prolong the catheter's shelf life.

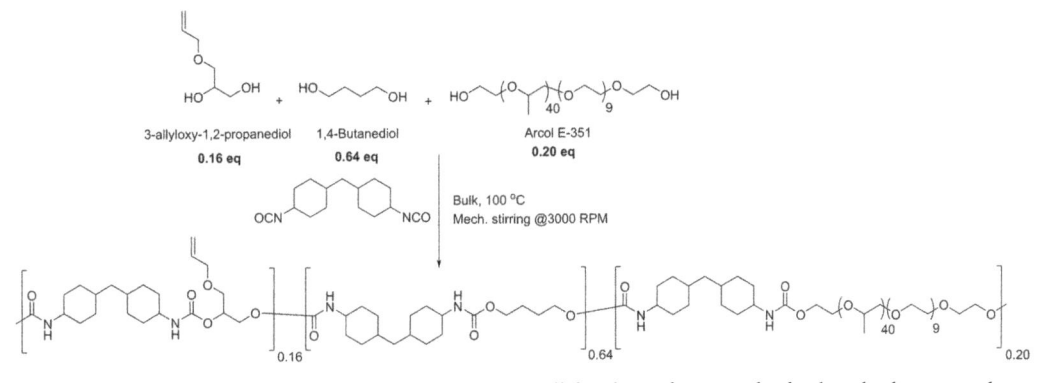

Figure 22. Synthetic procedure for TPU containing an allyl–ether side group for further thiol-ene coupling. Adapted with permission from reference (71). Copyright 2020 American Chemical Society.

A portion of memory shape foams, electronic devices, and artificial heart components such as pacemakers and tubes for hemodialysis are produced using PU ionomers (PUIs). These polymers present ionic groups in their backbones, which aids in improving thermal stability, conductibility, and mechanical properties because of stronger intermolecular interactions and better dispersibility of these polymers in aqueous media. Ionic groups can be introduced into the structure by forming a quaternary amine salt group. This salt can be produced by reacting the N atom in the urethane linkage with salt, followed by the ternization of the sulfur atom. An example from the literature that illustrates this scenario can be seen in Figure 23. A polydioxolane PUI was synthesized by introducing an ionic group using NaH and 1,3-propane sultone (72). The authors studied the effect of increasing ionization levels in the PU's backbone on thermal behavior, mechanical properties, and conductivity, with the latter being the property targeted for improvement.

The glass-transition temperature analysis in this study displayed an incremental increase concerning the ionization level, which went from around 215 to 230 °C. The dynamic mechanical analysis also demonstrated an increase in the modulus and a shift to a higher temperature to reach the plateau zone. The observed effects of increasing glass-transition temperature and the incremental change in mechanical properties in the temperature function were related to the appearance of strong intermolecular ion–ion or ion–dipole interactions after introducing ionic segments. Therefore, many HSs were formed because of the increase in these strong intermolecular interactions between the ionic groups of polymeric chains. Such interaction is also reflected with the increase in the microphase separation with the SSs. This phenomenon correlates to the PUI trend to present a two-phase character for tests, such as dynamic mechanical analysis and differential scanning calorimetry. The ionic conductibility increased up to 90 °C, reaching higher values for samples with a larger number of ionic groups in the chain. The reason for this is based on the mechanism for ion transportation for solid polymer electrolytes. The SSs present more mobility, and the free volume further increases when exposed to higher temperatures. It allows the ions to move more freely, hence improving the conductibility. Through these property analyses, it is notable that PUI can be used in electronic devices, allowing for different procedures and offering tunable properties.

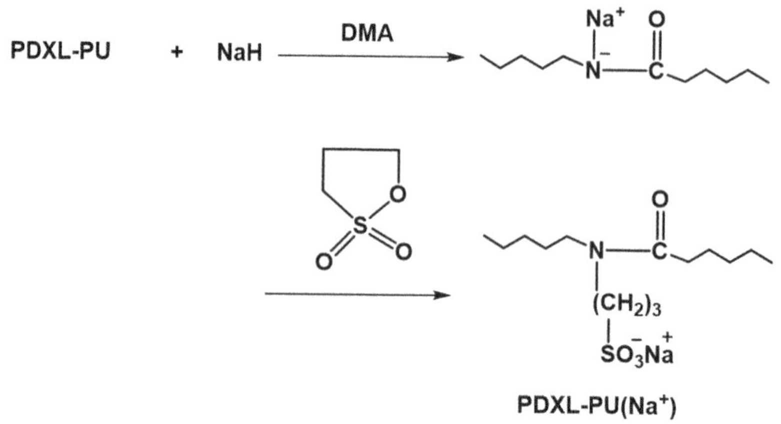

Figure 23. Synthesis of polydioxolane PUI. Adapted with permission from reference (72). Copyright 2001 Springer Nature.

Chen et al. used a similar approach by synthesizing PUI for thermal energy storage (73). A polyethylene glycol (PEG) PU was used because of its phase change capabilities. However, to further improve this property, higher levels of crystallinity in the SSs of PEG are required. The incorporation of ionic groups into the isocyanate's HSs results in a high microphase separation between the two domains. It allows the SSs to crystallize more easily because of lesser dissolution of asymmetric hard domains into the soft domains. Furthermore, the PEG PU is thermoplastic and can be easily reprocessed.

As mentioned previously, PUs are employed as components for medical or electronic devices. However, in these two sectors, they require various properties to become fully effective, such as conductibility, hydrophobicity, thermal stability, appreciable mechanical properties, and the self-healing factor. These properties are the main features that mimic human skin. Even though it is challenging to obtain these properties at once, Ying et. al. successfully synthesized a PU that satisfied those conditions and can be used as an electronic skin (e-skin) (74). This leads to several specific applications for PU, such as intelligent prosthetics, interactions between humans and computers, and medical rehab. High-performance materials should be used to introduce the desired properties adequately. Consequently, hydroxyl-terminated polybutadiene was used as the soft and hydrophobic segment by Ying et al. The self-healing property was introduced by integrating a dynamic disulfide bond, which can be broken through mechanical stress, cut, or tear and can be recombined by applying thermal curing. For this purpose, bis(4-hydroxyphenyl)-disulfide was used as a chain extender. The effect of self-healing arises from the ease of reversibility of disulfide bonds (S–S). When mechanical strain of some sort is exerted on the material, the S–S bond is disrupted. However, the disulfide bridge can be quickly reestablished after a thermal healing process is performed. The self-healing properties can also be introduced in other ways, such as Diels–Alder reactions, hydrogen bonding, the complex interaction between metal–organic ligand, and urea chemistry. (75–80). The last important factor in Ying et al.'s work was to ensure the microphase separation between the soft and rigid domains. This condition promotes mechanical properties, such as elasticity and toughness, in the PU, which is different from other homogenous polymers like polydimethylsiloxane. For the enhancement of PU matrix conduction, a conductive filler based on a Ga–In–Sn metal alloy was used. The PU composite synthesis was performed in two different ways, using the one-pot and the two-pot methods, for comparison. The overall schematics of the e-skin-based PU can be seen in Figure 24.

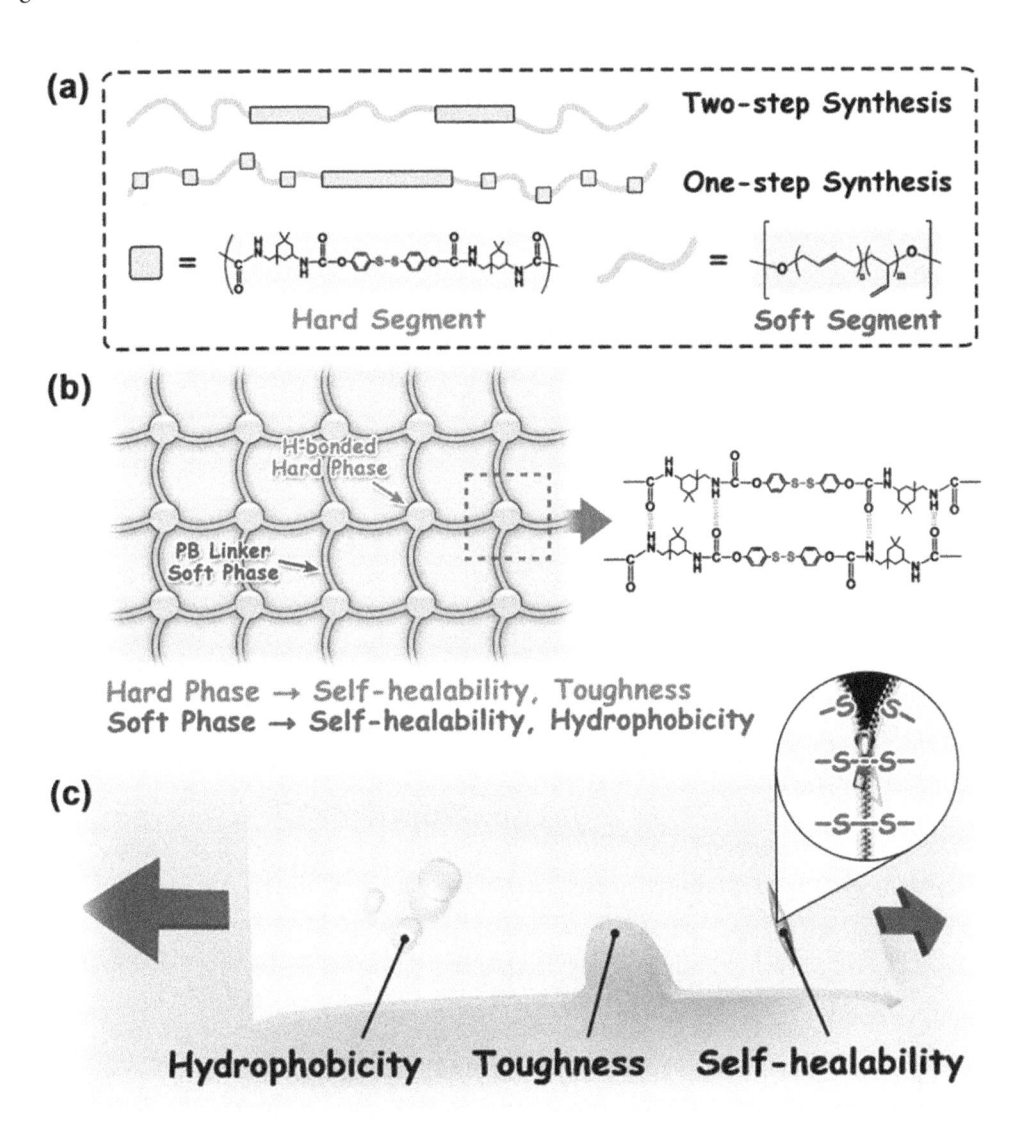

Figure 24. (a) Chemical structure of the PU matrix. (b) Representation of the network PU structure showing the interaction between the SSs and HSs. (c) Scheme for properties of the PU, such as hydrophobicity (left), arched notch that does not propagate (middle), and self-healing properties derived from the disulfide bond's dynamic (right). Adapted with permission from reference (74). Copyright 2020 American Chemical Society.

The conduction properties were analyzed through several tests and showed highly effective values, which suggest that this PU composite is suitable for e-skin applications. After cutting the sample and performing thermal healing, both the conductive and mechanical properties remained almost the same. The mechanical properties had a 93% recovery compared with the initial values, showing excellent self-healing properties. Figure 25a demonstrates the results obtained from the reestablished conductibility after cutting. Figure 25b shows that resistance was kept constant as a function of the soaking time. Figure 25c shows the decrease of electrical resistance as the PU underwent thermal healing and reached a resistance of around 7.5 Ω. Figure 25d displays the sample's electrical resistance before and after healing as a function of strain and reveals that it is slightly increased after the healing process.

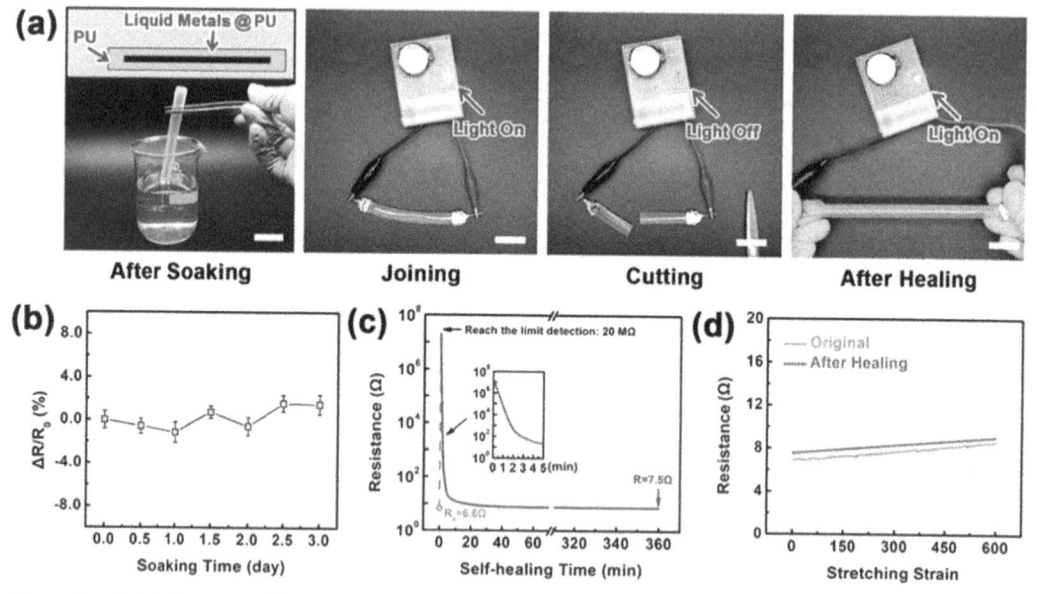

Figure 25. (a) Schematic of the PU composite and the reestablishment of conductibility after the sample was cut and thermally healed. (b) Effect of soaking time over resistance. (c) Conductibility of the sample after the cutting. (d) Effect of strain on the electrical resistance, comparing the original and the sample after the healing process. Adapted with permission from reference (74). Copyright 2020 American Chemical Society.

Another aspect that defines some of the PU properties is the control of hydrophilic and hydrophobic segments, which play an important role in applications related to the biomedical area, for instance. Materials used to transport body fluids such as blood and protein solutions may suffer from deposition of these biomaterials that eventually lead to coagulation, which is concerning in cardiovascular applications (81, 82). After long exposure to blood, PU can harden and lose its surface properties (83). To decrease the adhesion with blood, Lin et al. proposed the synthesis of a PU that contained hydrophilic and hydrophobic domains (84). The hydrophilic segments were derived from poly(tetramethylene glycol), whereas the hydrophobic domain was derived from polydimethylsiloxane. The combination of these polyols with different interactions with water led to microphase separation, which turned out to be an important factor for decreased adsorption of platelets. This occurs because a surface with hydrophilic and hydrophobic domains enables the protein adsorption and distribution while inhibiting the adsorption of platelets that could activate thrombosis (85, 86). Hence, controlling the microphase separation through hydrophobicity can lead to surface properties that prevent the adhesion of biological fluids.

Adhesives, sealants, binders, and coatings play a significant role in the PU market worldwide. Their uses in paints for cars, adhesives for wood composites, connections, and, most frequently, for anticorrosion coatings are extremely important for the economy of any country. The damage caused by corrosion in containers, ships, platforms, and any other marine or high-sea equipment accounts for an average of 3.5% of the gross domestic product values globally. This huge economic loss encourages the development of coatings to effectively protect a surface from corrosion and other external agents. This type of application requires proper adhesion, chemical resistance, effective drying, resistance to scratching, and flexibility even at lower temperatures (7, 87).

Waterborne PU (WPU) emerged as a promising candidate because of its high efficiency, relatively facile production, and ecofriendliness. International legislation banned the use of chlorofluorocarbons as blowing agents, and the amount of volatile organic compounds allowed for the formulation of polymers was drastically decreased in general. These new conditions led the industry and scientific community to reshape its experimental design to find new routes to obtain paints, adhesives, and coatings free of these components. Based on that, WPU was developed and had the advantage of a relatively constant viscosity regardless of the PU's molecular weight, which simplifies the processing and drying of a coating. The PU dispersions generally consist of a colloidal system in which carboxylic acids (anionic WPU) or quaternary amines (cationic WPU) are introduced into the PU's backbone. These groups are then neutralized with salts, promoting better interaction with water and yielding a stable suspension.

Generally, a stable WPU suspension has small particles. Various methods can be used to synthesize WPU, such as use of a prepolymer emulsifier, the acetone process, melt–dispersion, self-dispersion of solids, and the ketamine/ketazine process. Most of these methods emphasize the use of no or low quantities of organic solvents to perform the synthesis, which is beneficial not only from the environmental perspective but also economically. Changing the reaction environment from nonpolar (organic) to polar (aqueous) carries inherent challenges that can be handled by introducing internal emulsifiers, which are the monomers with polar pendant groups. They bond to the polymeric chain, which leads to an overall increase in hydrophilicity of the polymer and therefore makes it more dispersible in water. However, some parameters, such as type and percentage of internal emulsifier, must be taken into consideration. Even though a more hydrophilic WPU would form a smaller particle size and lead to coatings that are more homogenous, they sometimes cause premature oxidation and lower resistance to moisture. Thus, scientists should obtain the optimal amount of internal emulsifier to increase hydrophilicity and make it environmentally friendly while decreasing the cost by using less organic solvent without compromising the properties of WPU coatings. Wang et al. developed an ecofriendly and antimicrobial WPU (88). It was synthesized by forming a prepolymer first based on polycarbonate diol and isophorone diisocyanate. Then, 3-dimethylamino-1,2-propanediol was used as a chain extender, along with the introduction of a pending tertiary amine into the PU's backbone. Finally, two alkyl bromine-based compounds were used to perform the quaternization of the amine. This step is crucial for WPU applications because it makes the PU dispersible in water and also imparts antimicrobial properties. The positive charge of the amine salts can neutralize some bacteria's cellular walls and kills them. The scheme for this process is shown in Figure 26.

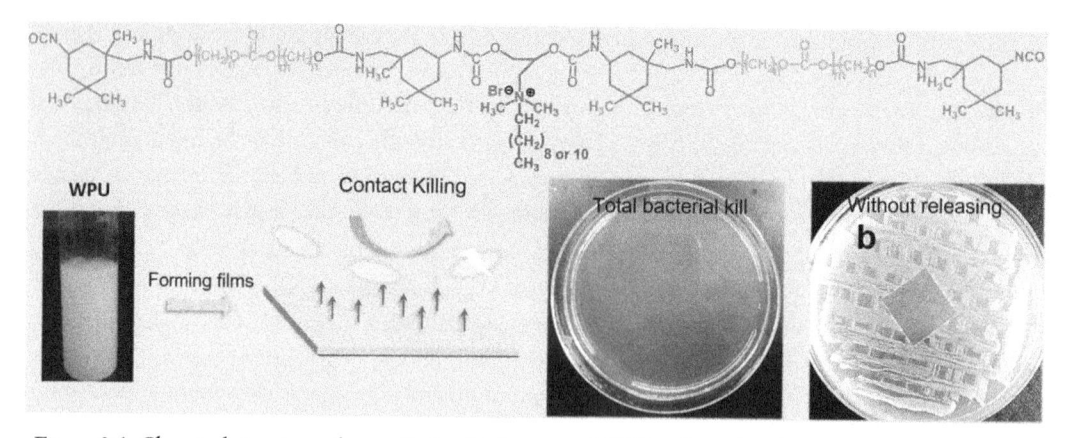

Figure 26. Chemical structure of an antimicrobial cationic WPU. Adapted with permission from reference (88). Copyright 2020 American Chemical Society.

To be suitable for large-scale application as coatings, the original polymer should provide satisfactory adhesion, hardness, tensile strength, and overall resistance against water, radiation, and chemical agents. The adhesion to a substrate relies on an effective drying process, which prevents swelling or peeling issues. Hardness and tensile strength are essential to protect the underlying material against mechanical forces. In some cases, the polymer also acts as the connective tissue between two substrates. The resistance against external agents can be introduced by creating a packed and tortuous structure that blocks the passage of moisture, salts, and radiation. All these properties can be effectively assimilated by the development of a cross-linked WPU structure. Hu et. al. (89) obtained drastic improvements in the synthesized cross-linked WPU compared to a conventional WPU, such as an increase in tensile strength from 0.43 to 6.47 MPa, an increase in hardness from 59 to 73 MPa, and a decrease in the absorption of water from 200 to around 20%. This behavior occurred due to the formation of a covalently connected structure, with the tortuous path in the WPU that prevented water from permeating through the polymer, thereby making it applicable for coatings or adhesives. The synthesis of the WPU was performed by using polypropylene glycol and isophorone diisocyanate. The structure with pending groups that can cross-link was obtained by synthesizing two chain extenders: one containing C–C triple bonds and the other diazide groups. The overall synthetic process and film-forming mechanism can be seen in Figures 27 and 28, respectively. Hu et al.'s study demonstrated that positively packed polymeric structures are effective for protective coatings. It also opens up many different possibilities for new synthetic routes and reagents that can be employed as anticorrosion coatings. Thus, there are many applications of PUs due to their wide range of properties obtained from the starting materials, synthetic routes, and processing.

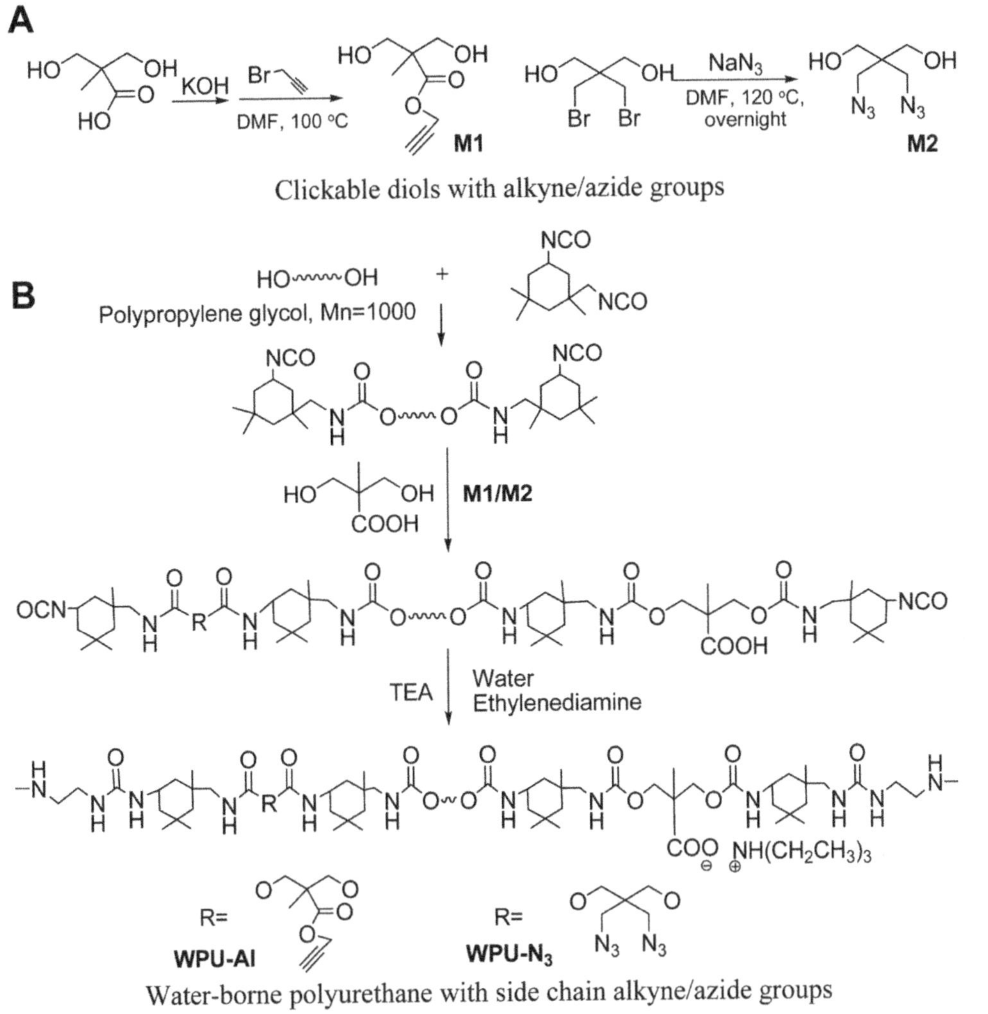

Figure 27. Synthetic process of (a) diol monomers for click reaction represented by M1 and M2. (b) Synthetic route for the synthesis of WPU. Adapted with permission from reference (89). Copyright 2016 American Chemical Society.

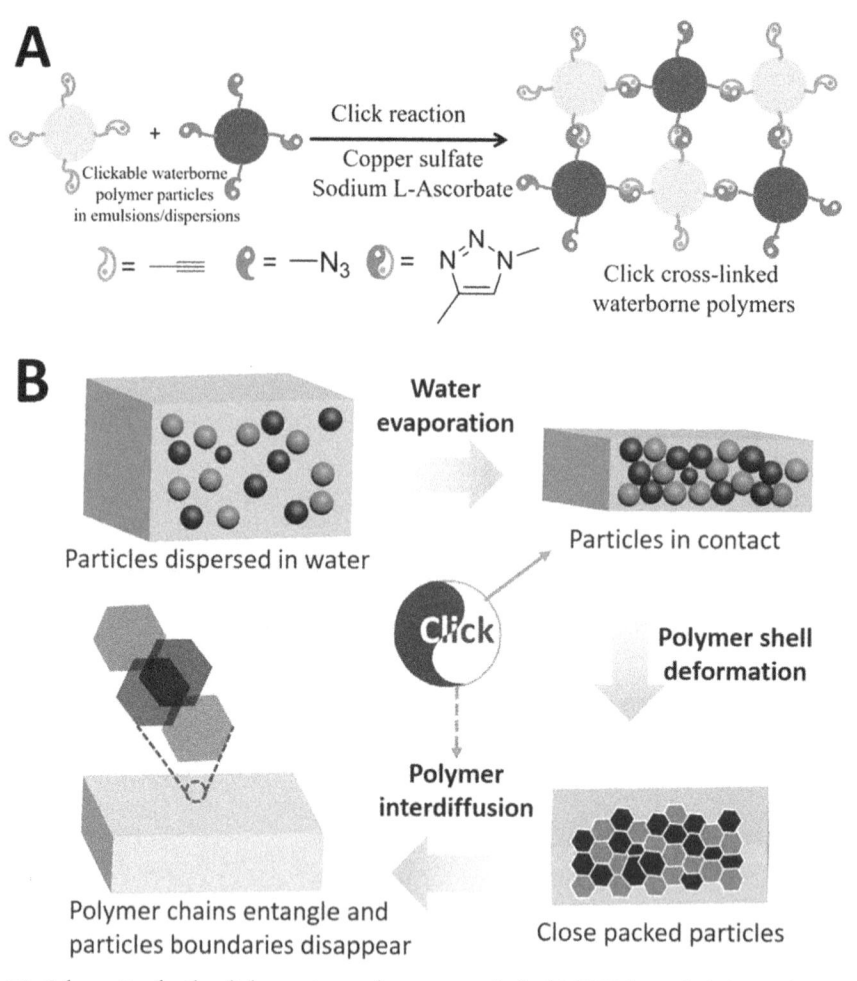

Figure 28. Schematics for the click reaction to form a cross-linked WPU through (a) acetylene and azo groups and (b) diffusion of microparticles for the formation of protective films. Adapted with permission from reference (89). Copyright 2016 American Chemical Society.

Conclusion and Future Aspects

The quick establishment of a solid PU industry made PU one of the most versatile polymers because of its wide range of properties. PU affords great comfort for humankind, since it is a core material in a wide range of products such as beds, furniture, dampers, and soles for shoes. It also helps save tremendous energy due to thermal insulation properties that drastically decrease the energy consumption of heaters and air conditioners. Similarly, the incorporation of PU in automobiles makes them efficient. It reduces fuel consumption and enhances safety features because of its capability to absorb mechanical energy. Despite these impressive properties and long-term use in our daily lives, the PU industry had to be reshaped due to increasing environmental concerns. This imposed new guidelines on the sectors that use PU, which led to novel biobased materials that are being challenged to surpass their petrochemical-based counterparts. Research has been showing that the implementation of both bio- and petrochemical-based materials is providing better results. The range of possible applications of PUs is incredibly vast due to the availability of different types of PU, such as rigid, flexible, thermoplastic, ionomer, and waterborne. Each type, despite their particularities and good performance, still has ample room for improvement. In conclusion, we expect that this book will provide readers with plenty of knowledge to tackle the current issues and help develop novel approaches and materials within PU chemistry.

Acknowledgments

The authors wish to thank Ms. Anjali Gupta from Pittsburg High School, Pittsburg, Kansas for their help in drawing some Figures.

References

1. Ionescu, M. *Chemistry and Technology of Polyols for Polyurethanes*; Ionescu, M., Ed.; Rapra Technology: Shawbury, UK, 2005.
2. Bayer, O. Das Di-Isocyanat-Polyadditionsverfahren-(Polyurethane). *Angew. Chemie* **1947**, *59*, 257–288.
3. Akindoyo, J. O.; Beg, M. D. H.; Ghazali, S.; Islam, M. R.; Jeyaratnam, N.; Yuvaraj, A. R. Polyurethane Types, Synthesis and Applications - A Review. *RSC Adv.* **2016**, *6*, 114453–114482.
4. Sharmin, E.; Zafar, F. *Polyurethane*; Sharmin, E., Zafar, F. , Eds.; IntechOpen Limited: London, 2012.
5. Singh, S. N. *Blowing Agents for Polyurethane Foams*; RAPRA Technology Limited; Rapra Technology Ltd.: Hamburg, Germany, 2001.
6. Moore, M. G. *The Economic Benefits of the U.S. Polyurethanes Industry*; American Chemistry Council: Washington, DC, 2020.
7. Szycher M. *Szycher's Handbook of Polyurethanes*; Szycher, M., Ed.; CRC Press: New York, 2012.
8. Fink, J. K. *Reactive Polymers: Fundamentals and Applications: A Concise Guide to Industrial Polymers*; Fink, J. K., Ed.; William Andrew/Elsevier: New York, 2017.
9. Babb, D. A. Polyurethanes from Renewable Resources. In *Synthetic Biodegradable Polymers*; Babb, D. A., Ed.; Springer, Verlag Berlin: Verlag, 2011; pp 315–360.
10. Li, Y. Y.; Luo, X.; Hu, S. Introduction to Bio-Based Polyols and Polyurethanes. In *Bio-Based Polyols and Polyurethanes*; Li, Y., Luo, X., Hu, S., Eds.; Springer: New York, 2015; pp 1–13.
11. Köhler, T.; Gutacker, A.; Mejiá, E. Industrial Synthesis of Reactive Silicones: Reaction Mechanisms and Processes. *Org. Chem. Front.* **2020**, *7*, 4108–4120.
12. Li, Y.; Sun, X. S. Synthesis and Characterization of Acrylic Polyols and Polymers from Soybean Oils for Pressure-Sensitive Adhesives. *RSC Adv.* **2015**, *5*, 44009–44017.
13. Palanisamy, A.; Rao, B. S.; Mehazabeen, S. Diethanolamides of Castor Oil as Polyols for the Development of Water-Blown Polyurethane Foam. *J. Polym. Environ.* **2011**, *19*, 698.
14. Challoner, K. R.; McCarron, M. M. Castor Bean Intoxication. *Ann. Emerg. Med.* **1990**, *19*, 1177–1183.
15. Bhoyate, S.; Ionescu, M.; Kahol, P. K.; Gupta, R. K. Castor-Oil Derived Nonhalogenated Reactive Flame-Retardant-Based Polyurethane Foams with Significant Reduced Heat Release Rate. *J. Appl. Polym. Sci.* **2019**, *136*, 1–7.
16. Yeadon, D. A.; McSherry, W. F.; Goldblatt, L. A. Preparations and Properties of Castor Oil Urethane Foams. *J. Am. Oil Chem. Soc.* **1959**, *36*, 16–20.

17. Ehrlich, A.; Smith, M. K.; Patton, T. C. Castor Polyols for Urethane Foams. *J. Am. Oil Chem. Soc.* **1959**, *36*, 149–154.
18. John, J.; Bhattacharya, M.; Turner, R. B. Characterization of Polyurethane Foams from Soybean Oil. *J. Appl. Polym. Sci.* **2002**, *86*, 3097–3107.
19. Gawryla, M. D.; Nezamzadeh, M.; Schiraldi, D. A. Foam-Like Materials Produced from Abundant Natural Resources. *Green Chem.* **2008**, *10*, 1078–1081.
20. Ramanujam, S.; Zequine, C.; Bhoyate, S.; Neria, B.; Kahol, P.; Gupta, R. Novel Biobased Polyol Using Corn Oil for Highly Flame-Retardant Polyurethane Foams. *C* **2019**, *5*, 13.
21. De Souza, V. H. R.; Silva, S. A.; Ramos, L. P.; Zawadzki, S. F. Synthesis and Characterization of Polyols Derived from Corn Oil by Epoxidation and Ozonolysis. *J. Am. Oil Chem. Soc.* **2012**, *89*, 1723–1731.
22. Xia, Y.; Zhang, Z.; Kessler, M. R.; Brehm-Stecher, B.; Larock, R. C. Antibacterial Soybean-Oil-Based Cationic Polyurethane Coatings Prepared from Different Amino Polyols. *ChemSusChem* **2012**, *5*, 2221–2227.
23. Yang, Z.; Feng, Y.; Liang, H.; Yang, Z.; Yuan, T.; Luo, Y.; Li, P.; Zhang, C. A Solvent-Free and Scalable Method to Prepare Soybean-Oil-Based Polyols by Thiol-Ene Photo-Click Reaction and Biobased Polyurethanes Therefrom. *ACS Sustain. Chem. Eng.* **2017**, *5*, 7365–7373.
24. Narine, S. S.; Kong, X.; Bouzidi, L.; Sporns, P. Physical Properties of Polyurethanes Produced from Polyols from Seed Oils: II. Foams. *J. Am. Oil Chem. Soc.* **2007**, *84*, 65–72.
25. Verlag, G. Cargill's BiOH Polyols Business Opens Manufacturing Site in Brasil. *PU Mag. Int.* **2007**, *26*, 12.
26. Effendi, A.; Gerhauser, H.; Bridgwater, A. V. Production of Renewable Phenolic Resins by Thermochemical Conversion of Biomass: A Review. *Renew. Sustain. Energy Rev.* **2008**, *12*, 2092–2116.
27. García, D.; Bustamante, F.; Villa, A. L.; Lapuerta, M.; Alarcón, E. Oxyfunctionalization of Turpentine for Fuel Applications. *Energy & Fuels* **2020**, *34*, 579–586.
28. Lora, J. Industrial Commercial Lignins: Sources, Properties and Applications. In *Monomers, Polymers and Composites from Renewable Resources*; Elsevier: New York, 2008; pp 201–224.
29. Josefsson, T.; Lennholm, H.; Gellerstedt, G. Steam Explosion of Aspen Wood. Characterisation of Reaction Products. *Holzforschung* **2002**, *56*, 289–297.
30. Tokay, B. A. Biomass Chemicals. In *Ullmann's Encyclopedia of Industrial Chemistry*; Major Reference Works; Wiley-VCH Verlag GmbH & Co. KGaA: Weinheim, Germany, 2000.
31. Suresh, K. I. Rigid Polyurethane Foams from Cardanol: Synthesis, Structural Characterization, and Evaluation of Polyol and Foam Properties. *ACS Sustain. Chem. Eng.* **2013**, *1*, 232–242.
32. Petrović, Z. S.; Zhang, W.; Javni, I. Structure and Properties of Polyurethanes Prepared from Triglyceride Polyols by Ozonolysis. *Biomacromolecules* **2005**, *6*, 713–719.
33. Petrović, Z. S. Polyurethanes from Vegetable Oils. *Polym. Rev.* **2008**, *48*, 109.
34. Petrović, Z. S.; Guo, A.; Javni, I.; Cvetković, I.; Hong, D. P. Polyurethane Networks from Polyols Obtained by Hydroformylation of Soybean Oil. *Polym. Int.* **2008**, *57*, 275–281.
35. Hoyle, C. E.; Lee, T. Y.; Roper, T. Thiol-Enes: Chemistry of the Past with Promise for the Future. *J. Polym. Sci., Part A: Polym. Chem.* **2004**, *42*, 5301–5338.

36. Ranaweera, C. K.; Ionescu, M.; Bilic, N.; Wan, X.; Kahol, P. K.; Gupta, R. K. Biobased Polyols Using Thiol-Ene Chemistry for Rigid Polyurethane Foams with Enhanced Flame-Retardant Properties. *J. Renew. Mater.* **2017**, *5*, 1–12.
37. Islam, M. R.; Beg, M. D. H.; Jamari, S. S. Development of Vegetable-Oil-Based Polymers. *J. Appl. Polym. Sci.* **2014**, *13*, 9016–9028.
38. Thomson, T. *Polyurethanes as Specialty Chemicals: Principles and Applications*; CRC Press: New York, United States, 2004.
39. Bähr, M.; Bitto, A.; Mülhaupt, R. Cyclic Limonene Dicarbonate as a New Monomer for Non-Isocyanate Oligo- and Polyurethanes (NIPU) Based upon Terpenes. *Green Chem.* **2012**, *14*, 1447–1454.
40. Jarfelt, U.; Ramnäs, O. Thermal Conductivity of Polyurethane Foam - Best Performance. In *10th International Symposium on District Heating and Cooling*; Chalmers University of Technology Goteborg: Goteborg, Sweden, 2006.
41. Harikrishnan, G.; Singh, S. N.; Kiesel, E.; Macosko, C. W. Nanodispersions of Carbon Nanofiber for Polyurethane Foaming. *Polymer (Guildf).* **2010**, *51*, 3349–3353.
42. Lorenzetti, A.; Roso, M.; Bruschetta, A.; Boaretti, C.; Modesti, M. Polyurethane-Graphene Nanocomposite Foams with Enhanced Thermal Insulating Properties. *Polym. Adv. Technol.* **2016**, *27*, 303–307.
43. Kuranska, M.; Prociak, A. Porous Polyurethane Composites with Natural Fibres. *Compos. Sci. Technol.* **2012**, *72*, 299–304.
44. Septevani, A. A.; Evans, D. A. C.; Annamalai, P. K.; Martin, D. J. The Use of Cellulose Nanocrystals to Enhance the Thermal Insulation Properties and Sustainability of Rigid Polyurethane Foam. *Ind. Crops Prod.* **2017**, *107*, 114–121.
45. Aloui, H.; Khwaldia, K.; Hamdi, M.; Fortunati, E.; Kenny, J. M.; Buonocore, G. G.; Lavorgna, M. Synergistic Effect of Halloysite and Cellulose Nanocrystals on the Functional Properties of PVA Based Nanocomposites. *ACS Sustain. Chem. Eng.* **2016**, *4*, 794–800.
46. Cinelli, P.; Anguillesi, I.; Lazzeri, A. Green Synthesis of Flexible Polyurethane Foams from Liquefied Lignin. *Eur. Polym. J.* **2013**, *49*, 1174–1184.
47. Yan, R.; Wang, R.; Lou, C.-W.; Huang, S.-Y.; Lin, J.-H. Quasi-Static and Dynamic Mechanical Responses of Hybrid Laminated Composites Based on High-Density Flexible Polyurethane Foam. *Compos. Part B Eng.* **2015**, *83*, 253–263.
48. Singhal, P.; Small, W.; Cosgriff-Hernandez, E.; Maitland, D. J.; Wilson, T. S. Low Density Biodegradable Shape Memory Polyurethane Foams for Embolic Biomedical Applications. *Acta Biomater.* **2014**, *10*, 67–76.
49. Hodlur, R. M.; Rabinal, M. K. Self Assembled Graphene Layers on Polyurethane Foam as a Highly Pressure Sensitive Conducting Composite. *Compos. Sci. Technol.* **2014**, *90*, 160–165.
50. Zhang, L.; Jeon, H. K.; Malsam, J.; Herrington, R.; Macosko, C. W. Substituting Soybean Oil-Based Polyol into Polyurethane Flexible Foams. *Polymer (Guildf).* **2007**, *48*, 6656–6667.
51. Guo, A.; Javni, I.; Petrovic, Z. Rigid Polyurethane Foams Based on Soybean Oil. *J. Appl. Polym. Sci.* **2000**, *77*, 467–473.
52. Hu, Y. H.; Gao, Y.; Wang, D. N.; Hu, C. P.; Zu, S.; Vanoverloop, L.; Randall, D. Rigid Polyurethane Foam Prepared from a Rape Seed Oil Based Polyol. *J. Appl. Polym. Sci.* **2002**, *84*, 591–597.

53. Chian, K. S.; Gan, L. H. Development of a Rigid Polyurethane Foam from Palm Oil. *J. Appl. Polym. Sci.* **1998**, *68*, 509–515.
54. Li, H.; Sun, J.-T.; Wang, C.; Liu, S.; Yuan, D.; Zhou, X.; Tan, J.; Stubbs, L.; He, C. High Modulus, Strength, and Toughness Polyurethane Elastomer Based on Unmodified Lignin. *ACS Sustain. Chem. Eng.* **2017**, *5*, 7942–7949.
55. Chang, K.-J.; Wang, Y.-Z.; Peng, K.-C.; Tsai, H.-S.; Chen, J.-R.; Huang, C.-T.; Ho, K.-S.; Lien, W.-F. Preparation of Silica Aerogel/Polyurethane Composites for the Application of Thermal Insulation. *J. Polym. Res.* **2014**, *21*, 338.
56. Nazeran, N.; Moghaddas, J. Synthesis and Characterization of Silica Aerogel Reinforced Rigid Polyurethane Foam for Thermal Insulation Application. *J. Non. Cryst. Solids* **2017**, *461*, 1–11.
57. M. de Souza, F.; Choi, J.; Bhoyate, S.; Kahol, P. K.; Gupta, R. K. Expendable Graphite as an Efficient Flame-Retardant for Novel Partial Bio-Based Rigid Polyurethane Foams. *C* **2020**, *6*, 27.
58. Weil, E. D.; Levchik, S. V. Phosphorus Flame Retardants. In *Kirk-Othmer Encyclopedia of Chemical Technology*; Major Reference Works; John Wiley & Sons: Hoboken, New Jersey, 2017; pp 1–34.
59. Bann, B.; Miller, S. A. Melamine and Derivatives of Melamine. *Chem. Rev.* **1958**, *58*, 131–172.
60. Wang, X.; Kalali, E. N.; Wan, J.-T.; Wang, D.-Y. Carbon-Family Materials for Flame Retardant Polymeric Materials. *Prog. Polym. Sci.* **2017**, *69*, 22–46.
61. Liu, Y.; Wang, Q. Melamine Cyanurate-Microencapsulated Red Phosphorus Flame Retardant Unreinforced and Glass Fiber Reinforced Polyamide 66. *Polym. Degrad. Stab.* **2006**, *91*, 3103–3109.
62. Chen, X.-Y.; Huang, Z.-H.; Xi, X.-Q.; Li, J.; Fan, X.-Y.; Wang, Z. Synergistic Effect of Carbon and Phosphorus Flame Retardants in Rigid Polyurethane Foams. *Fire Mater.* **2018**, *42*, 447–453.
63. Zheng, X.; Wang, G.; Xu, W. Roles of Organically-Modified Montmorillonite and Phosphorous Flame Retardant During the Combustion of Rigid Polyurethane Foam. *Polym. Degrad. Stab.* **2014**, *101*, 32–39.
64. Zhang, Y.; Yu, B.; Wang, B.; Liew, K. M.; Song, L.; Wang, C.; Hu, Y. Highly Effective P-P Synergy of a Novel DOPO-Based Flame Retardant for Epoxy Resin. *Ind. Eng. Chem. Res.* **2017**, *56*, 1245–1255.
65. Zhi, M.; Liu, Q.; Zhao, Y.; Gao, S.; Zhang, Z.; He, Y. Novel MoS2–DOPO Hybrid for Effective Enhancements on Flame Retardancy and Smoke Suppression of Flexible Polyurethane Foams. *ACS Omega* **2020**, *5*, 2734–2746.
66. Claeys, B.; Vervaeck, A.; Hillewaere, X. K. D.; Possemiers, S.; Hansen, L.; De Beer, T.; Remon, J. P.; Vervaet, C. Thermoplastic Polyurethanes for the Manufacturing of Highly Dosed Oral Sustained Release Matrices via Hot Melt Extrusion and Injection Molding. *Eur. J. Pharm. Biopharm.* **2015**, *90*, 44–52.
67. Datta, J.; Kasprzyk, P. Thermoplastic Polyurethanes Derived from Petrochemical or Renewable Resources: A Comprehensive Review. *Polym. Eng. Sci.* **2018**, *58*, E14–E35.
68. Drobny, J. G. *Handbook of Thermoplastic Elastomers*, 2nd ed.; Elsevier: New York, 2014.

69. Alagi, P.; Ghorpade, R.; Choi, Y. J.; Patil, U.; Kim, I.; Baik, J. H.; Hong, S. C. Carbon Dioxide-Based Polyols as Sustainable Feedstock of Thermoplastic Polyurethane for Corrosion-Resistant Metal Coating. *ACS Sustain. Chem. Eng.* **2017**, *5*, 3871–3881.
70. Cyriac, A.; Lee, S. H.; Varghese, J. K.; Park, J. H.; Jeon, J. Y.; Kim, S. J.; Lee, B. Y. Preparation of Flame-Retarding Poly(Propylene Carbonate). *Green Chem.* **2011**, *13*, 3469–3475.
71. Nikam, S. P.; Chen, P.; Nettleton, K.; Hsu, Y.-H.; Becker, M. L. Zwitterion Surface-Functionalized Thermoplastic Polyurethane for Antifouling Catheter Applications. *Biomacromolecules* **2020**, *21*, 2714–2725.
72. Zhu, W.; Wang, X.; Yang, B.; Wang, L.; Tang, X.; Yang, C. Synthesis and Characterization of Polydioxolane Polyurethane Ionomer. *J. Mater. Sci.* **2001**, *36*, 5137–5141.
73. Chen, K.; Liu, R.; Zou, C.; Shao, Q.; Lan, Y.; Cai, X.; Zhai, L. Linear Polyurethane Ionomers as Solid–Solid Phase Change Materials for Thermal Energy Storage. *Sol. Energy Mater. Sol. Cells* **2014**, *130*, 466–473.
74. Ying, W. Bin; Yu, Z.; Kim, D. H.; Lee, K. J.; Hu, H.; Liu, Y.; Kong, Z.; Wang, K.; Shang, J.; Zhang, R.; Zhu, J.; Li, R.-W. Waterproof, Highly Tough, and Fast Self-Healing Polyurethane for Durable Electronic Skin. *ACS Appl. Mater. Interfaces* **2020**, *12*, 11072–11083.
75. Chortos, A.; Liu, J.; Bao, Z. Pursuing Prosthetic Electronic Skin. *Nat. Mater.* **2016**, *15*, 937–950.
76. Wei, M.; Zhan, M.; Yu, D.; Xie, H.; He, M.; Yang, K.; Wang, Y. Novel Poly(Tetramethylene Ether)Glycol and Poly(ε-Caprolactone) Based Dynamic Network via Quadruple Hydrogen Bonding with Triple-Shape Effect and Self-Healing Capacity. *ACS Appl. Mater. Interfaces* **2015**, *7*, 2585–2596.
77. Pu, W.; Fu, D.; Wang, Z.; Gan, X.; Lu, X.; Yang, L.; Xia, H. Realizing Crack Diagnosing and Self-Healing by Electricity with a Dynamic Crosslinked Flexible Polyurethane Composite. *Adv. Sci.* **2018**, *5*, 1800101.
78. Wang, Z.; Xie, C.; Yu, C.; Fei, G.; Wang, Z.; Xia, H. A Facile Strategy for Self-Healing Polyurethanes Containing Multiple Metal–Ligand Bonds. *Macromol. Rapid Commun.* **2018**, *39*, 1700678.
79. Liu, W.-X.; Yang, Z.; Qiao, Z.; Zhang, L.; Zhao, N.; Luo, S.; Xu, J. Dynamic Multiphase Semi-Crystalline Polymers Based on Thermally Reversible Pyrazole-Urea Bonds. *Nat. Commun.* **2019**, *10*, 4753.
80. Wang, Z.; Gangarapu, S.; Escorihuela, J.; Fei, G.; Zuilhof, H.; Xia, H. Dynamic Covalent Urea Bonds and Their Potential for Development of Self-Healing Polymer Materials. *J. Mater. Chem. A* **2019**, *7*, 15933–15943.
81. Ishihara, K.; Miyazaki, H.; Kurosaki, T.; Nakabayashi, N. Improvement of Blood Compatibility on Cellulose Dialysis Membrane. III. Synthesis and Performance of Water-Soluble Cellulose Grafted with Phospholipid Polymer as Coating Material on Cellulose Dialysis Membrane. *J. Biomed. Mater. Res.* **1995**, *29*, 181–188.
82. Hilbert, S. L.; Ferrans, V. J.; Tornita, Y.; Eidbo, E. E.; Jones, M. Evaluation of Explanted Polyurethane Trileaflet Cardiac Valve Prostheses. *J. Thorac. Cardiovasc. Surg.* **1987**, *94*, 419–429.
83. Guidoin, R.; Sigot, M.; King, M.; Sigot-Luizard, M.-F. Biocompatibility of the Vascugraft®: Evaluation of a Novel Polyester Methane Vascular Substitute by an Organotypic Culture Technique. *Biomaterials* **1992**, *13*, 281–288.

84. Lin, Y.-H.; Chou, N.-K.; Chen, K.-F.; Ho, G.-H.; Chang, C.-H.; Wang, S.-S.; Chu, S.-H.; Hsieh, K.-H. Effect of Soft Segment Length on Properties of Hydrophilic/Hydrophobic Polyurethanes. *Polym. Int.* **2007**, *56*, 1415–1422.
85. Hsieh, K. H.; Liao, D. C.; Chen, C. Y.; Chiu, W. Y. Interpenetrating Polymer Networks of Polyurethane and Maleimide-Terminated Polyurethane for Biomedical Applications. *Polym. Adv. Technol.* **1996**, *7*, 265–272.
86. Okano, T.; Aoyagi, T.; Kataoka, K.; Abe, K.; Sakurai, Y.; Shimada, M.; Shinohara, I. Hydrophilic-Hydrophobic Microdomain Surfaces Having an Ability to Suppress Platelet Aggregation and Their in Vitro Antithrombogenicity. *J. Biomed. Mater. Res.* **1986**, *20*, 919–927.
87. Xu, Y.; Petrovic, Z.; Das, S.; Wilkes, G. L. Morphology and Properties of Thermoplastic Polyurethanes with Dangling Chains in Ricinoleate-Based Soft Segments. *Polymer (Guildf)*. **2008**, *49*, 4248–4258.
88. Wang, Y.; Chen, R.; Li, T.; Ma, P.; Zhang, H.; Du, M.; Chen, M.; Dong, W. Antimicrobial Waterborne Polyurethanes Based on Quaternary Ammonium Compounds. *Ind. Eng. Chem. Res.* **2020**, *59*, 458–463.
89. Hu, J.; Peng, K.; Guo, J.; Shan, D.; Kim, G. B.; Li, Q.; Gerhard, E.; Zhu, L.; Tu, W.; Lv, W.; Hickner, M. A.; Yang, J. Click Cross-Linking-Improved Waterborne Polymers for Environment-Friendly Coatings and Adhesives. *ACS Appl. Mater. Interfaces* **2016**, *8*, 17499–17510.

Chapter 2

Green Materials for the Synthesis of Polyurethanes

Ziwei Li,[1] Kaimin Chen,[*,1] and Mingwei Wang[*,2]

[1]College of Chemistry and Chemical Engineering,
Shanghai University of Engineering Science, Shanghai 201620, P. R. China
[2]State Key Laboratory of Chemical Engineering,
East China University of Science and Technology, Shanghai 200237, P. R. China
*Email: kmchen@sues.edu.cn
*Email: mingweiwang@ecust.edu.cn

Generally, isocyanate and polyol are two raw materials for the synthesis of polyurethane (PU). In the past, petroleum-based polyols have dominated the synthesis of PU because of their high hydroxyl content. However, biobased materials have attracted increasing attention because of a series of environmental problems and production safety caused by the depletion and overexploitation of fossil resources. Biobased materials possess many advantages, such as being economically reasonable, being environmentally friendly, and having a wide range of sources such as agricultural byproducts. Among all biobased polyols, vegetable oil is considered an environmentally friendly and renewable material that can replace conventional polyols from petroleum. Some isocyanate-free strategies have been also employed to replace the isocyanate and phosgene used in the production process. In this chapter, the green materials for the synthesis of PU materials are summarized in detail.

Introduction

Polyurethane (PU), containing repeated carbamate groups (−NH−COO−) on its main chain, is a common polymeric material, which has been widely used in many fields because of its unique properties, including light weight, high toughness, high elongation, high wear resistance, ease of processing, and low cost. PU was first synthesized in the 1930s by Heinrich Rinke and Otto Bayer by addition polymerization of diisocyanates (1). Since then, PU materials have continued to develop, through the development of their raw materials and the continuous innovation of the synthesis process. In the 21st century, PU has been widely used in practical applications. However, the toxicity of isocyanate itself and petroleum-based polyols do harm both to the environment and human health. It has become urgent to develop green raw materials for PU. PU is a kind of multipurpose synthetic resin with various product forms. PU can be processed to form foam plastics, elastomers, adhesives, coatings, synthetic leather, fibers, waterproof materials, and refractory materials, which can be used in aerospace, national defense, construction, transportation, and medical equipment, among other fields.

Traditional PU synthesis is mainly divided into a one-step method and a two-step method, as shown in Figure 1 and Figure 2. Isocyanates are extremely reactive and can react with other reagents containing reactive hydrogen—even with water at room temperature—to form carbamate, and with small amounts of excess $-NCO-$ to form urea bonds. Unstable isocyanate could react with carboxylic acid or water to create polyurea, which could enhance the PU properties to some degree. Carbamate readily breaks down into amines and carbon dioxide, and the resulting amines react with isocyanates to form urea, with the formation of biurets and allophanate as side reactions (2).

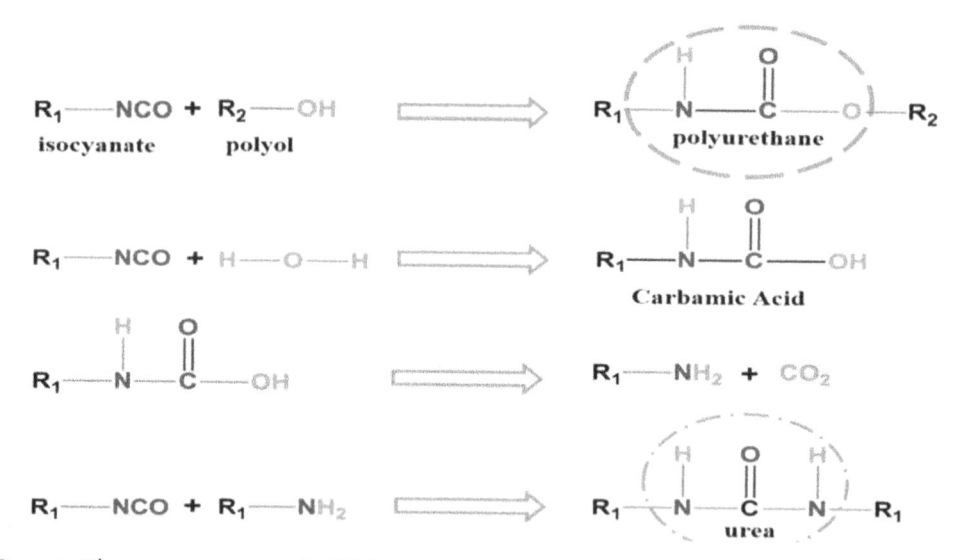

Figure 1. The primary reaction of a PU foaming system. Reproduced with permission from reference (2). Copyright 2020 John Wiley & Sons.

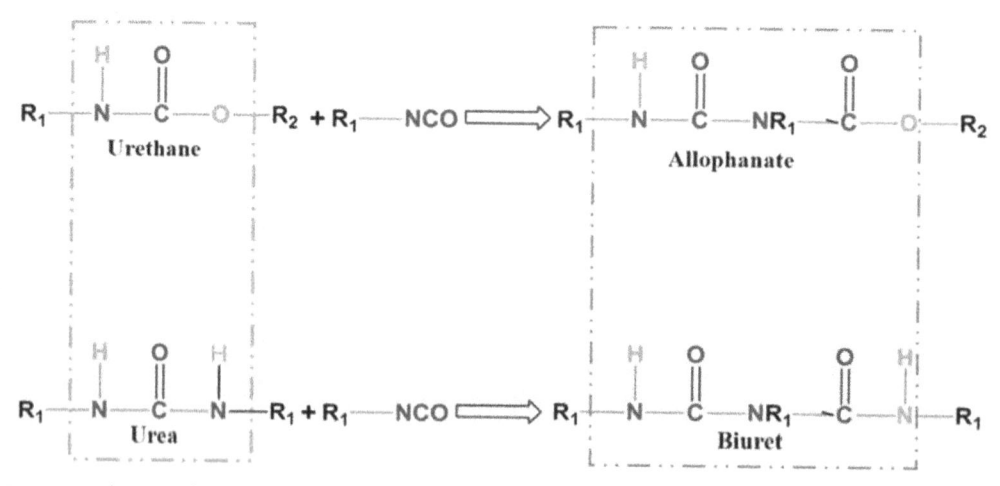

Figure 2. The secondary reaction of a PU foaming system. Reproduced with permission from reference (2). Copyright 2020 John Wiley & Sons.

Non-isocyanate PU (NIPU) has been developed as a more environmentally friendly material. NIPU was first synthesized in 1957 by Dyer and Scott. There are two ways to synthesize NIPU: polycondensation and polyaddition reactions (Figure 3) (3). Different raw materials have been developed, and different synthesis routes have been formulated in accordance with their various properties. The emergence of NIPU avoids toxic and harmful substances from raw materials, and this creates an alternative green route to synthesize PU.

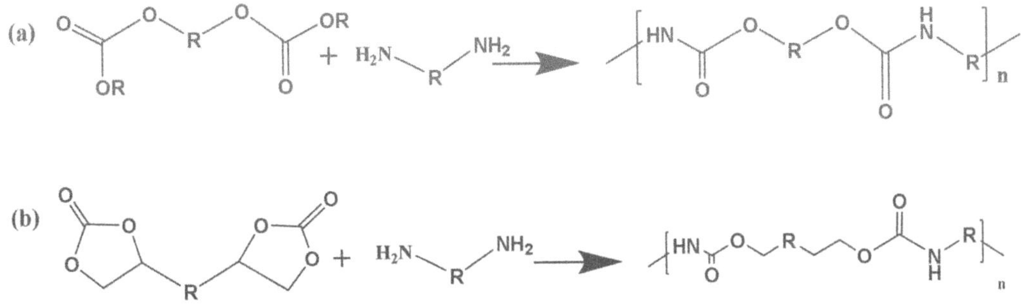

Figure 3. Synthetic scheme for NIPU via (a) polycondensation and (b) polyaddition. Reproduced with permission from reference (3). Copyright 2021 Elsevier.

PUs are mostly formed by the reactions of polyols such as polyester and polyether with isocyanates, chain extenders, or cross-linkers, whose properties ultimately depend on the properties and structures of the macromolecular chains. Especially for PU elastomer materials, the phase separation of soft segments and hard segments plays a decisive role in the properties of the final products. The composition of the soft and hard segments in PUs is illustrated in Figure 4, using thermoplastic PU as the example (4). Polyols with a low degree of polymerization, such as polyether and polyester, constitute the soft segment structures, which account for the majority of PUs. The polarity of soft segment structures mainly affects the mechanical properties of PUs. The crystallization of PUs is also affected by the soft segment structures. The hard segment is mainly composed of a diisocyanate after a reaction or a diisocyanate and a chain extender, which mainly controls the softening melting temperature and high temperature performance of PUs. The hard segment of PU will oxidize and degrade at a high temperature. The increase of hard segment content will increase the hardness and decrease the plasticity of PUs. According to the different content of blocks in PU structures, PU products can be divided into soft PU foam (PUF), rigid PUF (RPUF), and semirigid PUF, with different processing technologies.

Vegetable Oils

In the early stages of PU development, petroleum-based polyols were widely used in production. The decline of fossil reserves and the dangers of exploitation have forced the research community to accelerate the development of biochemicals and products. By changing the structure and composition of polymer materials, degradation can be achieved (5). By reducing the demand for nonrenewable energy, and thus reducing "greenhouse gas" emissions, the goal of reducing global warming is achieved. As renewable resources, biology-based products have become an ideal replacement for raw fossil materials and have successfully attracted great attention in the scientific community. Vegetable oils, which are kinds of triesters, are obtained from glycerol and fatty acids with various carbon atoms and carbon–carbon double bonds (6). Throughout the past few decades, researchers have synthesized PU materials from vegetable oil-based polyols, such as lignin, castor oil, soybean oil, and so forth.

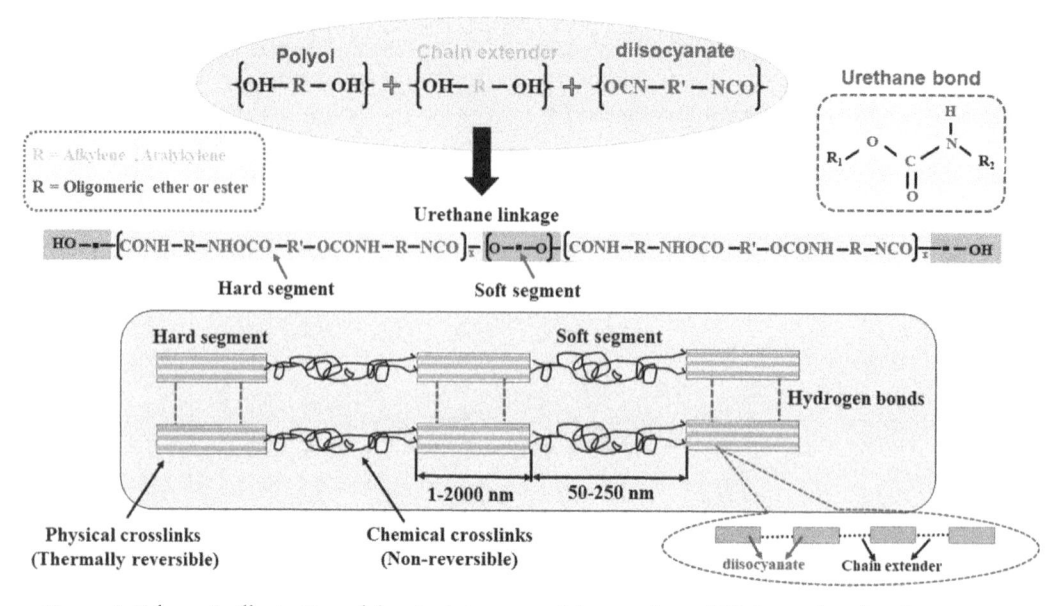

Figure 4. Schematic illustration of chemical structure of thermoplastic PU. Reproduced with permission from reference (4). Copyright 2020 Taylor & Francis.

Vegetable oils are also predominantly made up of triglyceride molecules. They are ideal replacement materials for the manufacture of biobased polymers, as they are renewable and can offer similar performance and low cost in comparison with petroleum-based polymer materials. In addition, petroleum-based monomers can be reacted with plant oils to produce modified macromonomers with good performance and properties (physical, mechanical, and thermal) for different applications. These modified macromolecules have environmental advantages over pure petroleum-based materials, which make them an attractive alternative. Biobased materials derived from natural sources and biomasses—such as castor oil, palm oil, canola oil, soybean oil, chitosan, and lignocellulose—have been used to synthesize natural polyols (7). The nontoxic and biodegradable properties of vegetable oils give them potential as functional polyols that can be used as ionizing groups. A synthesis of hyperbranched PUs based on vegetable oil can be found in Khanderay et al. (8).

The use of vegetable oil in the synthesis of PUs has been widely reported. Vegetable oil contains large amounts of triglycerides and various long-chain fatty acids, which create an ideal way to synthesize PUs. Synthetic PUs have special properties because of the presence of glycerides. Vegetable oil is also inexpensive and can reduce the consumption of fossil fuels. At the same time, the long-chain fatty acids in vegetable oil can be manipulated to produce side chains, which can be used as internal plasticizers to improve the plasticity of PUs (9). The use of biobased polyols to prepare PUs can also realize the reuse of waste resources. Examples include several types of unbleached screen residues from the pulp mill industry, waste from wastewater treatment and bark, wood chips from wood chip production, broken bleached cellulose fibers, sludge, and lignin particles. In addition, solid wood milling will produce bark, sawdust, small solid wood fragments, and wood chips. Other residues can be extracted from lignin and cellulose. These materials can be used in the synthesis of PU after treatment.

Different plant species and growing conditions result in different functional groups contained in plants themselves. The chemical general formula of vegetable oil is shown in Figure 5, where R_n represents the different functional groups that it carries. For example, the hydroxyl groups in castor oil and turtle/chrysanthemum oil contain active epoxy groups themselves (6). These vegetable oils can be directly synthesized into PUs without chemical modification. However, most oils require chemical modification of the reaction sites (esters and carbon–carbon double bonds) in triglycerides before they can be used in the synthesis of PUs. For example, in order to produce biobased polyols for water-based PU dispersions, multiple approaches have been developed to introduce functional groups at the reaction sites of fatty acid chains.

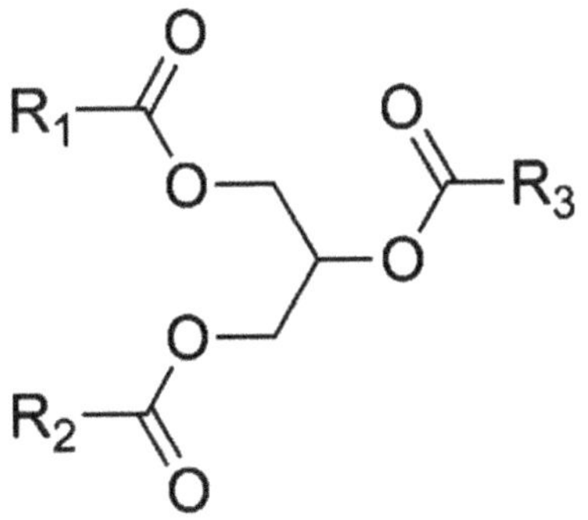

Figure 5. Triglyceride structure of vegetable oils (R_1, R_2, and R_3 represent fatty acid chains). Reproduced with permission from reference (10). Copyright 2010 Royal Society of Chemistry.

Vegetable oils are mainly triglycerides formed between glycerol and many fatty acids. The degree of unsaturation determines the comprehensive properties, which can be quantitatively characterized by the iodine value (IV). A larger IV value means more carbon–carbon double bonds. Thus, vegetable oils can be classified as drying oils (IV > 130), semidrying oils (100 < IV < 130), or nondrying oils (IV < 100) according to the IV data (Table 1) (10).

Table 1. Properties and Fatty Acid Compositions of the Most Common Vegetable Oils[a]

Vegetable Oil	Double Bonds	IV/mg per 100 g	Palmitic	Stearic	Oleic	Linoleic	Linolenic
Palm	1.7	44–58	42.8	4.2	40.5	10.1	
Olive	2.8	75–94	13.7	2.5	71.1	10.0	0.6
Groundnut	3.4	80–106	11.4	2.4	48.3	31.9	
Rapeseed	3.8	94–120	4.0	2.0	56.0	26.0	10.0
Sesame	3.9	103–116	9.0	6.0	41.0	43.0	1.0
Cottonseed	3.9	90–119	21.6	2.6	18.6	54.4	0.7
Corn	4.5	102–130	10.9	2.0	25.4	59.6	1.2
Soybean	4.6	117–143	11.0	4.0	23.4	53.3	7.8
Sunflower	4.7	110–143	5.2	2.7	37.2	53.8	1.0
Linseed	6.6	168–204	5.5	3.5	19.1	15.3	56.6

[a] Adapted with permission from reference (10). Copyright 2010 Royal Society of Chemistry.

Market Conditions

With the use of vegetable oil as the raw material of PU, in addition to saving resources and protecting the environment, the economic benefits are immeasurable. On example is palm oil, which is grown mainly in tropical and subtropical regions such as Malaysia, Indonesia, and Brazil. Palm farming is a controversial topic among tropical crops. There is no doubt that growing palm trees can bring significant economic benefits. However, cutting down forests to plant palm trees is environmentally harmful. In order to rationally plant palm trees, an influential international management system—the Roundtable on Sustainable Palm Oil—has been established to make a certain plan for relevant aspects (*11*). Palm oil production reached about 73.5 million tons per year in 2018, and it is increasing yearly. In Malaysia, palm oil accounts for about 43% of the country's agricultural output (*12*). There is no denying that the development of the palm oil industry not only provides cheap and green raw materials, but it also plays a positive role in regional economic development. More aspects should be taken into consideration to evaluate the advantages and disadvantages of using petroleum-based materials and biobased materials to make a comprehensive definition. A life-cycle assessment (LCA) is a rough standard to make an assessment for decision-makers. Related researchers described an LCA by saying: that "real-world users of LCA understand the trade-offs and the 'associated learnings' involved in using LCA for decision-making" (*13*). At the present time, LCAs have been widely recognized.

When LCA analysis was used for rapeseed oil (RO), the raw material source was the Latvian State Institute of Wood Chemistry. Two polyols were obtained by combining RO with diethanolamine and transforming it with melamine, respectively. Figure 6 shows the processing flow of RO under ideal conditions. Compared with petroleum-based PUs, especially regarding the impact on the environment (global warming), it was found that the biobased materials had significant advantages over the petroleum-based materials in terms of nonrenewable energy use, greenhouse gas emissions, and water consumption. However, this is not a simple conclusion, as the material can lead to land use, marine eutrophication, and ecotoxicity. In addition, the distribution method chosen by the RO producers can greatly impact economic outcomes. The use of biobased PUs will bring environmental benefits, whereas the comprehensive effect is impacted by many factors, and some aspects might lead to negative effects. Another LCA study on lignin also confirmed this point of view. The study cited various sources of lignin, such as black wine produced during winemaking and side streams in the pulp and paper industry. By reviewing 42 LCAs on lignin and lignin products, the findings showed that lignin's use can effectively reduce greenhouse gas emissions. However, land acidification and water eutrophication were both potentially negative consequences (*14*). Researchers should not only look at the good side (sustainable development, etc.) of using plant-based oils, but they should also consider a series of possible ecological and economic consequences.

Typical Green Polyol Materials

Depending on the type and composition of a raw material, there are various routes to synthesize PUs from vegetable oil-based polyols: epoxidation and ring opening of ethylene oxide (*16*), ozone decomposition (*17*), hydroformylation and hydrogenation (*4*), transesterification (*18*), and thiolene (*19*). The five different ways for synthesis are shown in Table 2. Epoxy ring opening is the most commonly used method for epoxidizing carbon–carbon double bonds of vegetable oils to produce polyols. This method has been applied in the processing of sunflower oil, soybean oil, cottonseed oil, and so forth. The type of product obtained after ring opening depends on the nature of the polyols and fatty acids contained in the plant. In addition, conditions such as the type of ring-opening reagent used in the processing also impact the final product. The ozone reaction usually takes place in two steps. The pros and cons of this approach are obvious. Ozonization eliminates half of the adipose chain and prohibits the addition of plasticizers in the process, which results in a higher hardness of the synthesized product. However, the molecular weight of the polyols obtained by this method is nearly 40% lower than that obtained by the other four methods, and the viscosity is lower after melting. Hydroformylation is an important synthetic route, and the products obtained by this method have higher reactivity to isocyanates. They are characterized by higher curing ability and shorter gel time, mainly for soybean and linseed. This method has high requirements for vegetable oil, and the properties of the products after the reaction mainly depend on the structures of the raw materials. Transesterification is mainly realized by using an ester mechanism, and a catalyst is introduced in most cases. The product properties of the reaction also depend on the properties of raw materials. The last one is the thiolene. This reaction is a single-step reaction under mild conditions with a high efficiency and yield (*18*).

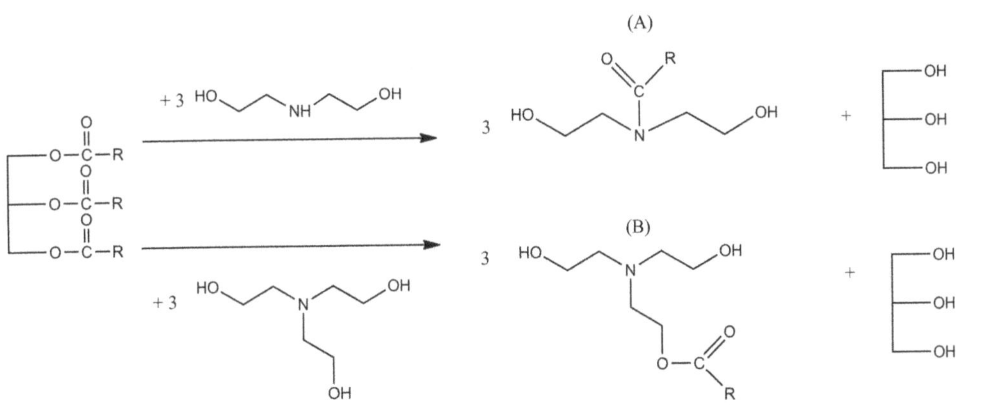

Figure 6. Idealized synthesis scheme for RO-based polyols: (A) RO/diethanolamine polyol, (B) RO/melamine polyol. Reproduced with permission from reference (15). Copyright 2020 Elsevier.

In addition to the five methods mentioned previously, the UV curing method is also a novel and efficient synthetic method. When isophorone diisocyanate and 2-hydroxyethyl methacrylate were used as precursors, renewable dimer fatty acid-based PU acrylate resin was synthesized from dimer acid. The synthesis route is shown in Figure 7. This study introduced an unconventional method for the synthesis of PU with good mechanical properties, thermal properties, hydrophobicity, and volumetric shrinkage resistance (*20*). The study also inspired researchers to actively explore new synthetic pathways in the future and reduce the pollution generated in the production process through more environmentally friendly synthetic pathways. When it comes to dimer acid, it is necessary to discuss a series of saturated branched fatty acids obtained from vegetable oils. Next, this chapter will introduce some typical vegetable oils and take those vegetables as examples to introduce a series of breakthroughs in the improvement of raw materials in PUs.

Table 2. Comparison of Different Methods for Vegetable Oil-Based Polyols[a]

	Epoxidation/Ring Opening	Ozonolysis/Reduction	Hydroformylation/Hydrogenation	Transesterification/Amidation	Thiolene
Number of steps	2	2	2	1	1
Functionality	Secondary, tunable	Primary and terminal, 3	Primary, tunable	Primary and terminal, 2–3	Primary, tunable
Hydroxyl number	70–340	200–260	140–210	150–400	190–330
Molecular weight	>1000	400–700	900–1150	350–550	1000–1500
Viscosity	High	Low	Medium	Low	Medium
Reaction temperature	50–190 °C	Room temperature	Approx. 120 °C	120–220 °C	Room temperature
Reaction time	Long	Medium	-	-	Medium
Reaction time	Yes	Yes	No	Yes or no	No
Catalysis	Cheap	Cheap	Expensive	Cheap	Cheap

[a] Adapted with permission from reference (6). Copyright 2018 Elsevier.

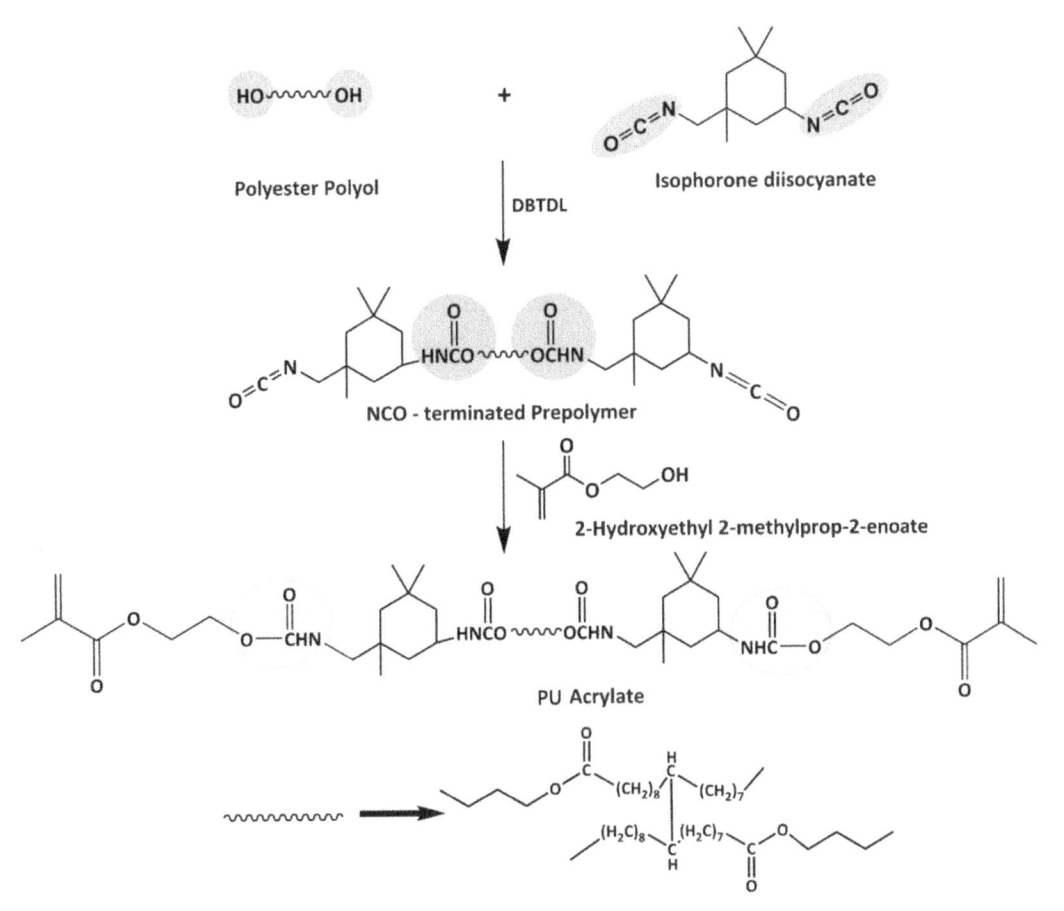

Figure 7. Synthesis of dimer fatty acid-based PU acrylate resin oligomer. DBTDL: dibutyltin dilaurate. Reproduced with permission from reference (20). Copyright 2020 Elsevier.

Lignin

Wood contains three main polymers: cellulose, lignin, and hemicellulose, in descending order of content. Lignin is a cross-linked natural material from oxygen radical polymerization of three monomers (i.e., coniferyl alcohol, P-coumaryl alcohol, and sinapyl alcohol). Different sources and separation methods of lignin will lead to different contents of each monomer. These three monomers, because of oxygen-free radicals with different structures, are coupled to each other to obtain complex cross-links, thus generating huge and complex network structures (Figure 8). The formed bonds vary depending on the amounts of monomers. The parameters of preparation of lignin—like pH, temperature, and solubility—have a significant influence on the product structure. Typically, the commercially obtained lignin mainly include five types: lignin sulfonate, kraft lignin, organosol lignin, alkaline lignin, and enzymatic lignin (21). It is worth noting that the unmodified lignin can be directly put into use. Although this may lead to poor performance of the product because of the poor activity of lignin, its advantages of green environmental protection and strong feasibility are still attractive. By contrast, although modified lignin increases the activity of the reaction, it will produce more waste and have a higher cost compared with unmodified lignin. How to choose between these points in the actual production depends on the specific situation.

The unmodified lignin mentioned previously may have some defects, but in some ways, it still has relatively excellent performance. Lignin-based PU elastomers (LPUes) can be obtained by combining hard segments and soft segments. The hard segments are cross-linked unfunctionalized lignins, and the soft segments are poly(propylene glycol), tolylene 2,4-diisocyanate terminated. The prepared LPUes have a high amount of stiffness, strength, and toughness, as indicated in the literature (22). In this work, the influence of molecular weight (3600 and 600 g mol^{-1}) and mass fraction (5–40 wt %) of lignin on the thermal and mechanical properties of LPUes was studied. The results demonstrated that the lignin content had a positive correlation with the thermal stability of LPUes. This phenomenon was most obvious in LPUes prepared with 600 g/mol^{-1} lignin. The excellent dispersion of low molecular weight lignin led to good mechanical properties under the optimal recipe. This research result fully indicated the potential of unmodified lignin as a green raw material of PU.

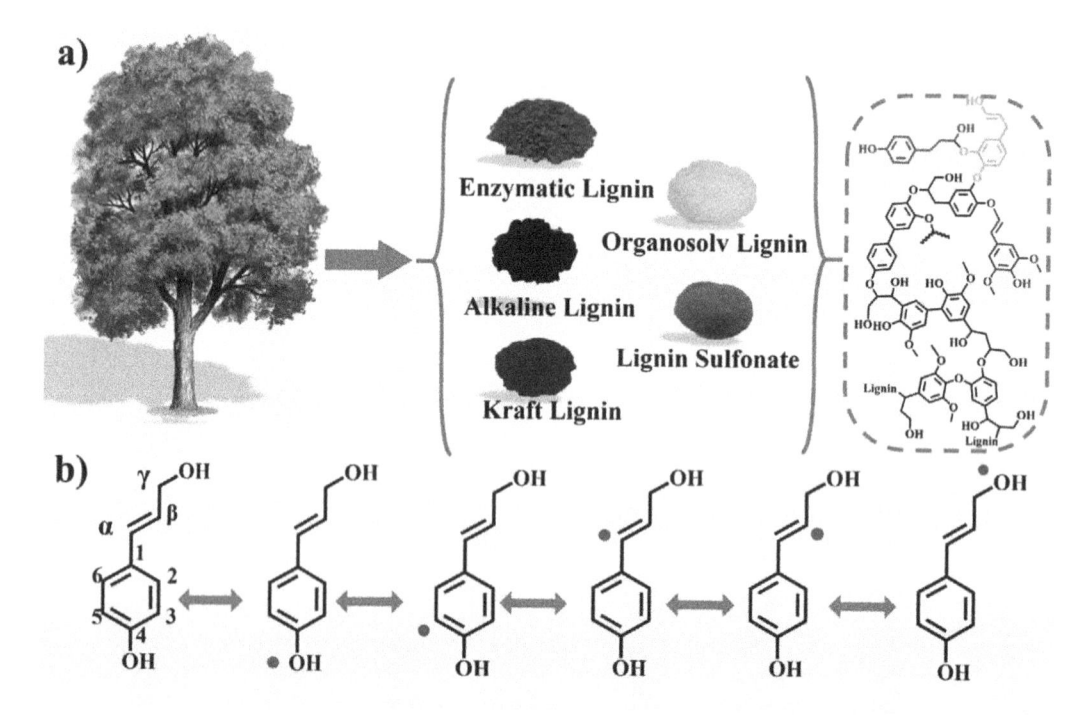

Figure 8. (a) Lignin and its structure. (b) Lignin unit resonant type. Reproduced with permission from reference (21). Copyright 2020 John Wiley & Sons.

Modification can provide better properties of lignin. Researchers have synthesized mechanically strong PUs with very high lignin content (23). When the modification of lignin by reacting it with formaldehyde in mild conditions was carried out, the obtained product had a high molecular weight with a high isolated yield and improved glass-transition temperature. Lignin hydroxyls could also be converted to isocyanate, which was then applied to react with polyols to prepare elastomers with excellent mechanical properties and good thermal stability. This demonstrated that the lignin PU properties could be largely fine-tuned by adjusting the molecular weight and glass-transition temperature of the lignin and rubber segment. This research demonstrated that both modified and unmodified lignin have excellent properties. Lignin has bright prospects and significant space for development in the future.

Cottonseed Oil

Cotton is a mallow plant with a large amount of plant protein and is the ninth largest oil-producing crop. Cottonseed oil occupies an extremely important position in the textile industry (24). The unsaturated degree and composition of fatty acids in cottonseed oil are shown in Table 3. The main fatty acid in cottonseed oil is linoleic acid. The ratio of polyunsaturated fatty acids (linoleic acid) to saturated fatty acids (palmitic acid/stearic acid) is 2:1. Cottonseed contains a large number of plant fibers from which cottonseed oil can be extracted. Cottonseed production is high, but its utilization rate is low (25). The fibrous material crop in cotton has been widely used in the textile industry, while other parts, such as cottonseed, have received less attention. The development and utilization of cottonseed can reduce pollutant emission while obtaining a higher utilization rate, which plays a significant role in promoting economic development.

Table 3. Degree of Unsaturation in Terms of Fatty Acid and Fatty Acid Composition in Cottonseed Oil Measured by Gas Chromatography Analysis[a]

Fatty Acids in Cottonseed Oil	Double Bonds	Fatty Acid Composition(%)	Molar Masses (g/mol)	Moles of Double Bonds per 100 g of Oil	Moles of Fatty Acids per 100 g of Oil
Linoleic acid	2	47.77	280.45	0.340	0.170
Palmitic acid	0	24.85	256.42	0.00	0.096
Oleic acid	1	20.61	282.46	0.072	0.072
Stearic acid	0	3.08	284.48	0.00	0.010
Palmitolic acid	1	0.58	254.41	0.002	0.002
Unknown fatty acid		3.11	292.00	0.00	0.010
Total		100		0.414	0.360

[a] Adapted with permission from reference (24). Copyright 2015 Springer.

The synthesis of PU from cottonseed can be divided into one-step and two-step methods. Various RPUFs can be obtained by a one-step process (Figure 9). Cottonseed oil was epoxidized first, and then polyols were prepared by biobased chain extenders such as ethylene glycol and lactic acid. (Cottonseed oil was epoxidized first and then was used to prepare polyol with biobased chain extenders such as ethylene glycol and lactic acid.) RPUFs can also be obtained from as-prepared polyols and methylene diisocyanate (MDI) assisted by suitable blowing agents. The thermal properties based on cottonseed oil exhibited no obvious difference compared with petroleum-based PUFs, which indicated that cottonseed oil-based polyols were promising candidates for green materials of PUs. A two-step process could also be applied to get PUs from cottonseed oil. The fatty acids were converted into amides, first followed by further conversion to amide ester to get the final polyols (24). It was reported that the gloss of PUs made from cottonseed oil turned out even better than that made from petroleum-based materials.

Castor Oil

Castor oleic acid is the main component of castor oil, accounting for about 90% of the total fatty acid content. Castor oleic acid contains a hydroxyl structure (as shown in Figure 10), which can be used as the raw material of PU (26). The properties of PU synthesized from castor oil are relatively stable and not easily deformed. The rich hydroxyl content enables castor oil-based products to perform as well as or better than petroleum-based PUs. At the same time, because castor oil is not used as the main edible oil and its price is low, it is an ideal candidate for PU raw materials.

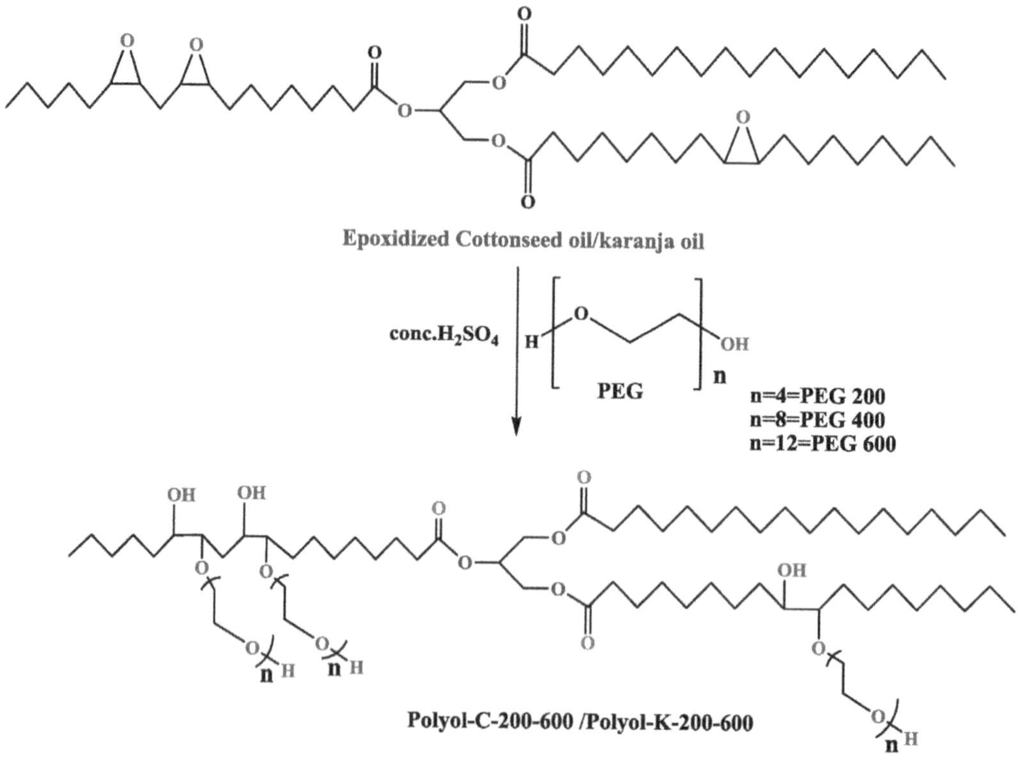

Figure 9. Preparation of polyols from epoxidized cottonseed oil. PEG: polyethylene glycol. Reproduced with permission from reference (25). Copyright 2015 Elsevier.

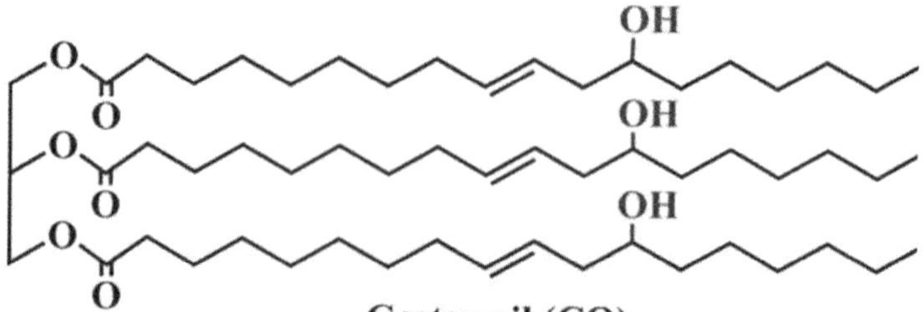

Chart 10. Chemical structures of castor oil. Reproduced with permission from reference (26). Copyright 2018 John Wiley & Sons.

The hydroxyl content will have a great impact on the performance of the final product. Some novel progress has been made in the development of castor oil. Researchers have used itaconic acid to cross-link castor oil. The plant oil-based raw materials were treated with an amide method and then combined with itaconic dihydroxate to produce polyamide polyols. The synthesis route is shown in Figure 11. The cross-linking effects of castor oil, linseed oil, and Karanja oil with itaconic acid were compared (27). The results demonstrated that the influence of different structures of hexamethylene isocyanate and polyols on the cross-linking degree and rigidity of PU was mainly reflected in the mechanical, chemical, thermal, and corrosion resistance aspects of the PU coating. The castor oil and itaconic acid PU had higher cross-linking density and thermal stability because of the presence of hydroxyl groups in castor oil. However, PU synthesized from linseed oil and Karanja oil with lower hydroxyl content had poor performance. This study fully showed the effect of fatty acid content in vegetable oil on synthetic products, and it also proved that castor oil is a kind of vegetable oil with good prospects.

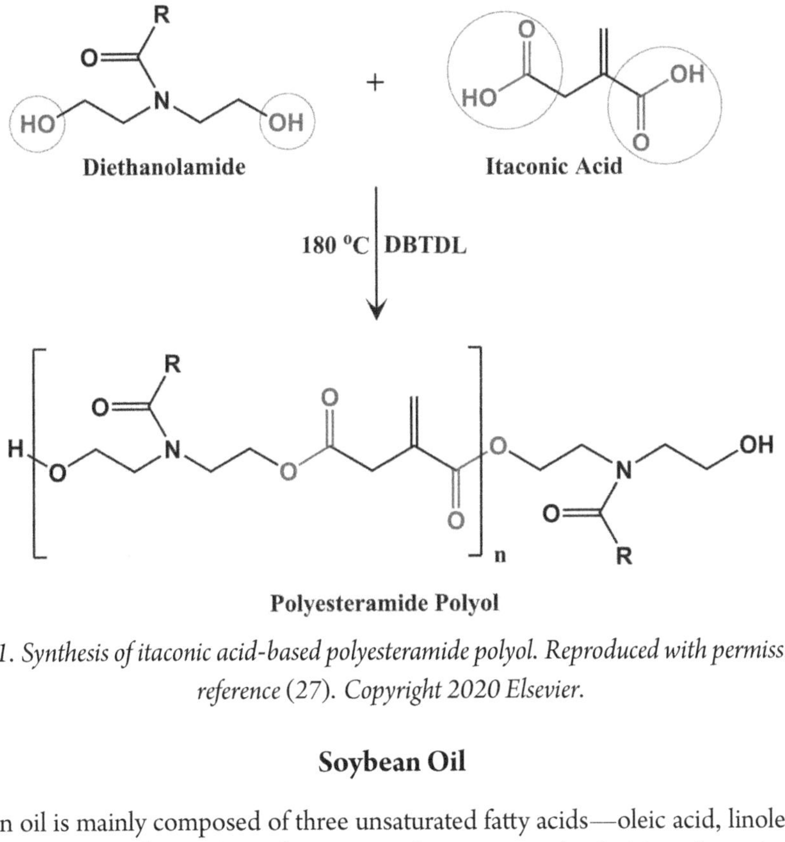

Figure 11. Synthesis of itaconic acid-based polyesteramide polyol. Reproduced with permission from reference (27). Copyright 2020 Elsevier.

Soybean Oil

Soybean oil is mainly composed of three unsaturated fatty acids—oleic acid, linoleic acid, and linolenic acid—along with a triglyceride mixture of two saturated palmitic and stearic acids. Only saturated fatty acids remain stable and unaffected when processed with oils to convert them into PUs. Excluding the possibility of a positional isomerism, there are still about 35 different triacylglyceride combinations in soybean oil. The triglyceride in soybean oil has some double bonds, ranging from 0 to 9, and the average number is 4.5. The triglycerides' structure may even vary the number of double bonds, as the combination of components could be different (28).

Using soybean oil to synthesize PU is of significant interest. Researchers have used oleic acid to create an open-loop treatment for most plants, as shown in Figure 12 (29). The existence of long carbon chains and unsaturated fatty acid in oleic acid creates synthetic PU coatings with better properties. Compared with PUs synthesized from petroleum-based polyols, bio-PU has a higher cross-linking degree and better thermal stability than petroleum-based PU. In addition, bio-PU has better corrosion resistance and a shorter cure time because of the presence of unsaturated oleic acid. The results have shown that the PU synthesized from vegetable oil has better properties than that synthesized from petroleum. On this basis, soybean oil is expected to replace petroleum as a raw material for the synthesis of PU.

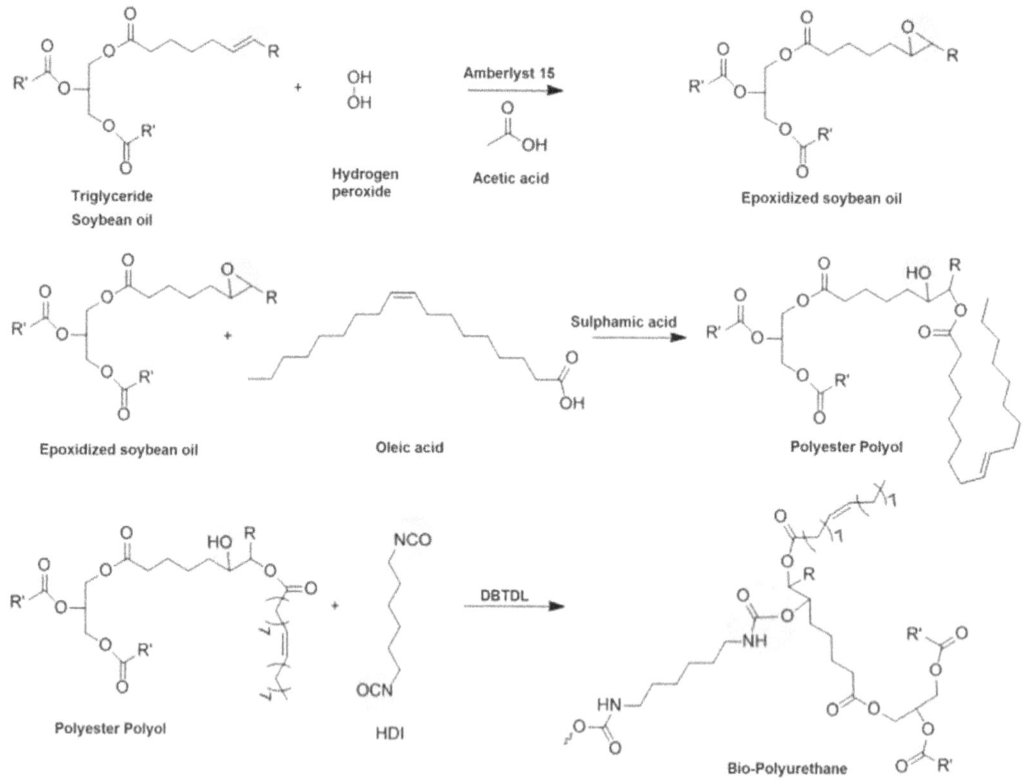

Figure 12. Reaction scheme for synthesis of epoxidized soybean oil, polyester polyol, and PU. Reproduced with permission from reference (29). Copyright 2020 Springer.

Soybean oil-based polyols can also be prepared by a thiolene click reaction, which is solvent-free and can be scalable in a homemade photochemical reactor. The obtained soybean oil-based polyol is highly reactive with diisocyanates because of primary hydroxyl groups in its structure. The results demonstrate that the obtained PU from soybean oil has improved thermal and mechanical properties (30). The microwave method is another route to improve the properties of soybean oil-based PU. A better performance PU could be obtained by increasing reaction yield, shortening reaction time, and improving ring-opening rate (31).

Biobased Chain Extender

Chain extender is a typical additive in PU synthesis. Chain extenders are usually small molecules such as ethanolamine and diols that contain two functional groups, mainly combined with diisocyanate, and mostly impact the hard segment. When the solid content remains unchanged, more chain extender means harder segment content. At present, there are few studies on chain extenders, but the improvement of chain extenders with cleaner raw materials is also a concern for scientists.

It has been reported that green PUF was prepared from biological malonic acid and trimethylamine (32). The effect of malonate content on PU performance was observed by changing the content of malonate. The results were compared with conventional 1,4-butanediol. The results showed that using biological malonate as a chain extender not only increased the rate of viscosity during the reaction process, but it also gave the produced PU more of a closed-cell structure. This was different from the traditional chain extender. Although the chain extender broke the hydrogen bond, it maintained the normal ordered form of the foam. In addition, the chain extender produced additional carbon dioxide for foaming and could therefore be used as a cross-linking agent. The development of this chain extender showed a biologically based material with the same, or in some ways better, performance than traditional chain extenders.

Non- and Green Isocyanate Materials

Although PU synthesized from isocyanates is a stable product, the synthesis of isocyanates is associated with severe toxicity, which is the most important disadvantage of the PU synthesis circuit. In addition, the reaction involving the use of phosgene and amine salts to synthesize isocyanates at around 200 °C is a major factor, as it is extremely harmful to human health (33). Long-term exposure to phosgene causes proteins in the alveoli to mutate and disrupt the blood–air barrier, which is one of the three barriers in the human body, causing asphyxia. Toluene diisocyanate and MDI are the two most frequently used monomers for the synthesis of PU, but they are very harmful. Toluene diisocyanate and MDI vapor can cause severe irritation to the skin, eyes, and respiratory tract, leading to asthmatic breathing or difficulty breathing (34). Scientists have become increasingly aware of the importance of reducing toxic substances produced in chemical products and processes. At the same time, the development of more environmentally friendly and recyclable products with better economic benefits and less potential for pollution has been high on their agenda.

NIPU has received significant interest. Non-isocyanates can also be obtained from biobased materials. The use of nontoxic green non-isocyanates instead of isocyanates in production is a very interesting approach. NIPUs, which are fully conformed to the "green" concept, are also being studied. This method avoids the use of isocyanates and thus reduces a series of hazards caused by isocyanates at the source. Generally, three main synthetic pathways have been explored: acyl azide, transurethanization, and aminolysis (Figure 13) (35). The first method is simple and controllable, and the generated A–B monomer is also reactive. The acyl azide group is converted to isocyanate by Curtis rearrangement. (Figure 13a). However, the disadvantage of this method is also very obvious: the formation of isocyanate as an intermediate of the reaction. The second method involves a two-step reaction. First, diurethane (or carbamate) is polycondensation with diol at high temperature to form isocyanate structure. (Figure 13b). Next, the method requires continuous removal of intermediates and yields can be increased to 89%. The ammonolysis reaction has received the most attention. It involves the reaction between cyclic carbonate and amine to form urethane bonds (Figure 13c). The pentacarbonate system is obtained through a process involving carbon dioxide consumption. The molar ratio of dicyclic carbonate to diamine in these polymers is close to 1:1. The resulting NIPU is polyhydroxyurea, which normally has a primary/secondary hydroxyl ratio of 30/70. Urethanes bearing secondary hydroxyl groups are more stable. In this process, the ammonolysis reaction also has some problems and shortcomings. The main drawbacks of the aminolysis reaction are the slow kinetics and low molar masses. These are mainly because of side reactions, which are promoted by heat or catalysts. However, in the comprehensive evaluation, this system can bring significant benefits for ecology and the economy, and it has a bright future.

The synthesis of non-isocyanates can be carried out by utilizing the reaction of carbon dioxide gas and cyclic carbonate functional materials from epoxides. The biobased NIPU membrane obtained from the formed cyclic carbonate has good physical, mechanical, and biological properties (36). Castor oil can also be converted to NIPU. Toxic reactions of Hofmann and Curtius rearrangements are also alternative methods for synthesizing NIPU. In addition, in the production process, the use of halogens such as bromine and chlorine will cause certain amounts of harm to the human body and the environment. However, some researchers have successfully developed a way to synthesize PU by using a nontoxic route of loose rearrangement. They have used dimethyl carbonate (DMC) as an activator and tertiary amine as a catalyst, dispersed in an alcohol medium. The use of DMC and tertiary amine can greatly reduce the toxicity of a rearrangement reaction. This method can be regarded as a nontoxic substitute for the rearrangement reaction. Unverferth et al. tried to synthesize renewable NIPU from biobased materials. Dicarboxylic acids were converted to DMC monomers via a base-catalyzed Lossen rearrangement. Diols and dicarbamates were obtained from the ricinoleic acid in castor oil after suitable functionalization. Renewable NIPU could be synthesized by polycondensation of the castor oil–derived DMC and diols. PUs with high molecular weights and improved thermal properties could be obtained under optimized reaction conditions (37).

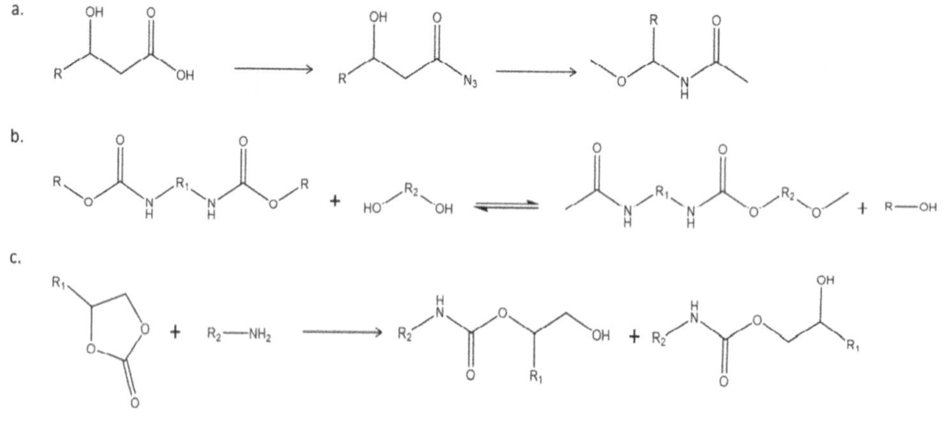

Figure 13. Main chemical routes to synthetized NIPU (a) with acyl azide, (b) with transurethanization, or (c) with aminolysis. Reproduced with permission from reference (35). Copyright 2018 Royal Society of Chemistry.

For the NIPU system, the low reactivity of some amine–carbonate systems might be a problem. This problem has been improved by Stachak et al. In the process, dangerous chemical intermediates have been gradually eliminated and replaced by more environmentally friendly, greener, and healthier raw materials and processes. At the same time, in the research process, a wider range of processing methods for the synthesis of PU has been developed, such as electrospinning and three-dimensional printing (38). At the present time, there is a great deal of literature and many patents about NIPU. It is believed that in the near future, NIPU will show more interesting properties.

Other

Biobased polyphenols are another source of PU materials in addition to biobased polyols because of their phenolic hydroxyl groups. Cashew nutshell liquid (CNSL) is extracted from the fruits of the cashew tree, which is native to Brazil (39). Between the hard inner shell and outer shell of the nut is a honeycomb-like structure from which the oil of CNSL of about 20–25% of the total weight of the nut can be extracted. CNSL is aromatic oil composed of natural phenolic compounds, which contain about 90% anacardic acid and 10% cardol. Anacardic acid is a natural phenol, which has a carboxyl and a long carbon chain in the ortho- and metaposition, respectively. Cardanol can be obtained from anacardic acid by decarboxylation (Figure 14). The results show that the PU synthesized from a cashew shell has good properties. In the reaction process, the small adjustment of the biological base unit will have a great impact on its performance. The process can be controlled by controlling the amount of CNSL. In addition, the synthesized PUs show intrinsic ultrafast self-healing behavior through the segmental mobility of the soft segment, along with excellent corrosion resistance properties. Most interestingly, the synthesized PUs undergo degradation under microbial exposure (*40*). Therefore, CNSL is a potential renewable resource and is expected to be the main raw material for the synthesis of PU with excellent properties in the future.

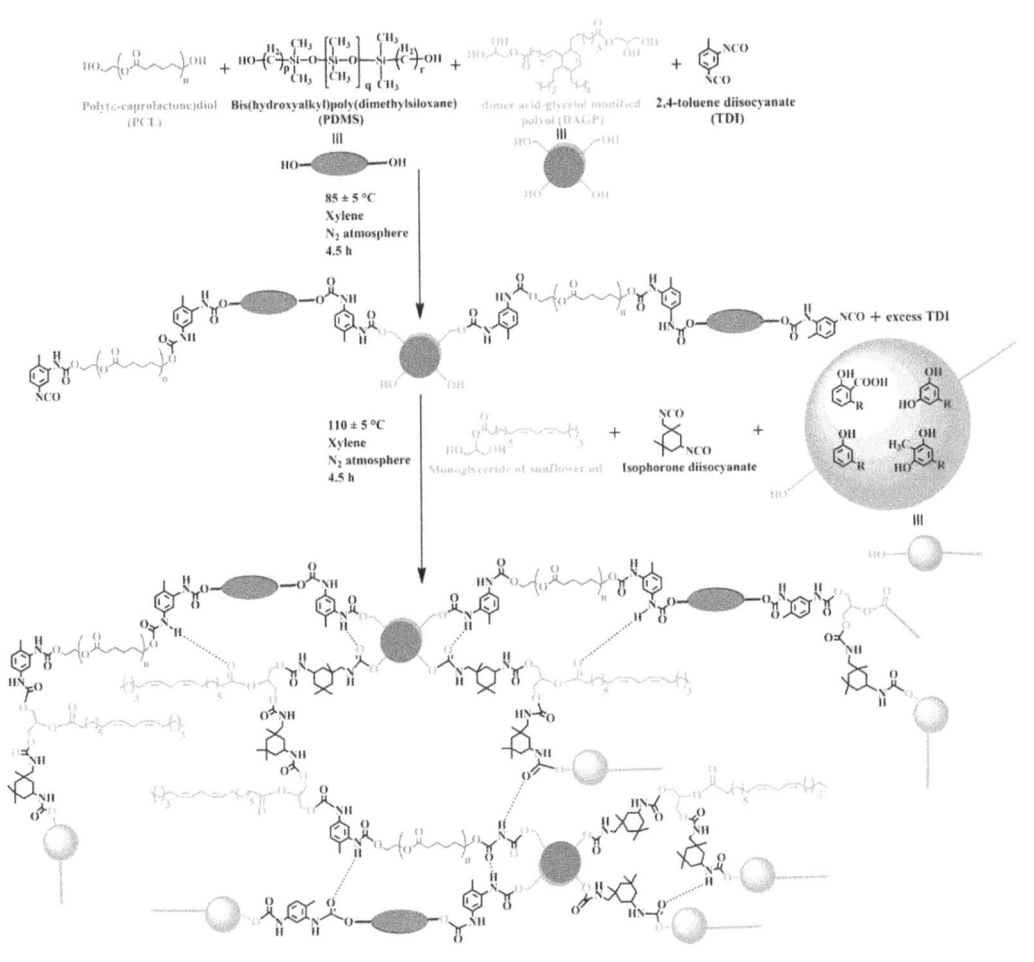

Figure 14. Synthesis route of PU–CNSL. Reproduced with permission from reference (40). Copyright 2020 Elsevier.

In addition to the aforementioned green raw materials, there are some other materials that can be used as green raw materials for the synthesis of PU. Food residues often contain large amounts of alkanes that can be used as PU raw materials, and the produced PU products also manifest excellent performance (*41*). Sea buckthorn oil-based fatty acid methyl ester, which is a byproduct of tocopherol extracted from sea buckthorn oil, was obtained after a series of pretreatments with sea buckthorn oil-based fatty acid methyl ester-polyesteramide polyols (SBTPEPs). SBTPEP can be used to synthesize PUs (Figure 15). The results showed that the PUs synthesized by SBTPEP had good surface properties, mechanical properties, and thermal stability (*42*). This study demonstrated the great potential of biobased raw materials by rationally utilizing the waste products produced in the process of using biobased raw materials.

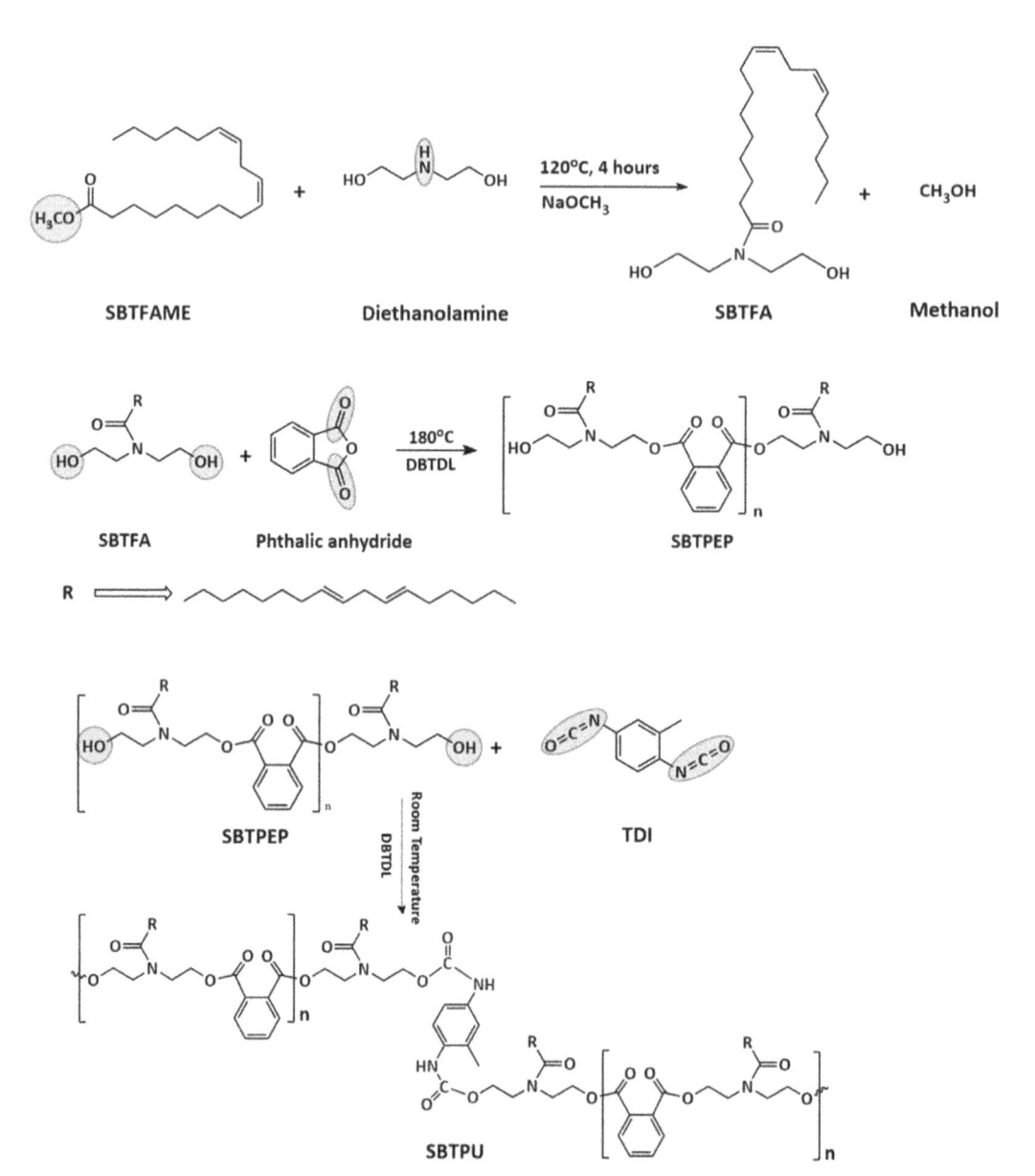

Figure 15. Reaction scheme for synthesis of fatty amide, SBTPEP, and PU. SBTFAME: sea buckthorn oil fatty acid methyl ester, SBTFA: sea buckthorn fatty acid methyl ester-based fatty amide, and SBTPU: sea buckthorn oil-based PU. Reproduced with permission from reference (42). Copyright 2020 John Wiley & Sons.

Fatty acid dimer is also an ideal candidate to produce polyols and PUs. It was used in the study as an alternative to vegetable oil. The unique structure of long-chain fatty acids provides excellent properties such as high impact strength, high stability, and less membrane shrinkage. The strategy of "green plus green" can be realized through the preparation of biobased coatings by a UV-curable solution. The results showed that the physical properties of PU based on dimeric acid structure were significantly improved compared with the previously explored biobased PU in terms of flexibility, hydrophobicity, and impact resistance. This study showed that excellent PU products could also be synthesized using green materials from abiotic sources. It seems to be a sensible choice without using biobased materials (43). This option may become an important research direction in the future, because it avoids many problems associated with using biobased materials. In addition, the development of a "greener" route is urgent. The issue of how to combine green raw materials with a green synthesis method is a problem worth thinking about thoroughly.

Conclusion

In the 21st century, people have become deeply aware of the importance of environmental protection and sustainable development. The excessive exploitation of fossil materials has brought irreversible damage to the environment, such as climate warming and melting glaciers. The processing of petroleum-based raw materials might cause harm to human health. Therefore, biobased raw materials have been widely studied because of their environmentally friendly and renewable characteristics. Their use greatly reduces the environmental pollution caused by the exploitation and use of petroleum-based raw materials. Biobased raw materials come from a wide range of sources, even including the waste left after crop production and processing. PUs produced from biobased materials have good properties and are expected to replace fossil materials as the main sources of some effective functional groups in the future. In addition, the large-scale use of biobased materials could also boost the development of industries such as agriculture. However, it should be noted that despite the advantages of biobased raw materials, there are also certain risks (water eutrophication, overexploitation, etc.). When using biobased raw materials as raw materials for PUs, various factors should be taken into account, and a comprehensive evaluation should be made after a comprehensive consideration of cost and other factors. Biobased materials provide a reasonable solution to solve a series of problems caused by petroleum-based materials.

References

1. Jiang, D.; Wang, Y.; Li, B.; Sun, C.; Guo, Z. Environmentally Friendly Alternative to Polyester Polyol by Corn Straw on Preparation of Rigid Polyurethane Composite. *Compos. Commun.* **2020**, *17*, 109–114. https://doi.org/10.1016/j.coco.2019.11.007.
2. Singh, I.; Samal, S. K.; Mohanty, S.; Nayak, S. K. Recent Advancement in Plant Oil Derived Polyol-Based Polyurethane Foam for Future Perspective: A Review. *Eur. J. Lipid Sci. Tech.* **2020**, *122* (3), 1900225. https://doi.org/10.1002/ejlt.201900225.
3. Khatoon, H.; Iqbal, S.; Irfan, M.; Darda, A.; Rawat, N. K. A Review on the Production, Properties and Applications of Non-Isocyanate Polyurethane: A Greener Perspective. *Prog. Org. Coatings* **2021**, *154*, 106124. https://doi.org/10.1016/j.porgcoat.2020.106124.
4. Khalifa, M.; Anandhan, S.; Wuzella, G.; Lammer, H.; Mahendran, A. R. Thermoplastic Polyurethane Composites Reinforced with Renewable and Sustainable Fillers–a Review. *Polym-Plast. Technol.* **2020**, *59* (16), 1751–1769. https://doi.org/10.1080/25740881.2020.1768544.

5. Somisetti, V.; Allauddin, S.; Narayan, R.; Raju, K. V. S. N. Flexible, Hard, and Tough Biobased Polyurethane Thermosets from Renewable Materials: Glycerol and 10-Undecenoic Acid. *J. Coatings Technol. Res.* **2018**, *15* (1), 199–210. https://doi.org/10.1007/s11998-017-9998-2.
6. Liang, H.; Feng, Y.; Lu, J.; Liu, L.; Yang, Z.; Luo, Y.; Zhang, Y.; Zhang, C. Bio-Based Cationic Waterborne Polyurethanes Dispersions Prepared from Different Vegetable Oils. *Ind. Crops Prod.* **2018**, *122*, 448–455. https://doi.org/10.1016/j.indcrop.2018.06.006.
7. Moghadam, P. N.; Yarmohamadi, M.; Hasanzadeh, R.; Nuri, S. Preparation of Polyurethane Wood Adhesives by Polyols Formulated with Polyester Polyols Based on Castor Oil. *Int. J. Adhes. Adhes.* **2016**, *68*, 273–282. https://doi.org/10.1016/j.ijadhadh.2016.04.004.
8. Khanderay, J. C.; Gite, V. V. Vegetable Oil-Based Polyurethane Coatings: Recent Developments in India. *Green Mater.* **2017**, *5* (3), 109–122. https://doi.org/10.1680/jgrma.17.00009.
9. Ahmad Hazmi, A. S.; Nik Pauzi, N. N. P.; Abd. Maurad, Z.; Abdullah, L. C.; Aung, M. M.; Ahmad, A.; Salleh, M. Z.; Tajau, R.; Mahmood, M. H.; Saniman, S. E. Understanding Intrinsic Plasticizer in Vegetable Oil-Based Polyurethane Elastomer as Enhanced Biomaterial. *J. Therm. Anal. Calorim* **2017**, *130* (2), 919–933. https://doi.org/10.1007/s10973-017-6459-1.
10. Xia, Y.; Larock, R. C. Vegetable Oil-Based Polymeric Materials: Synthesis, Properties, and Applications. *Green Chem.* **2010**, *12* (11), 1893–1909. https://doi.org/10.1039/c0gc00264j.
11. Brandão, F.; Schoneveld, G.; Pacheco, P.; Vieira, I.; Piraux, M.; Mota, D. The Challenge of Reconciling Conservation and Development in the Tropics: Lessons from Brazil's Oil Palm Governance Model. *World Dev.* **2021**, *139*, 105268. https://doi.org/10.1016/j.worlddev.2020.105268.
12. Adam, N. I.; Hanibah, H.; Subban, R. H. Y.; Kassim, M.; Mobarak, N. N.; Ahmad, A.; Badri, K. H.; Su'ait, M. S. Palm-Based Cationic Polyurethane Membranes for Solid Polymer Electrolytes Application: A Physico-Chemical Characteristics Studies of Chain-Extended Cationic Polyurethane. *Ind. Crops Prod.* **2020**, *155*, 112757. https://doi.org/10.1016/j.indcrop.2020.112757.
13. Pryshlakivsky, J.; Searcy, C. Life Cycle Assessment as a Decision-Making Tool: Practitioner and Managerial Considerations. *J. Clean. Prod.* **2021**, *309*, 127344. https://doi.org/10.1016/j.jclepro.2021.127344.
14. Moretti, C.; Corona, B.; Hoefnagels, R.; Vural-Gürsel, I.; Gosselink, R.; Junginger, M. Review of Life Cycle Assessments of Lignin and Derived Products: Lessons Learned. *Sci. Total Environ.* **2021**, *770*, 144656. https://doi.org/10.1016/j.scitotenv.2020.144656.
15. Fridrihsone, A.; Romagnoli, F.; Kirsanovs, V. Cabulis, U. Life Cycle Assessment of Vegetable Oil Based Polyols for Polyurethane Production. *J. Clean. Prod.* **2020**, *266*, 121403. https://doi.org/10.1016/j.jclepro.2020.121403.
16. Sinadinović-Fišer, S.; Janković, M.; Petrović, Z. S. Kinetics of in Situ Epoxidation of Soybean Oil in Bulk Catalyzed by Ion Exchange Resin. *J. Am. Oil Chem. Soc.* **2001**, *78* (7), 725–731. https://doi.org/10.1007/s11746-001-0333-9.
17. Petrović, Z. S.; Zhang, W.; Javni, I. Structure and Properties of Polyurethanes Prepared from Triglyceride Polyols by Ozonolysis. *Biomacromolecules* **2005**, *6* (2), 713–719. https://doi.org/10.1021/bm049451s.

18. Paraskar, P. M.; Prabhudesai, M. S.; Hatkar, V. M.; Kulkarni, R. D. Vegetable Oil Based Polyurethane Coatings – A Sustainable Approach: A Review. *Prog. Org. Coatings* **2021**, *156*, 106267. https://doi.org/10.1016/j.porgcoat.2021.106267.
19. Chuayjuljit, S.; Maungcharoen, A.; Saravari, O. Preparation and Properties of Palm Oil-Based Rigid Polyurethane Nanocomposite Foams. *J. Reinf. Plast. Compos.* **2010**, *29* (2), 218–225. https://doi.org/10.1177/0731684408096949.
20. Paraskar, P. M.; Hatkar, V. M.; Kulkarni, R. D. Facile Synthesis and Characterization of Renewable Dimer Acid-Based Urethane Acrylate Oligomer and Its Utilization in UV-Curable Coatings. *Prog. Org. Coatings* **2020**, *149*, 105946. https://doi.org/10.1016/j.porgcoat.2020.105946.
21. Ma, X.; Chen, J.; Zhu, J.; Yan, N. Lignin-Based Polyurethane: Recent Advances and Future Perspectives. *Macromol. Rapid Commun.* **2020**, *42* (3), 2000492. https://doi.org/10.1002/marc.202000492.
22. Li, H.; Sun, J. T.; Wang, C.; Liu, S.; Yuan, D.; Zhou, X.; Tan, J.; Stubbs, L.; He, C. High Modulus, Strength, and Toughness Polyurethane Elastomer Based on Unmodified Lignin. *ACS Sustain. Chem. Eng.* **2017**, *5* (9), 7942–7949. https://doi.org/10.1021/acssuschemeng.7b01481.
23. Saito, T.; Perkins, J. H.; Jackson, D. C.; Trammel, N. E.; Hunt, M. A.; Naskar, A. K. Development of Lignin-Based Polyurethane Thermoplastics. *RSC Adv.* **2013**, *3* (44), 21832–21840. https://doi.org/10.1039/c3ra44794d.
24. Pawar, M. S.; Kadam, A. S.; Singh, P. C.; Kusumkar, V. V.; Yemul, O. S. Rigid Polyurethane Foams from Cottonseed Oil Using Bio-Based Chain Extenders: A Renewable Approach. *Iran. Polym. J.*, *25* (1), 59–68. https://doi.org/10.1007/s13726-015-0401-9.
25. Gaikwad, M. S.; Gite, V. V.; Mahulikar, P. P.; Hundiwale, D. G.; Yemul, O. S. Eco-Friendly Polyurethane Coatings from Cottonseed and Karanja Oil. *Prog. Org. Coatings* **2015**, *86*, 164–172. https://doi.org/10.1016/j.porgcoat.2015.05.014.
26. Wang, C.; Xu, F.; He, M.; Ding, L.; Li, S.; Wei, J. Castor Oil-Based Polyurethane/Silica Nanocomposites: Morphology, Thermal and Mechanical Properties. *Polym. Compos.* **2018**, *39*, E1800–E1806. https://doi.org/10.1002/pc.24798.
27. Paraskar, P. M.; Prabhudesai, M. S.; Kulkarni, R. D. Synthesis and Characterizations of Air-Cured Polyurethane Coatings from Vegetable Oils and Itaconic Acid. *React. Funct. Polym.* **2020**, *156*, 104734. https://doi.org/10.1016/j.reactfunctpolym.2020.104734.
28. SantosMiranda, M. E.; Marcolla, C.; Rodriguez, C. A.; Wilhelm, H. M.; Sierakowski, M. R.; BelleBresolin, T. M.; Alves de Freitas, R. I. The Role of N-Carboxymethylation of Chitosan in the Thermal Stability and Dynamic. *Polym Int* **2006**, *55*, 961–969. https://doi.org/10.1002/pi.
29. Paraskar, P. M.; Prabhudesai, M. S.; Deshpande, P. S.; Kulkarni, R. D. Utilization of Oleic Acid in Synthesis of Epoxidized Soybean Oil Based Green Polyurethane Coating and Its Comparative Study with Petrochemical Based Polyurethane. *J. Polym. Res.* **2020**, *27* (8), 242. https://doi.org/10.1007/s10965-020-02170-w.
30. Yang, Z.; Feng, Y.; Liang, H.; Yang, Z.; Yuan, T.; Luo, Y.; Li, P.; Zhang, C. A Solvent-Free and Scalable Method to Prepare Soybean-Oil-Based Polyols by Thiol-Ene Photo-Click Reaction and Biobased Polyurethanes Therefrom. *ACS Sustain. Chem. Eng.* **2017**, *5* (8), 7365–7373. https://doi.org/10.1021/acssuschemeng.7b01672.

31. Favero, D.; Marcon, V. R. R.; Barcellos, T.; Gómez, C. M.; Sanchis, M. J.; Carsí, M.; Figueroa, C. A.; Bianchi, O. Renewable Polyol Obtained by Microwave-Assisted Alcoholysis of Epoxidized Soybean Oil: Preparation, Thermal Properties and Relaxation Process. *J. Mol. Liq.* **2019**, *285*, 136–145. https://doi.org/10.1016/j.molliq.2019.04.078.
32. Zhao, W.; Nolan, B.; Bermudez, H.; Hsu, S. L.; Choudhary, U.; van Walsem, J. Spectroscopic Study of the Morphology Development of Closed-Cell Polyurethane Foam Using Bio-Based Malonic Acid as Chain Extender. *Polymer* **2020**, *193*, 122344. https://doi.org/10.1016/j.polymer.2020.122344.
33. Loepp, G.; Vollmer, S.; Witte, G.; Wöll, C. Adsorption of Heptanethiol on Cu(110). *Langmuir* **1999**, *15* (11), 3767–3772. https://doi.org/10.1021/la981510y.
34. Baur, X.; Marek, W.; Ammon, J.; Czuppon, A. B.; Marczynski, B.; Raulf-Heimsoth, M.; Roemmelt, H.; Fruhmann, G. Respiratory and Other Hazards of Isocyanates. *Int. Arch. Occup. Environ. Health* **1994**, *66* (3), 141–152. https://doi.org/10.1007/BF00380772.
35. Furtwengler, P.; Avérous, L. Renewable Polyols for Advanced Polyurethane Foams From Diverse Biomass Resources. *Polym. Chem.* **2018**, *9*, 4258–4287. https://doi.org/10.1039/C8PY00827B
36. Gholami, H.; Yeganeh, H. Soybean Oil-Derived Non-Isocyanate Polyurethanes Containing Azetidinium Groups as Antibacterial Wound Dressing Membranes. *Eur. Polym. J.* **2021**, *142*, 110142. https://doi.org/10.1016/j.eurpolymj.2020.110142.
37. Unverferth, M.; Kreye, O.; Prohammer, A.; Meier, M. A. R. Renewable Non-Isocyanate Based Thermoplastic Polyurethanes via Polycondensation of Dimethyl Carbamate Monomers with Diols. *Macromol. Rapid Commun.* **2013**, *34* (19), 1569–1574. https://doi.org/10.1002/marc.201300503.
38. Stachak, P.; Łukaszewska, I.; Hebda, E.; Pielichowski, K. Recent Advances in Fabrication of Non-Isocyanate Polyurethane-Based Composite Materials. *Materials* **2021**, *14* (13), 3497. https://doi.org/10.3390/ma14133497
39. Ionescu, M.; Wan, X.; Bilić, N.; Petrović, Z. S. Polyols and Rigid Polyurethane Foams from Cashew Nut Shell Liquid. *J. Polym. Environ.* **2012**, *20* (3), 647–658. https://doi.org/10.1007/s10924-012-0467-9.
40. Ghosh, T.; Karak, N. Cashew Nut Shell Liquid Terminated Self-Healable Polyurethane as an Effective Anticorrosive Coating with Biodegradable Attribute. *Prog. Org. Coatings* **2020** (139), 105472. https://doi.org/10.1016/j.porgcoat.2019.105472.
41. Augu, M.; Prociak, A.; Ryszkowska, I. Open-Cell Polyurethane Foams of Very Low Density Modi Fi Ed with Various Palm Oil-Based Bio-Polyols in Accordance with Cleaner Production. *J. Clean. Prod.* **2021**, *290*, 125875. https://doi.org/10.1016/j.jclepro.2021.125875.
42. Prabhudesai, M. S.; Paraskar, P. M.; Kedar, R.; Kulkarni, R. D. Sea Buckthorn Oil Tocopherol Extraction's By-Product Utilization in Green Synthesis of Polyurethane Coating. *Eur. J. Lipid Sci. Technol.* **2020**, *122* (4), 1900387. https://doi.org/10.1002/ejlt.201900387.
43. Paraskar, P. M.; Kulkarni, R. D. Synthesis of Isostearic Acid / Dimer Fatty Acid - Based Polyesteramide Polyol for the Development of Green Polyurethane Coatings. *J. Polym. Environ.* **2021**, *29*, 54–70. https://doi.org/10.1007/s10924-020-01849-x.

Chapter 3

Overview on Classification of Flame-Retardant Additives for Polymeric Matrix

Mattia Bartoli,[*,1,2] Giulio Malucelli,[2,3] and Alberto Tagliaferro[2,4]

[1]Center for Sustainable Future Technologies @POLITO, Fondazione Istituto Italiano di Tecnologia, Via Livorno 60, 10144 Turin, Italy

[2]National Interuniversity Consortium of Materials Science and Technology (INSTM), Via G. Giusti 9, Florence 50121, Italy

[3]Department of Applied Science and Technology, Viale Teresa Michel 5, Alessandria 15121, Italy

[4]Department of Applied Science and Technology, Politecnico di Torino, Corso Duca degli Abruzzi 24, Torino 10129, Italy

*Email: mattia.bartoli@iit.it

Plastic additives are among the most relevant assets in the polymer industry for improving the properties of polymeric materials. The improvement of flame resistance is a major issue for the safe use of plastics for industrial and civilian purposes. For this reason, flame-retardant (FR) additives represent a very relevant class of compounds that are gaining further importance year by year. In this chapter, we report a general overview of the main FR additives, focusing on their action mechanisms together with the most significant achievements reported in the literature. We aim to provide a comprehensive description of the wide realm of FR additives that will represent a reference point for newcomers and a guide for experts in the field.

Introduction

The polymer production industry is driven by massive production of different plastic and composite materials in an attempt to fulfill marketplace demand.

Within all commodities related to plastic production, flame retardants (FRs) are among the most used additives for various plastic and composite products in sectors ranging from automotive to high-tech industries. These materials can be classified into two main families: inorganic and organic FRs.

Inorganic species include a wide range of materials such as metal oxides and hydroxides (*1*), phosphates (*2*), and silicates (*3*).

Organic FRs are represented by halogenated or phosphorus/nitrogen–containing organic compounds, biomacromolecules, and carbon-based materials (4).

Inorganic materials are the oldest species, and a great deal of work has been done to replace or engineer them because of their poor efficiency and compatibility. Compatibility is an issue in the use of inorganic species because it can lead to poor dispersion into the polymeric matrix with an appreciable drop in mechanical properties. An exception is glass fiber composites, in which both mechanical and flame resistance are related to the fibers rather than to the matrix (5).

Organic FRs usually require low concentrations to achieve the same performance as that of inorganic counterparts and can be easily tailored to improve their properties and compatibility with the polymer matrix. Among organic additives, the halogenated ones have represented the best compromise between performance and cost for many years. Their use is being reduced due to environmental issues related to their manufacturing, disposal, and degradation processes. Accordingly, several alternative materials have been developed that aim to reach the best FR performances without environmental pollution.

FRs act mainly through three mechanisms (Scheme 1): inducing the formation of a protective char layer, capturing radical species, and inducing intumescence (i.e., a great volume increment).

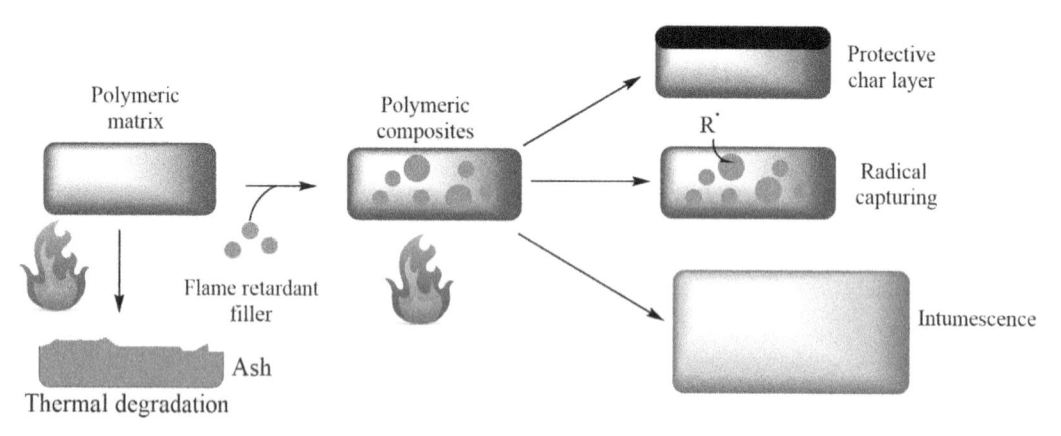

Scheme 1. Main mechanisms occurring during burning of a polymeric matrix with and without FR addition.

In this chapter, we are going to give a general overview on both inorganic and organic FRs, focusing our attention on developments. We also briefly report some insights on organic halogenated FRs because they are among the most used additives, although there are new policies and regulations to discourage their use.

Inorganic FR Additives

Hydroxides and Oxides

One of the most interesting FR classes is represented by metal hydroxides and oxides (1, 6). These species have gained a great deal of relevance because of useful properties such as low cost, anticorrosion effects (7), and low emission of smoke during burning processes (8).

Metal hydroxides act as FRs by exploiting an endothermic decomposition mechanism, with the formation of the corresponding oxides and of water. Accordingly, metal hydroxide dilutes flammable gases and removes a part of the heat.

Metal oxides display a different FR mechanism based on intumescence (i.e., producing char, a carbonaceous residue layer that swells upon the application of a flame or the exposure to a heat flux, hence providing an improved thermal insulation).

Metal oxides and hydroxides exhibit relevant disadvantages such as a relatively low FR efficiency and a remarkable decrease of both chemical and mechanical properties of the polymer matrix (9).

Among these FRs, magnesium oxide and hydroxide are very popular and effective FR additives. Since early systematic and rigorous studies in the 1980s (10, 11), magnesium oxide and hydroxide have been proven to be extremely efficient smoke-suppressant additives for plastic composites without decreasing flammability.

Xu and coworkers (12) prepared a platelet-shaped and superfine magnesium hydroxide (Figure 1).

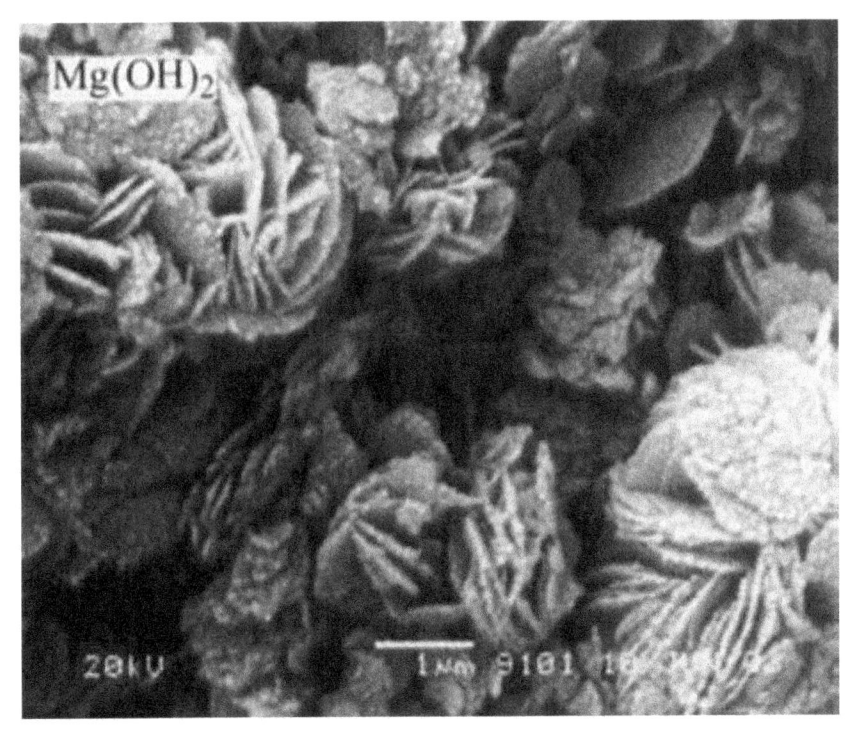

Figure 1. Scanning electron microscope capture of ultrafine magnesium hydroxide particles. Reproduced with permission from reference (12). Copyright 2006 Elsevier.

The authors directly converted and modified a bischofite mineral collected from Cha'erhan saline through a simple dissolution process using ammonia and a thermal treatment at a relatively low temperature, below 180 °C. The recovered materials consisted of platelet-shaped lamellae with thicknesses of up to 230 nm, reaching a specific surface area <48 m^2/g. A further modification with sodium stearate changed the contact angle of magnesium, allowing a better interaction with such an apolar polymer matrix as poly(ethylene), reaching a limiting oxygen index (LOI) of 32.5% in the presence of 60 wt % of magnesium hydroxide.

Qiu et al. (13) improved the dispersibility of magnesium hydroxide by tuning its morphology and producing rodlike nanocrystalline particles. The particles were efficiently dispersed in ethylene–vinyl acetate copolymer with a filler loading up to 50 wt %, achieving an LOI as high as 38.3% (i.e., about 60% higher with respect to the unfilled copolymer matrix).

Fang and coworkers (*14*) further modified the magnesium hydroxide morphology, obtaining nanosized hexagonal particles with an average size of up to 174 nm and a specific surface area not exceeding 51 m²/g, through a template synthesis using poly(vinyl pyrrolidone). The resulting poly(vinyl pyrrolidone)/magnesium hydroxide composites showed an appreciable increment of thermal resistance as assessed through thermogravimetric analysis carried out in air atmosphere.

A surface tailoring of magnesium hydroxide was reported by Cao et al. (*15*) by treating neat magnesium hydroxide with glycine in a basic environment. After a further treatment at 150 °C, a highly hydrophobic material (water contact angle beyond 150°) was recovered and mixed with an acrylonitrile–butadiene–styrene (ABS) copolymer at low concentrations (namely 1 and 5 wt %). As clearly reported in Figure 2, the presence of 5 wt % of hydrophobic magnesium hydroxide remarkably decreased the heat release rate (HRR). The authors did not observe any decrement of mechanical properties in the presence of the inorganic filler.

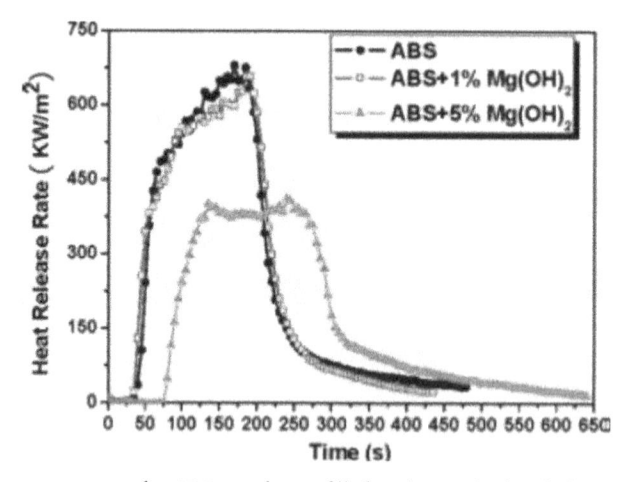

Figure 2. HRR versus time curves for ABS copolymer filled with superhydrophobic magnesium hydroxide. Reproduced with permission from reference (15). Copyright 2010 American Chemical Society.

Wang et al. (*16*) produced nanosized magnesium hydroxide by using a rotating packed bed reactor. The nanofiller was dispersed into a 10 wt % solution of poly(vinyl alcohol), giving rise to transparent, flexible, and stretchable composites. The authors achieved a very promising increase of the LOI (+45%) compared with the unfilled polymer. The peak of HRR observed by using microscale combustion calorimetry was lowered from 453 to 332 W/g. The authors also reported the formation of compact and smooth residue that acted as a protective layer during combustion.

The method of producing magnesium hydroxide seemed to be a very crucial issue for the final FR properties of the resulting composites. In particular, Ren et al. (*17*) reported a very effective hydrothermal method able to produce size-controlled magnesium hydroxide with a large surface area by a simple hydrothermal treatment of a slurry of magnesium oxide and sodium hydroxide.

Rigolo et al. (*18*) improved the simple use of magnesium hydroxide by adding magnesium carbonate. By assessing different ratios of magnesium hydroxide and carbonate, the authors were able to prove the beneficial effect of carbonate addition using a 1:1 ratio and a filler loading approaching 60 wt % in poly(propylene) (PP), reaching an LOI of 28.2%.

Li et al. (*19*) synthesized magnesium oxide microcapsules through an in situ emulsion polymerization for obtaining cellulose fibers. The materials produced through electrospinning were also compared with those containing pure magnesium hydroxide. Microcapsules-based materials showed a significantly lower HRR with respect to the others. The results proved that the morphology of magnesium hydroxide played a relevant role in the fire behavior. The thermogravimetric curves showed that the generated residues increased from 13.79% (cellulose fiber) to 30.26% (cellulose–magnesium oxide fiber).

Another common metal hydroxide used as an FR is aluminum hydroxide, which shows interesting properties as discussed in the next paragraphs.

Hiremath and coworkers (20) mixed aluminum hydroxide with an epoxy matrix and investigated the fire and mechanical performances of carbon-fiber-reinforced composites. The authors clearly proved that the use of filler loadings between 0.3 and 6 wt % did not compromise the mechanical behavior of the composite but significantly improved its fire retardance. Similarly, Laachachi et al. (21) used boehmite for improving the fire behavior of poly(methyl methacrylate).

As shown in Figure 3, composites containing up to 20 wt % of the filler showed a flameout time that significantly increased (up to 879 s) as compared with the unfilled matrix, suggesting a strong effect of the filler on the combustion kinetics of the polymer.

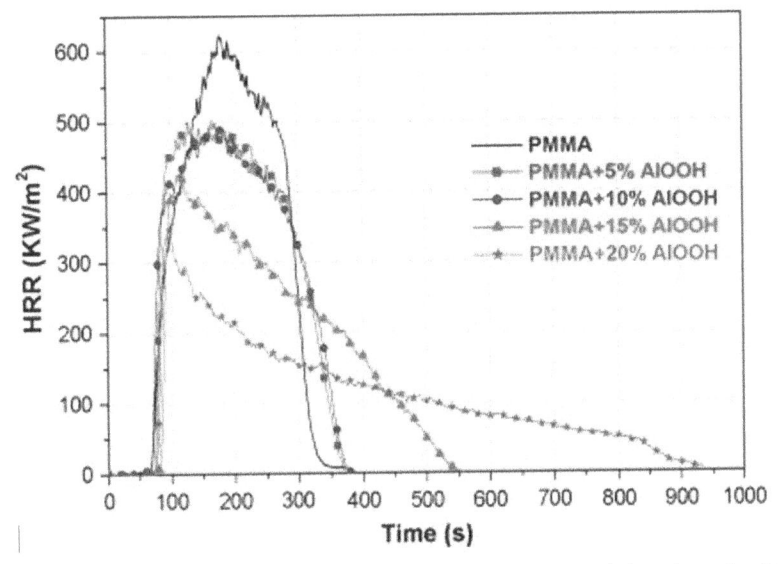

Figure 3. The effect of boehmite (AlOOH) content on the HRR of poly(methyl methacrylate) (PMMA) at 35 kW/m². Reproduced with permission from reference (21). Copyright 2009 Elsevier.

Ahmad et al. (22) combined alumina and boron nitride to enhance the fire behavior of epoxy resin coatings. The authors reported a massive char formation when 0.5 wt % of filler loading was employed.

Boron-based materials can also be used solely as additives, exploiting their activity in the condensed phase by driving the matrix decomposition to the creation of char layer over the formation of carbon monoxide and dioxide (23). Additionally, boron-containing additives reduce the afterglow process in halogenated FR systems (24). The most popular boron-based FR additive is represented by zinc borate and its analogs.

Wang et al. (25) developed a new zinc borate—$Zn_2Al(OH)_6[B_4O_5(OH)_4]_{0.5}$ and Zn_2AlBO_4—and employed it as filler for the preparation of PP composites; the filler loading ranged from 1 up to 30 wt %. The micrometric particles of zinc borate showed an average size of up to 10 μm and were well dispersed into the polymer matrix. The cone calorimetry results shown in Figure 4 clearly demonstrated the optimal performances of 15 wt % filler loading with a decrease of the peak of HRR by about 64% with only negligible further improvement moving up to higher filler loadings.

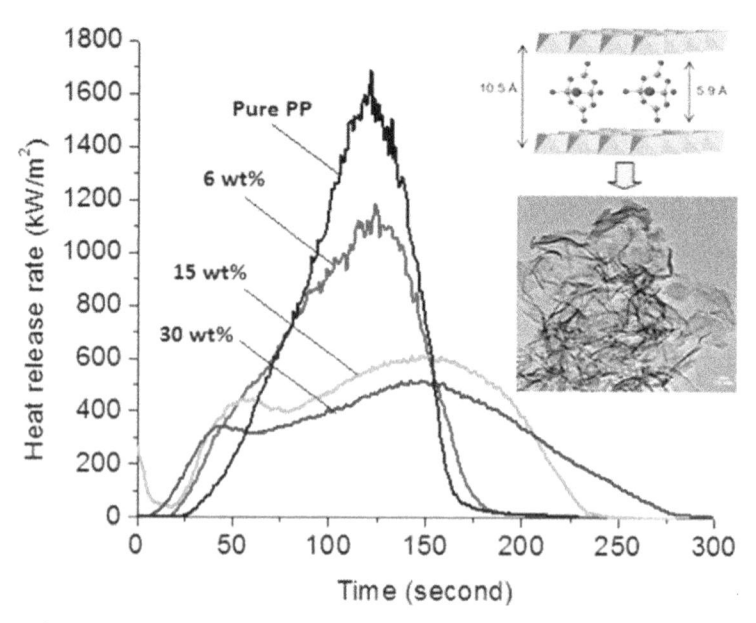

Figure 4. HRR of zinc borate containing PP. Reproduced with permission from reference (25). Copyright 2013 American Chemical Society.

Another widely used metal oxide is antimony oxide, an FR additive that has found a wide application as a stand-alone additive (26) or in mixture with other agents.

Li et al. (27) developed a multicomponent FR system mixing antimony oxide with ammonium polyphosphate and pentaerythritol and tested it on PP. Antimony oxide below 2 wt % loading displayed a synergistic effect with the other additives, increasing the LOI up to 37% and producing a very stable char layer that slowed down the heat and oxygen transfer.

Li and coworkers (28) designed an FR system mixing antimony oxide with micrometric mica particles and compared it with a chlorinated product. The authors proved that antimony oxide mixed with mica displays the same properties of the halogenated FR, providing self-extinction to the polymer.

Despite the high cost compared with other metals, molybdenum oxides are among the best FRs and smoke suppressants for a great variety of polymers. As an example, the unique overall fire performances of molybdenum oxides when added to poly(vinyl chloride) are caused by the char formation directly on the polymer backbone (29, 30).

Nabipour et al. (31) provided a complete and comprehensive overview of all metal-based materials. It was observed that transition metal species are very complex to rationalize in a unique way. The complex interaction between metal center and polymer is affected not only by the metal type but also by its chemical state (i.e., oxide, organometallic species, salt). Among them, layered hydroxides have gained remarkable attention because of the high effectiveness and wide range of applications to several polymeric matrices (32).

Compounds Containing Silicone

Silicone-based FR materials induce both char formation in the condensed phase and radicals trapping in the gas phase (33).

Simple systems based on sepiolite were used by Liu et al. (34) with remarkable results, but further modification and conjugation are generally required over the neat silica use.

Wang et al. (*35*) described a new route to combine silica nanoparticles with expandable graphite (EG) for improving the FR properties of a PP matrix (Figure 5).

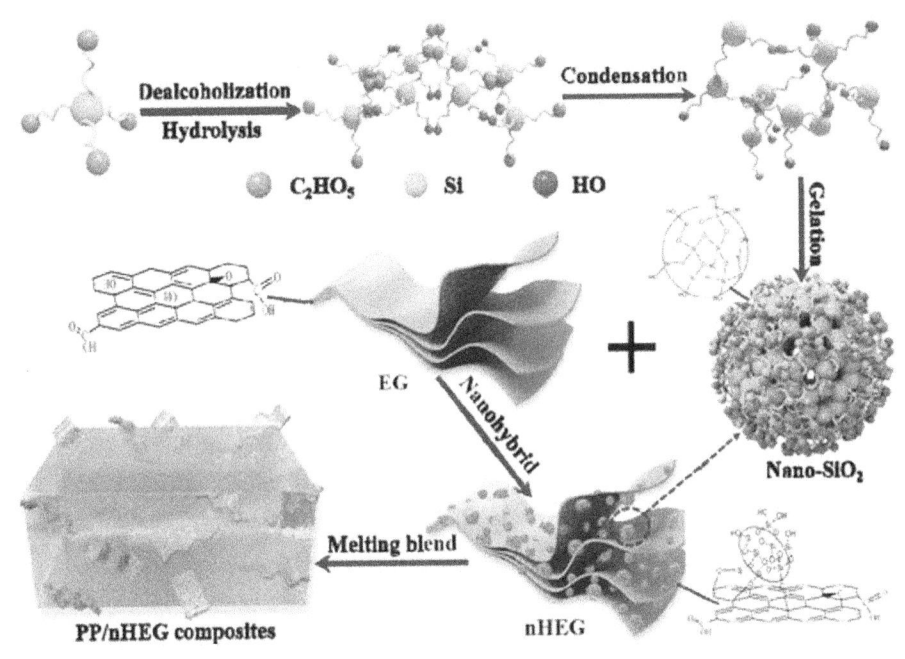

Figure 5. Synthetic route for the production of silica-nanohybrid-based PP composites. nHEG: nanohybrid EG. Reproduced with permission from reference (35). Copyright 2021 American Chemical Society.

The authors designed an in situ one-pot synthesis for dispersing the silica particles on the surface and between planes of graphite lamellae as PP additive. The authors ascribed the enhancements of the fire behavior to the formation of continuous and tight char during the combustion.

Kawahara and coworkers (*36*) further modified a silica-based additive by tailoring the particle with cyclotriphosphazene and bis(4-aminophenoxy)phenyl phosphine oxide. This way, the LOI increased in a remarkable way. A grafting process was used also by Gu et al. (*37*) to link the silica particles to poly(aniline) by using phosphoric acid. This system was then dispersed into an epoxy matrix, for which a significant decrease of the peak of HRR was observed.

This research field is vital, and every year, new systems based on tailoring the silica surface are produced by using hydrophobic or inorganic new materials for improving the FR performances of polymer composites (*38–41*).

Brannum and coworkers (*42*) developed an unconventional method, finding inspiration in biological systems. The authors reproduced a silica additive based on silica diatoms of algae to improve the flame retardance of polyurethane (PU) foams. The use of this bioinspired silica lowered the peak of HRR from 560 to 262 kW/m^2.

A further well-established route for the preparation of silica-based FRs is based on the sol–gel synthetic route. As reported by Zhu and coworkers (*43*), the creation of a silica network can be achieved using a template sol–gel route directly on cotton, polyester, or polyester–cotton fabrics. The adopted coating procedure allowed the formation of a nanostructured layer that increased the LOI. This approach opened the way toward a further development of FR additives for textile applications. Fan et al. (*44*) used tetraethyl orthosilicate as a precursor for the production of a silica-coated cotton fabric with superior flame resistance. The LOI of the treated fabrics was 24%. Similarly, Ren et al. (45) used the same approach with poly(aniline) fibers, reaching an LOI as high as 42%.

An improvement of the sol–gel approach is represented by the production of aerogels that display higher surface area and better dispersibility. As reported by Sun et al. (46), silica aerogel represents a very promising FR and smoke-suppressant additive for isocyanate-based polyimide foams. The authors described the easy mixing process: working with a polymer precursor slurry containing silica aerogel, an LOI of 33.0% was achieved; the total smoke production (TSP) was as low as 0.37 m^2/m^2, as assessed by forced-combustion tests.

Nechyporchuk et al. (47) created a complex network between cellulose fibers and silica gel through the interfacial complexation of the FR additive. The authors reported a very simple procedure for producing highly coated material with superior FR properties. Similarly, Fanglong and coworkers (48) proved the synergistic effect occurring in intumescent hybrid cellulose/silica treated cotton fabrics, using only a limited amount of FR additive (i.e., 4 wt %).

Phytic acid is another bioderived material that boosted the efficiency of silica materials, as reported by Cheng et al. (38, 49). The combination of tetraethoxysilane with phytic acid was exploited for treating wool fabrics. A self-extinguishing and smoke-suppressant material with enhanced thermal stability was obtained. The durability of the designed sol–gel treatment was very high; the conferred FR features were maintained even after 30 washing cycles.

Compounds Containing Phosphorous

Phosphorous-based compounds are very common and widely used inorganic FR additives. Their action mechanism is quite complex, and it is based on formation of polyphosphoric acid acting as a dehydrating agent, which favors the formation of a stable char and heat removal through vaporization of phosphorus compounds, which also determines a dilution of released gases and a decrease of viscosity of the melted polymer matrix (2).

Among phosphorous-containing compounds, ammonium polyphosphate is the most used (50) and studied (51–56). Ammonium polyphosphate and its derivatives are the common FRs employed for the production of PP-based FR composites.

Qin et al. (57) modified the original structure of ammonium polyphosphate by using a simple precipitation method, achieving a rougher surface (Figure 6).

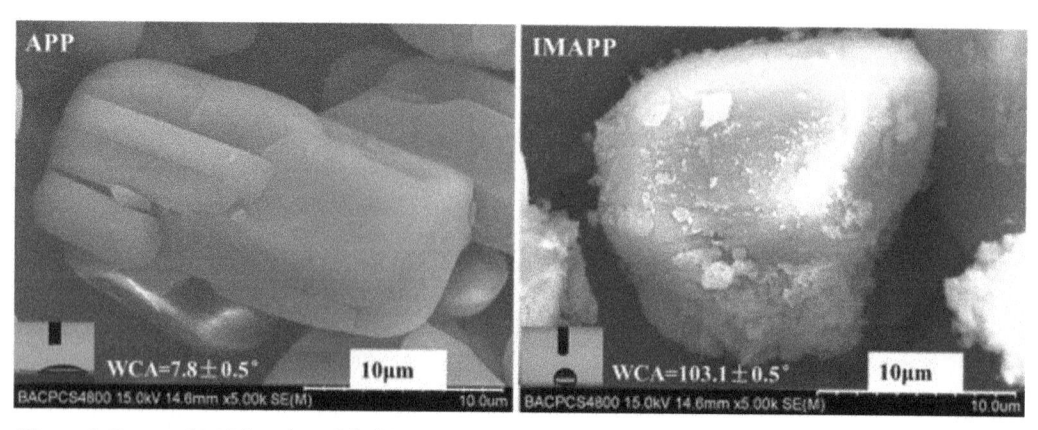

Figure 6. Pristine (AAP) and modified (IMAPP) ammonium polyphosphate; WCA: water contact angle. Reproduced with permission from reference (57). Copyright 2016 Elsevier.

Compounding ammonium polyphosphates and dipentaerythritol with PP improved the LOI from 25 to 32%. The same research group also studied a combination of aluminum phosphate and ammonium polyphosphate, highlighting the beneficial effect of the aluminum salt. Scanning electron microscopy showed a tight crystallized aluminum metaphosphate deriving from the phase transition of bare aluminum phosphate that compacted the char surface.

Similarly, Xu et al. (58) investigated the synergistic effect of ammonium polyphosphate and hydrotalcite. The designed FR system was able to increase the LOI up to 28%, achieving the V-0 rating in vertical flame spread tests. Furthermore, as assessed by cone calorimetry tests, the HRR, the total heat release (THR), and the total smoke release decreased by 44, 21, and 75%, respectively, as compared with the unfilled polymer.

Further modifications of ammonium polyphosphate systems were proposed by Guan et al. (59) by using ethanolamine to tailor the particle surface through an ion exchange reaction. This new additive led to an increase of the LOI of a maze flour/PP composite up to 43%, along with an increment of flexural strength of about 30%. Duan et al. (60) conjugated ammonium polyphosphate with a hyperbranched polymer synthesized by using tris(2-hydrooxyethyl) isocyanurate and 2-carboxyethyl(phenyl)phosphinic acid. V-0 rating was achieved for the FR system in vertical flame spread tests.

Liu and coworkers (61) used a noncovalent functionalization by combining ammonium polyphosphate with a dye as cointercalating agent for the preparation of low density poly(ethylene)/PP compounds. As assessed by forced-combustion tests, the peak of HRR decreased by 63% with respect to the unfilled compound.

Feng et al. (62) flame retarded low-density poly(ethylene) by using a combination of ammonium polyphosphate and an aromatic char-promoting agent based on triazine and aromatic moieties. The authors demonstrated that the hybrid material altered the decomposition behavior of the polymeric matrix, forming an intumescent char layer that lowered the flammability of the polymer. Smoke production rate, HRR, THR, TSP, mass loss, CO production, and CO_2 production were remarkably decreased. The analysis of the morphology of the residues after cone calorimetry tests by scanning electron microscopy demonstrated the formation of a continuous and compact intumescent char layer. The results gathered from Fourier transform infrared spectroscopy and laser Raman spectroscopy analyses revealed the achievement of an appropriate graphitization degree and network structure of the char residue, with improved strength and shield properties, hence enhancing the FR properties of the polyolefin.

Chen et al. (63) used a combination of ammonium polyphosphate and ferrite as filler of PU; the so-obtained system reduced the smoke produced with or without flame, HRR, THR, and total smoke release, as observed in cone calorimetry tests.

Shi et al. (64) proved that ammonium polyphosphate can be easily combined with graphitic carbon nitride for producing an effective FR for poly(styrene). The authors reported an increased thermal stability and a significant decrease of both HRR and THR; these findings were ascribed to the formation of a stable protective char layer.

Zhao et al. (65) combined ammonium polyphosphate with a layered double hydroxide; the resulting FR was incorporated into poly(vinyl alcohol). An LOI value of 33% was achieved with 25 wt % of ammonium polyphosphate mixed with 1 wt % of layered double hydroxide.

Even better results were obtained by Luo et al. (66) by replacing metal hydroxide with α-zirconium phosphate up to 13 wt % loading. The authors claimed a decrease of peak of HRR and of THR by 83 and 76%, respectively, as compared with the unfilled polymer.

Ammonium polyphosphate has found many applications as an FR additive for poly(lactic acid). Xue et al. (67) proved that 2 wt % of ammonium polyphosphate was sufficient to reach the V-0 rating in vertical flame spread tests without compromising the mechanical behavior of the polymer. Further improvements could be achieved by using particular morphologies such as leaf-structured ammonium polyphosphate, as reported by Zhao et al. (68), or by mixing ammonium polyphosphate with aluminum phosphate in a shape of microcapsules as described by Mao et al. (69).

Moving away from other phosphorous compounds, it is worthy to note the use of phosphate glasses as FRs for epoxy resins (70).

Phosphates were also combined with transition metal species and melamine (71) or with alkaline-earth elements such as magnesium and calcium (72, 73) for achieving high performances as FR additives in polyolefins.

Zhang and coworkers (74) proved that phosphate recovered from wastewater streams could be used as a valuable and sustainable FR for paper.

Organic FR Additives

Halogenated Organic Compounds

Halogen-containing organic (mostly chlorinated and brominated) FR compounds are among the most used additives for commercial applications (75), even though they are highly toxic and not environmentally friendly (76). These compounds act through a complex series of mechanisms that deplete the ignition and flame spread and remove heat from the combustion environment (77).

Halogen-containing organic compounds exploit their FR properties in the gas phase, reacting with hydrogenyl and hydroxyl radicals formed during the degradation process of the polymer, suppressing combustion. This mechanism is closely related to the lability of halogen–carbon bonds: the formed chloridyl and bromidyl radicals are able to slow down the radical propagation during combustion (78).

The halogen–carbon bond energy is also a very crucial parameter: linear aliphatic residues induce better performances than do aryl counterparts because of the lower carbon–halogen bond energies with more effective release of halogen radicals (79). Although these compounds are highly effective, their use has been discouraged with severe limitations and policy regulations that aim to preserve and protect the environment (80–82).

Halogenated compounds such as polyhalogenated phenyl ethers are decreasing in importance in the research world because their production and use have become a matter for serious environmental studies, which have traced them in atmosphere and water (83) as sources of dangerous pollution.

Compounds Containing Phosphorous and Nitrogen

The growing concerns about the use of halogenated FRs have boosted the use and development of organic compounds containing phosphorus and nitrogen as substitutes. Organic phosphorous compounds exploit their FR properties mainly in the condensed phase rather than in the gas phase (84). These compounds reduce the formation of combustible species while at the same time promoting charring. Furthermore, phosphorous FRs can promote the orientation and the strength of carbon layers (85).

Among phosphorous organic materials, one of the most used is 9,10-dihydro-9-oxa-10-phosphaphenanthrene-10-oxide (DOPO). DOPO has undergone extensive modification studies for improving its FR ability together with its dispersibility in different polymer matrices.

Ao et al. (86) modified DOPO by creating a polymer with phosphorus-linked groups through a two-step esterification/polycondensation synthesis. The so-obtained product was utilized as an additive for poly(ethylene terephthalate), increasing the LOI up to 31%.

Wang et al. (87) reported an interesting strategy based on a multicomponent reaction for tailoring DOPO systems that aims to improve the FR behavior of a polyester diol. The authors claimed a reduction of THR of 20% by using a limited amount of additive (i.e., 0.8 wt %), without affecting the mechanical properties.

DOPO is not the only widely used system, and other systems based on phosphine oxides have been designed and developed. In particular, Liu et al. (88) synthesized 2,3-dicarboxy propyl diphenyl phosphine oxide as an additive for polyamide 6. The authors achieved an LOI of 32% because of the quenching effect of phosphorus-containing radicals in the gas phase. They observed a positive effect of caprolactam formed through back-biting during combustion of the polymeric matrix.

Nazir and coworkers (89) proposed a more complex phosphorous-based FR bonding alkylated sulphones to various complex organic phosphate and phosphine oxides such as 6,6'-(sulfonylbis(ethane-2,1-diyl))bis(dibenzo[c,e][1,2]oxaphosphinine 6-oxide), sulfonylbis(ethane-2,1-diyl)bis(diphenylphosphine oxide), and tetraphenyl (sulfonyl bis(ethane-2,1-diyl)) bis(phosphonate). These systems increased the thermal stability of PP, decreasing the HRR by 48%. Simpler approaches also could be effective as proved by the use of bis-phenoxy (3-hydroxy) phenyl phosphine oxide (90) or phosphonamide (91). The latter was able to increase the LOI of poly(lactic acid) up to 38% in the presence of 2.5 wt % of FR. The conjugation of phosphorous-based additive with bioderived materials was reported by Nazir et al. (92), who bonded trivinyl phosphine oxide and cyclic amines with acetyl cellulose. As a result, a supramolecular network was obtained. This system was used with very good results as a coating agent for cotton fabrics, retaining up to 95 wt % of phosphorus content after 50 laundry cycles and reaching an LOI beyond 27%.

Heteroatoms can help improve the flame retardance induced by phosphorous compounds as reported by Varbanov and coworkers (93) for halogenated residues and by Tao et al. (94) for nitrogen-rich organic compounds.

Nitrogen-containing organic additives mainly comprise melamine and its derivatives, potentially in combination with phosphorus additives or alone.

A melamine-based additive FR mechanism was proposed by Yang et al. (95). The authors suggested that melamine in combination with phosphates act as acidic species and gas-releasing sources, inducing the formation of an intumescent char layer on the burning surface of the polymeric matrix. Xu et al. (96) supported this hypothesis further, observing the occurrence of cross-linking reactions with the formation of polyphosphate compounds at temperatures of less than 300 °C.

Several authors (97, 98) reported the synergistic effect of melamine with DOPO that allowed decreasing HRR up to 80%, as assessed in cone calorimetry tests.

Członka et al. (99) combined melamine with silica and an ionic liquid, namely methylimidazolium chloride. The obtained system was dispersed into PU foams. A remarkable decrease of the peak of HRR (−84%) was observed in the treated polymer with respect to the neat counterpart. The authors ascribed this improvement to the formation of a tight char layer, as observed by scanning electron microscopy analyses.

Melamine could also be combined with inorganic salts such as copper phosphate (71), zinc borate (100), and aluminum salts (101, 102) or organic derivatives such as phytic acid.

The combination with phytic acid represents a very attractive matter for exploiting the FR properties of melamine as reported by Shang et al. (103). The authors described the production of a layered material composed of nanoflakes assembled though a supramolecular process used as a filler for PP. Shang and coworkers documented a decrease of peak HRR of 22% when only 2 wt % of the filler was added to the polyolefin.

Similarly, Li et al. (104) produced melamine phytate nanosheets doped with manganese salts. In this case, the fire behavior of PP loaded with 13.5 wt % of the filler was greatly improved, achieving an LOI of 32% and lowering the HRR by 56%.

Carbon-Based Compounds

Carbon-containing FRs create a physical barrier to burning and slow down the degradation of the unburned zone of materials. Carbon-based additives can induce intumescence, creating a foamed cellular structure that preserves the underlying material from combustion (105).

Among carbonaceous materials, carbon black is undoubtedly the most utilized additive for polymer composite production (106). Carbon black has many applications as an FR additive in various polymeric matrices (107). Yang et al. (108) reported the use of carbon black as an additive for carbon-fiber-reinforced PP composites. The authors achieved a 70% decrease of the peak of both HRR and THR. These findings were attributed to the formation of a tough filler three-dimensional network with carbon fibers acting as bridges that connected individual carbon black particles. Liu et al. (109) showed that carbon black acted also as a smoke suppressant in PU composites, depleting the amount of ammonia released during combustion. Carbon black was also combined with melamine and phosphates by Yan et al. (110) in an epoxy resin. The obtained results showed a beneficial and synergistic effect of the presence of 15 wt % of carbon black that reduced the peaks of both HRR and TSP by 67 and 54%, respectively, as compared with the unfilled polymer.

Another carbon additive widely used as an FR is graphite. Higginbotham and coworkers (111) reported a comprehensive study on the effect of graphite oxide on the flame retardance of several polymer matrices (Figure 7).

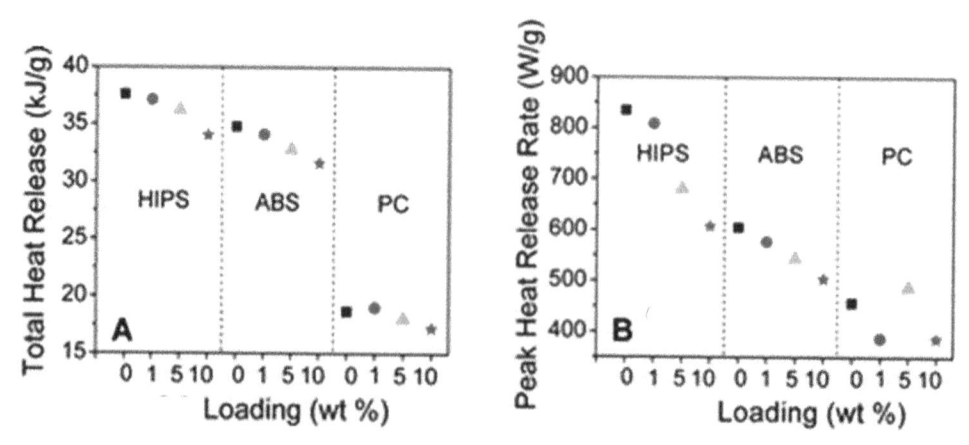

Figure 7. The effect of graphite oxide on the THR (A) and peak HRR (B) of high-impact poly(styrene) (HIPS), ABS polymer, and polycarbonate (PC). Reproduced with permission from reference (111). Copyright 2009 American Chemical Society.

The addition of graphite oxide (up to 10 wt % loading) induced an overall improvement of the flame retardance of each selected matrix together with an increased thermal stability. Similarly, Qiu et al. (*112*) used graphene oxide to reduce the peak of HRR of poly(vinyl alcohol) by 63% as compared with the unfilled counterpart.

Laachachi et al. (*113*) investigated the use of expanded graphite in epoxy matrix composites, which showed a decrease of HRR together with a great delay of the time to ignition.

The FR effect of graphite is strongly related to its particle size distribution as reported by Wang et al. (*114*), who showed that graphite with an average particle size of 1500 mesh provided the best performances in a protective coating layer on plywood. The authors claimed a decrease of the peak of HRR by 75% and a fire growth index of only 0.06 kW m^{-2}s^{-1}. Further improvements of graphite performances could be achieved by adding inorganic salts (*115*, *116*) that led to the formation of a protective char layer during combustion. Lee et al. (*117*) showed the beneficial effect of the addition of a small quantity of carbon nanotubes to graphite; this allowed the formation of a network between carbon filler particles that simultaneously inhibited the polymer degradation and the evaporation of volatile organic matter during combustion.

Another carbon source that looks very promising as an FR is biochar, the carbon recovered from pyrolysis of biomass. Bhattacharyya and coworkers have worked on biochar-based PP composites (*118–121*), documenting the effectiveness of this bioderived carbon material as an FR. The effect of neat biochar is not much different from that of any of other carbon-based FR even if there are some unique features to be considered. Depending on its origin, biochar has different amounts of inorganic components; Matta et al. (*122*) exploited this feature to prove a correlation between the amount and composition of inorganic fraction with the resulting FR properties. The authors proved that an average amount of 20 wt % of inorganic content induced the best fire performances in cone calorimetry tests. The combination or carbon-based and inorganic components are among the most interesting points in the use of biochar as an FR.

Biochar was used in several polymer matrixes such as poly(ethylene) (*123*) and ethylene-vinyl acetate (*122*) because of its easy dispersibility, as reported by many authors, and also as a coating component for cotton textiles (*124*).

Biomacromolecules and Biosourced Products

Biomacromolecules and biosourced products are emerging materials in flame retardance. Because of environmental concerns regarding the use of traditional FRs, materials produced from biomasses have become the central pillar of advanced research in the field.

Cellulose-based materials have been mixed with EG by Zheng et al. (*125*) for producing a halogen-free FR for PP. The original cellulose scaffold was functionalized with phosphates and melamine; this way, the LOI of PP was 32% in the presence of 30 wt % of filler. As reported by Tong et al. (*126*), cellulose could also be used without other additives for producing polymeric films with intrinsic FR properties because of cellulose nanofibrils.

Lignin was also used in a very effective way to improve the fire behavior of PP as reported by De Chirico et al. (*127*). Liang and coworkers used a lignin derivative in combination with phosphates and melamine for improving the LOI of an epoxy resin up to 36% (at 20 wt % of filler loading) (*128*). The authors claimed a simultaneous reduction of TSP (−43%) and of THR (−25%). Further modification of lignin could boost the smoke suppression as in the case of its combination with layered double hydroxides (*129*).

Chitosan is another valuable source for the production of bioderived FRs (*130*). Chen et al. (*131*) modified chitosan with DOPO, inducing a decrease of both THR and TSP by 39 and 72%, respectively, in epoxy systems. The authors proved that chitosan acts as an intumescent FR. Similar findings were discussed by Kundu et al. (*132*) in terms of the effect of chitosan on polyamide systems and by Yang and coworkers (*133*) on poly(vinyl alcohol).

Chitosan is also widely used as a coating for cotton fabrics (*134*) and synthetic textiles (*135*). Deoxyribonucleic acid has been used as an alternative FR. Rajczak et al. (*136*) tested the effect of a clay modified with deoxyribonucleic acid on the fire behavior of an ethylene-vinyl acetate copolymer. In particular, it was found that the modified clay promoted a remarkable lowering of HRR and peak of HRR and TSP, as assessed by cone calorimetry tests. Deoxyribonucleic acid also has proven to be a good intumescent coating system for different fabrics (*134, 137*).

Another biosourced molecule that has attracted great research interest is phytic acid, the principal organic molecule used by plants to store phosphates in tissues. In particular, the high phosphorous content together with the renewable origin make the use of phytic acid very suitable for FR purposes. Several authors (*49, 138–144*) have used phytic acid as components for both coatings formulations and bulk additives with good FR results.

Conclusion

In this chapter, we briefly overviewed the main classes of FR compounds and their applications, providing an incomplete but comprehensive guide to the field. We wish to encourage the reader to approach this immense field, keeping in mind the key points discussed, such as the main mechanisms involved in the action of each class of FRs together with the almost infinite material combinations available.

Finally, we wish to point out that there is not the perfect and universal FR system but that the choice of an FR system should be balanced according to the final application, the environmental impact, and, of course, the overall fire performance.

We firmly believe that sustainable materials such as biomacromolecules and biochar will at least partially replace the traditional FR additives, promoting a greener mindset and a more accountable society.

References

1. Morgan, A. B. A Review of Transition Metal-Based Flame Retardants: Transition Metal Oxide/Salts, and Complexes. In *Fire and Polymers V*; American Chemical Society: Washington, DC, 2009; Vol. 1013, pp 312–328,
2. Green, J. A review of phosphorus-containing flame retardants. *Journal of Fire Sciences* **1992**, *10*, 470–487.
3. Kiliaris, P.; Papaspyrides, C. Polymer/layered silicate (clay) nanocomposites: an overview of flame retardancy. *Progress in Polymer Science* **2010**, *35*, 902–958.
4. Chen, L.; Wang, Y. Z. A review on flame retardant technology in China. Part I: development of flame retardants. *Polymers for Advanced Technologies* **2010**, *21*, 1–26.
5. Sathishkumar, T.; Satheeshkumar, S.; Naveen, J. Glass fiber-reinforced polymer composites–a review. *Journal of Reinforced Plastics and Composites* **2014**, *33*, 1258–1275.
6. Gao, Y.; Wu, J.; Wang, Q.; Wilkie, C. A.; O'Hare, D. Flame retardant polymer/layered double hydroxide nanocomposites. *Journal of Materials Chemistry A* **2014**, *2*, 10996–11016.

7. McMahon, M. E.; Santucci, R. J., Jr; Glover, C. F.; Kannan, B.; Walsh, Z. R.; Scully, J. R. A Review of Modern Assessment Methods for Metal and Metal-Oxide Based Primers for Substrate Corrosion Protection. *Frontiers in Materials* **2019**, *6*, 190.
8. Cusack, P.; Monk, A.; Pearce, J.; Reynolds, S. J. An investigation of inorganic tin flame retardants which suppress smoke and carbon monoxide emission from burning brominated polyester resins. *Fire and Materials* **1989**, *14*, 23–29.
9. Liu, S.-P. Flame retardant and mechanical properties of polyethylene/magnesium hydroxide/montmorillonite nanocomposites. *Journal of Industrial and Engineering Chemistry* **2014**, *20*, 2401–2408.
10. Hirschler, M.; Thevaranjan, T. Effects of magnesium oxide/hydroxide on flammability and smoke production tendency of polystyrene. *European Polymer Journal* **1985**, *21*, 371–375.
11. Alfonso, G. C.; Costa, G.; Pasolini, M.; Russo, S.; Ballistreri, A.; Montaudo, G.; Puglisi, C. Flame-resistant polycaproamide by anionic polymerization of ε-caprolactam in the presence of suitable flame-retardant agents. *Journal of Applied Polymer Science* **1986**, *31*, 1373–1382.
12. Xu, H.; Deng, X.-r. Preparation and properties of superfine $Mg(OH)_2$ flame retardant. *Transactions of Nonferrous Metals Society of China* **2006**, *16*, 488–492.
13. Qiu, L.; Xie, R.; Ding, P.; Qu, B. Preparation and characterization of $Mg(OH)_2$ nanoparticles and flame-retardant property of its nanocomposites with EVA. *Composite Structures* **2003**, *62*, 391–395.
14. Fang, H.; Zhou, T.; Chen, X.; Li, S.; Shen, G.; Liao, X. Controlled preparation and characterization of nano-sized hexagonal $Mg(OH)_2$ flame retardant. *Particuology* **2014**, *14*, 51–56.
15. Cao, H.; Zheng, H.; Yin, J.; Lu, Y.; Wu, S.; Wu, X.; Li, B. $Mg(OH)_2$ Complex Nanostructures with Superhydrophobicity and Flame Retardant Effects. *Journal of Physical Chemistry C* **2010**, *114*, 17362–17368.
16. Wang, M.; Han, X.-W.; Liu, L.; Zeng, X.-F.; Zou, H.-K.; Wang, J.-X.; Chen, J.-F. Transparent Aqueous $Mg(OH)_2$ Nanodispersion for Transparent and Flexible Polymer Film with Enhanced Flame-Retardant Property. *Industrial & Engineering Chemistry Research* **2015**, *54*, 12805–12812.
17. Ren, M.; Yang, M.; Li, S.; Chen, G.; Yuan, Q. High throughput preparation of magnesium hydroxide flame retardant via microreaction technology. *RSC Advances* **2016**, *6*, 92670–92681.
18. Rigolo, M.; Woodhams, R. Basic magnesium carbonate flame retardants for polypropylene. *Polymer Engineering & Science* **1992**, *32*, 327–334.
19. Li, X.; Zhang, K.; Shi, R.; Ma, X.; Tan, L.; Ji, Q.; Xia, Y. Enhanced flame-retardant properties of cellulose fibers by incorporation of acid-resistant magnesium-oxide microcapsules. *Carbohydrate Polymers* **2017**, *176*, 246–256.
20. Hiremath, P.; Arunkumar, H. S.; Shettar, M. Investigation on Effect of Aluminium Hydroxide on Mechanical and Fire Retardant Properties of GFRP- Hybrid Composites. *Materials Today: Proceedings* **2017**, *4*, 10952–10956.
21. Laachachi, A.; Ferriol, M.; Cochez, M.; Cuesta, J.-M. L.; Ruch, D. A comparison of the role of boehmite (AlOOH) and alumina (Al_2O_3) in the thermal stability and flammability of poly (methyl methacrylate). *Polymer Degradation and Stability* **2009**, *94*, 1373–1378.

22. Ahmad, F.; Zulkurnain, E. S. B.; Ullah, S.; Al-Sehemi, A. G.; Raza, M. R. Improved fire resistance of boron nitride/epoxy intumescent coating upon minor addition of nano-alumina. *Materials Chemistry and Physics* **2020**, *256*, 123634.
23. Dogan, M.; Unlu, S. M. Flame retardant effect of boron compounds on red phosphorus containing epoxy resins. *Polymer Degradation and Stability* **2014**, *99*, 12–17.
24. Shen, K. K.; Kochesfahani, S.; Jouffret, F. Zinc borates as multifunctional polymer additives. *Polymers for Advanced Technologies* **2008**, *19*, 469–474.
25. Wang, Q.; Undrell, J. P.; Gao, Y.; Cai, G.; Buffet, J.-C.; Wilkie, C. A.; O'Hare, D. Synthesis of Flame-Retardant Polypropylene/LDH-Borate Nanocomposites. *Macromolecules* **2013**, *46*, 6145–6150.
26. Riyazuddin; Bano, S.; Husain, F. M.; Khan, R. A.; Alsalme, A.; Siddique, J. A. Influence of Antimony Oxide on Epoxy Based Intumescent Flame Retardation Coating System. *Polymers* **2020**, *12*, 2721.
27. Li, N.; Xia, Y.; Mao, Z.; Wang, L.; Guan, Y.; Zheng, A. Influence of antimony oxide on flammability of polypropylene/intumescent flame retardant system. *Polymer Degradation and Stability* **2012**, *97*, 1737–1744.
28. Li, F.; Jianhuai, W.; Jiongtian, L.; BingguoO, W.; Shuojiang, S. Preparation and Fire Retardancy of Antimony Oxide Nanoparticles/Mica Composition. *Journal of Composite Materials* **2007**, *41*, 1487–1497.
29. Edelson, D.; Kuck, V.; Lum, R.; Scalco, E.; Starnes, W., Jr; Kaufman, S. Anomalous behavior of molybdenum oxide as a fire retardant for polyvinyl chloride. *Combustion and Flame* **1980**, *38*, 271–283.
30. Zhang, M.; Wu, W.; He, S.; Wang, X.; Jiao, Y.; Qu, H.; Xu, J. Synergistic flame retardant effects of activated carbon and molybdenum oxide in poly (vinyl chloride). *Polymer International* **2018**, *67*, 445–452.
31. Nabipour, H.; Wang, X.; Song, L.; Hu, Y. Metal-organic frameworks for flame retardant polymers application: A critical review. *Composites Part A: Applied Science and Manufacturing* **2020**, *139*, 106113–106123.
32. Nalawade, P.; Aware, B.; Kadam, V.; Hirlekar, R. Layered double hydroxides: A review. *Journal of Scientific & Industrial Research* **2009**, *68*, 262–272.
33. Mercado, L.; Galia, M.; Reina, J. Silicon-containing flame retardant epoxy resins: Synthesis, characterization and properties. *Polymer Degradation and Stability* **2006**, *91*, 2588–2594.
34. Liu, Y.; Zhao, J.; Deng, C.-L.; Chen, L.; Wang, D.-Y.; Wang, Y.-Z. Flame-Retardant Effect of Sepiolite on an Intumescent Flame-Retardant Polypropylene System. *Industrial & Engineering Chemistry Research* **2011**, *50*, 2047–2054.
35. Wang, N.; Chen, S.; Li, L.; Bai, Z.; Guo, J.; Qin, J.; Chen, X.; Zhao, R.; Zhang, K.; Wu, H. An Environmentally Friendly Nanohybrid Flame Retardant with Outstanding Flame-Retardant Efficiency for Polypropylene. *Journal of Physical Chemistry C* **2021**, *125*, 5185–5196.
36. Kawahara, T.; Yuuki, A.; Hashimoto, K.; Fujiki, K.; Yamauchi, T.; Tsubokawa, N. Immobilization of flame-retardant onto silica nanoparticle surface and properties of epoxy resin filled with the flame-retardant-immobilized silica (2). *Reactive and Functional Polymers* **2013**, *73*, 613–618.

37. Gu, H.; Guo, J.; He, Q.; Tadakamalla, S.; Zhang, X.; Yan, X.; Huang, Y.; Colorado, H. A.; Wei, S.; Guo, Z. Flame-Retardant Epoxy Resin Nanocomposites Reinforced with Polyaniline-Stabilized Silica Nanoparticles. *Industrial & Engineering Chemistry Research* **2013**, *52*, 7718–7728.
38. Cheng, X.-W.; Tang, R.-C.; Guan, J.-P.; Zhou, S.-Q. An eco-friendly and effective flame retardant coating for cotton fabric based on phytic acid doped silica sol approach. *Progress in Organic Coatings* **2020**, *141*, 105539.
39. Nageswara Rao, T.; Hussain, I.; Heun Koo, B. Enhanced thermal properties of silica nanoparticles and chitosan bio-based intumescent flame retardant Polyurethane coatings. *Materials Today: Proceedings* **2020**, *27*, 369–375.
40. Wang, Y.; Zhao, J. Preliminary study on decanoic/palmitic eutectic mixture modified silica fume geopolymer-based coating for flame retardant plywood. *Construction and Building Materials* **2018**, *189*, 1–7.
41. Xu, S.; Zhou, C.; Li, J.; Shen, L.; Lin, H. Simultaneously improving mechanical strength, hydrophobic property and flame retardancy of ethylene vinyl acetate copolymer/intumescent flame retardant/FeOOH by introducing modified fumed silica. *Materials Today Communications* **2021**, *26*, 102114.
42. Brannum, D. J.; Price, E. J.; Villamil, D.; Kozawa, S.; Brannum, M.; Berry, C.; Semco, R.; Wnek, G. E. Flame-Retardant Polyurethane Foams: One-Pot, Bioinspired Silica Nanoparticle Coating. *ACS Applied Polymer Materials* **2019**, *1*, 2015–2022.
43. Zhu, Y.-s.; You, F.; Zhou, H.-t.; Huangfu, W.-h.; Wang, Z.-h. Pyrolysis Properties and Flame Retardant Effects of Fabrics Finished by Hybrid Silica-based Sols. *Procedia Engineering* **2018**, *211*, 1091–1101.
44. Fan, D.-d.; You, F.; Zhang, Y.; Huang, Z. Flame Retardant Effects of Fabrics Finished by Hybrid Nano-Micro Silica-based Sols. *Procedia Engineering* **2018**, *211*, 160–168.
45. Ren, Y.; Zhang, Y.; Gu, Y.; Zeng, Q. Flame retardant polyacrylonitrile fabrics prepared by organic-inorganic hybrid silica coating via sol-gel technique. *Progress in Organic Coatings* **2017**, *112*, 225–233.
46. Sun, G.; Duan, T.; Liu, C.; Zhang, L.; Chen, R.; Wang, J.; Han, S. Fabrication of flame-retardant and smoke-suppressant isocyanate-based polyimide foam modified by silica aerogel thermal insulation and flame protection layers. *Polymer Testing* **2020**, *91*, 106738.
47. Nechyporchuk, O.; Bordes, R.; Köhnke, T. Wet Spinning of Flame-Retardant Cellulosic Fibers Supported by Interfacial Complexation of Cellulose Nanofibrils with Silica Nanoparticles. *ACS Applied Materials & Interfaces* **2017**, *9*, 39069–39077.
48. Fanglong, Z.; Qun, X.; Qianqian, F.; Rangtong, L.; Kejing, L. Influence of nano-silica on flame resistance behavior of intumescent flame retardant cellulosic textiles: Remarkable synergistic effect? *Surface and Coatings Technology* **2016**, *294*, 90–94.
49. Cheng, X.-W.; Guan, J.-P.; Yang, X.-H.; Tang, R.-C.; Fan, Y. Phytic acid/silica organic-inorganic hybrid sol system: a novel and durable flame retardant approach for wool fabric. *Journal of Materials Research and Technology* **2020**, *9*, 700–708.
50. Schacker, O.; Wanzke, W. Compounding with ammonium polyphosphate-based flame retardants. *Plastics, Additives and Compounding* **2002**, *4*, 28–33.

51. Montaudo, G.; Scamporrino, E.; Vitalini, D. Intumescent flame retardants for polymers. II. The polypropylene-ammonium polyphosphate-polyurea system. *Journal of Polymer Science: Polymer Chemistry Edition* **1983**, *21*, 3361–3371.
52. Ballistreri, A.; Montaudo, G.; Puglisi, C.; Scamporrino, E.; Vitalini, D. Intumescent flame retardants for polymers. I. The poly (acrylonitrile)–ammonium polyphosphate–hexabromocyclododecane system. *Journal of Applied Polymer Science* **1983**, *28*, 1743–1750.
53. Camino, G.; Costa, L.; Trossarelli, L. Study of the mechanism of intumescence in fire retardant polymers: Part I—Thermal degradation of ammonium polyphosphate-pentaerythritol mixtures. *Polymer Degradation and Stability* **1984**, *6*, 243–252.
54. Camino, G.; Costa, L.; Trossarelli, L. Study of the mechanism of intumescence in fire retardant polymers: Part V—Mechanism of formation of gaseous products in the thermal degradation of ammonium polyphosphate. *Polymer Degradation and Stability* **1985**, *12*, 203–211.
55. Camino, G.; Costa, L.; Trossarelli, L.; Costanzi, F.; Pagliari, A. Study of the mechanism of intumescence in fire retardant polymers: Part VI—Mechanism of ester formation in ammonium polyphosphate-pentaerythritol mixtures. *Polymer Degradation and Stability* **1985**, *12*, 213–228.
56. Montaudo, G.; Scamporrino, E.; Puglisi, C.; Vitalini, D. Intumescent flame retardant for polymers. III. The polypropylene–ammonium polyphosphate–polyurethane system. *Journal of Applied Polymer Science* **1985**, *30*, 1449–1460.
57. Qin, Z.; Li, D.; Yang, R. Study on inorganic modified ammonium polyphosphate with precipitation method and its effect in flame retardant polypropylene. *Polymer Degradation and Stability* **2016**, *126*, 117–124.
58. Xu, S.; Zhang, M.; Li, S.-Y.; Zeng, H.-Y.; Du, J.-Z.; Chen, C.-R.; Wu, K.; Tian, X.-Y.; Pan, Y. The effect of ammonium polyphosphate on the mechanism of phosphorous-containing hydrotalcite synergism of flame retardation of polypropylene. *Applied Clay Science* **2020**, *185*, 105348.
59. Guan, Y.-H.; Huang, J.-Q.; Yang, J.-C.; Shao, Z.-B.; Wang, Y.-Z. An Effective Way To Flame-Retard Biocomposite with Ethanolamine Modified Ammonium Polyphosphate and Its Flame Retardant Mechanisms. *Industrial & Engineering Chemistry Research* **2015**, *54*, 3524–3531.
60. Duan, L.; Yang, H.; Song, L.; Hou, Y.; Wang, W.; Gui, Z.; Hu, Y. Hyperbranched phosphorus/nitrogen-containing polymer in combination with ammonium polyphosphate as a novel flame retardant system for polypropylene. *Polymer Degradation and Stability* **2016**, *134*, 179–185.
61. Liu, Y.; Gao, Y.; Zhang, Z.; Wang, Q. Preparation of ammonium polyphosphate and dye co-intercalated LDH/polypropylene composites with enhanced flame retardant and UV resistance properties. *Chemosphere* **2021**, *277*, 130370.
62. Feng, C.; Liang, M.; Jiang, J.; Huang, J.; Liu, H. Preparation and characterization of oligomeric char forming agent and its effect on the thermal degradation and flame retardant properties of LDPE with ammonium polyphosphate. *Journal of Analytical and Applied Pyrolysis* **2016**, *119*, 75–86.

63. Chen, M.-J.; Shao, Z.-B.; Wang, X.-L.; Chen, L.; Wang, Y.-Z. Halogen-Free Flame-Retardant Flexible Polyurethane Foam with a Novel Nitrogen–Phosphorus Flame Retardant. *Industrial & Engineering Chemistry Research* **2012**, *51*, 9769–9776.
64. Shi, Y.; Xing, W.; Wang, B.; Hong, N.; Zhu, Y.; Wang, C.; Gui, Z.; Yuen, R. K. K.; Hu, Y. Synergistic effect of graphitic carbon nitride and ammonium polyphosphate for enhanced thermal and flame retardant properties of polystyrene. *Materials Chemistry and Physics* **2016**, *177*, 283–292.
65. Zhao, C.-X.; Liu, Y.; Wang, D.-Y.; Wang, D.-L.; Wang, Y.-Z. Synergistic effect of ammonium polyphosphate and layered double hydroxide on flame retardant properties of poly(vinyl alcohol). *Polymer Degradation and Stability* **2008**, *93*, 1323–1331.
66. Luo, Y.; Xie, D.; Chen, Y.; Han, T.; Chen, R.; Sheng, X.; Mei, Y. Synergistic effect of ammonium polyphosphate and α-zirconium phosphate in flame-retardant poly(vinyl alcohol) aerogels. *Polymer Degradation and Stability* **2019**, *170*, 109019.
67. Xue, Y.; Zuo, X.; Wang, L.; Zhou, Y.; Pan, Y.; Li, J.; Yin, Y.; Li, D.; Yang, R.; Rafailovich, M. H.; Guo, Y. Enhanced flame retardancy of poly(lactic acid) with ultra-low loading of ammonium polyphosphate. *Composites Part B: Engineering* **2020**, *196*, 108124.
68. Zhao, X.; Chen, L.; Li, D.-F.; Fu, T.; He, L.; Wang, X.-L.; Wang, Y.-Z. Biomimetic construction peanut-leaf structure on ammonium polyphosphate surface: Improving its compatibility with poly(lactic acid) and flame-retardant efficiency simultaneously. *Chemical Engineering Journal* **2021**, *412*, 128737.
69. Mao, N.; Jiang, L.; Li, X.; Gao, Y.; Zang, Z.; Peng, S.; Ji, L.; Lv, C.; Guo, J.; Wang, H.; Niu, E.; Zhai, Y. Core-shell ammonium polyphosphate@nanoscopic aluminum hydroxide microcapsules: Preparation, characterization, and its flame retardancy performance on wood pulp paper. *Chemical Engineering Journal Advances* **2021**, *6*, 100096.
70. Liu, W.; Pan, Y.-T.; Zhang, J.; Zhang, L.; Moya, J. S.; Cabal, B.; Wang, D.-Y. Low-melting phosphate glasses as flame-retardant synergists to epoxy: Barrier effects vs flame retardancy. *Polymer Degradation and Stability* **2021**, *185*, 109495.
71. Salasinska, K.; Mizera, K.; Celiński, M.; Kozikowski, P.; Borucka, M.; Gajek, A. Thermal properties and fire behavior of polyethylene with a mixture of copper phosphate and melamine phosphate as a novel flame retardant. *Fire Safety Journal* **2020**, *115*, 103137.
72. Yan, L.; Xu, Z.; Liu, D. Synthesis and application of novel magnesium phosphate ester flame retardants for transparent intumescent fire-retardant coatings applied on wood substrates. *Progress in Organic Coatings* **2019**, *129*, 327–337.
73. Zhang, Q.; Zhang, X.; Cheng, W.; Li, Z.; Li, Q. In situ-synthesis of calcium alginate nano-silver phosphate hybrid material with high flame retardant and antibacterial properties. *International Journal of Biological Macromolecules* **2020**, *165*, 1615–1625.
74. Zhang, X.; Shen, J.; Ma, Y.; Liu, L.; Meng, R.; Yao, J. Highly efficient adsorption and recycle of phosphate from wastewater using flower-like layered double oxides and their potential as synergistic flame retardants. *Journal of Colloid and Interface Science* **2020**, *562*, 578–588.
75. Green, J. An overview of the fire retardant chemicals industry, past—present—future. *Fire and Materials* **1995**, *19*, 197–204.
76. Pearce, E.; Liepins, R. Public health implications of components of plastics manufacture. Flame retardants. *Environmental Health Perspectives* **1975**, *11*, 59–69.

77. Weil, E. D.; Zhu, W.; Patel, N.; Mukhopadhyay, S. M. A systems approach to flame retardancy and comments on modes of action. *Polymer Degradation and Stability* **1996**, *54*, 125–136.

78. Eljarrat, E.; Barceló, D. *Brominated Flame Retardants*; Springer Science & Business Media: New York, 2011; Vol. 16.

79. Camino, G.; Costa, L.; Di Cortemiglia, M. L. Overview of fire retardant mechanisms. *Polymer Degradation and Stability* **1991**, *33*, 131–154.

80. Shaw, S. Halogenated flame retardants: do the fire safety benefits justify the risks? *Reviews on Environmental Health* **2010**, *25*, 261–306.

81. Zaikov, G.; Lomakin, S. Ecological issue of polymer flame retardancy. *Journal of Applied Polymer Science* **2002**, *86*, 2449–2462.

82. Hull, T.; Law, R.; Bergman, Å. Environmental drivers for replacement of halogenated flame retardants. *Polymer Green Flame Retardants* **2014**, 119–179.

83. Ma, S.; Yu, Y.; Yang, Y.; Li, G.; An, T. A new advance in the potential exposure to "old" and "new" halogenated flame retardants in the atmospheric environments and biota: From occurrence to transformation products and metabolites. *Critical Reviews in Environmental Science and Technology* **2020**, *50*, 1935–1983.

84. Jianjun, L.; Yuxiang, O. The flame-retardation mechanism of organic phosphorus flame retardants. In *Theory of Flame Retardation of Polymeric Materials*; De Gruyter: Boston, 2019; pp 69–94,

85. Chen, X. Y.; Huang, Z. H.; Xi, X. Q.; Li, J.; Fan, X. Y.; Wang, Z. Synergistic effect of carbon and phosphorus flame retardants in rigid polyurethane foams. *Fire and Materials* **2018**, *42*, 447–453.

86. Ao, X.; Du, Y.; Yu, D.; Wang, W.; Yang, W.; Sun, B.; Zhu, M. Synthesis, characterization of a DOPO-based polymeric flame retardant and its application in polyethylene terephthalate. *Progress in Natural Science: Materials International* **2020**, *30*, 200–207.

87. Wang, H.; Liu, Q.; Zhao, X.; Jin, Z. Synthesis of reactive DOPO-based flame retardant and its application in polyurethane elastomers. *Polymer Degradation and Stability* **2021**, *183*, 109440.

88. Liu, K.; Li, Y.; Tao, L.; Liu, C.; Xiao, R. Synthesis and characterization of inherently flame retardant polyamide 6 based on a phosphine oxide derivative. *Polymer Degradation and Stability* **2019**, *163*, 151–160.

89. Nazir, R.; Gooneie, A.; Lehner, S.; Jovic, M.; Rupper, P.; Ott, N.; Hufenus, R.; Gaan, S. Alkyl sulfone bridged phosphorus flame-retardants for polypropylene. *Materials & Design* **2021**, *200*, 109459.

90. Ren, H.; Sun, J.; Wu, B.; Zhou, Q. Synthesis and properties of a phosphorus-containing flame retardant epoxy resin based on bis-phenoxy (3-hydroxy) phenyl phosphine oxide. *Polymer Degradation and Stability* **2007**, *92*, 956–961.

91. Liu, L.; Xu, Y.; Pan, Y.; Xu, M.; Di, Y.; Li, B. Facile synthesis of an efficient phosphonamide flame retardant for simultaneous enhancement of fire safety and crystallization rate of poly (lactic acid). *Chemical Engineering Journal* **2020**, 127761.

92. Nazir, R.; Parida, D.; Borgstädt, J.; Lehner, S.; Jovic, M.; Rentsch, D.; Bülbül, E.; Huch, A.; Altenried, S.; Ren, Q.; Rupper, P.; Annaheim, S.; Gaan, S. In-situ phosphine oxide physical networks: A facile strategy to achieve durable flame retardant and antimicrobial treatments of cellulose. *Chemical Engineering Journal* **2020**, 128028.

93. Varbanov, S.; Vasileva, V.; Vaseva, I.; Borisov, G. Flame-retardant properties of epoxide and unsaturated polyester resins modified with tertiary phosphine oxides containing chlorinated phenoxy groups. *European Polymer Journal* **1986**, *22*, 211–215.

94. Tao, W.; Hu, X.; Sun, J.; Qian, L.; Li, J. Effects of P–N flame retardants based on cytosine on flame retardancy and mechanical properties of polyamide 6. *Polymer Degradation and Stability* **2020**, *174*, 109092.

95. Yang, H.; Song, L.; Tai, Q.; Wang, X.; Yu, B.; Yuan, Y.; Hu, Y.; Yuen, R. K. K. Comparative study on the flame retarded efficiency of melamine phosphate, melamine phosphite and melamine hypophosphite on poly(butylene succinate) composites. *Polymer Degradation and Stability* **2014**, *105*, 248–256.

96. Xu, D.; Lu, H.; Huang, Q.; Deng, B.; Li, L. Flame-retardant effect and mechanism of melamine phosphate on silicone thermoplastic elastomer. *RSC Advances* **2018**, *8*, 5034–5041.

97. Bifulco, A.; Parida, D.; Salmeia, K. A.; Lehner, S.; Stämpfli, R.; Markus, H.; Malucelli, G.; Branda, F.; Gaan, S. Improving flame retardancy of in-situ silica-epoxy nanocomposites cured with aliphatic hardener: Combined effect of DOPO-based flame-retardant and melamine. *Composites Part C: Open Access* **2020**, *2*, 100022.

98. Koedel, J.; Callsen, C.; Weise, M.; Puchtler, F.; Weidinger, A.; Altstaedt, V.; Schobert, R.; Biersack, B. Investigation of melamine and DOPO-derived flame retardants for the bioplastic cellulose acetate. *Polymer Testing* **2020**, *90*, 106702.

99. Członka, S.; Strąkowska, A.; Strzelec, K.; Kairytė, A.; Kremensas, A. Melamine, silica, and ionic liquid as a novel flame retardant for rigid polyurethane foams with enhanced flame retardancy and mechanical properties. *Polymer Testing* **2020**, *87*, 106511.

100. Riyazuddin; Nageswara Rao, T.; Hussain, I.; Heun Koo, B. Effect of aluminum tri-hydroxide/zinc borate and aluminium tri-hydroxide/melamine flame retardant systems synergies on epoxy resin. *Materials Today: Proceedings* **2020**, *27*, 2269–2272.

101. Xu, S.; Li, J.; Ye, Q.; Shen, L.; Lin, H. Flame-retardant ethylene vinyl acetate composite materials by combining additions of aluminum hydroxide and melamine cyanurate: Preparation and characteristic evaluations. *Journal of Colloid and Interface Science* **2021**, *589*, 525–531.

102. Zhou, R.; Mu, J.; Sun, X.; Ding, Y.; Jiang, J. Application of intumescent flame retardant containing aluminum diethyphosphinate, neopentyl glycol, and melamine for polyethylene. *Safety Science* **2020**, *131*, 104849.

103. Shang, S.; Yuan, B.; Sun, Y.; Chen, G.; Huang, C.; Yu, B.; He, S.; Dai, H.; Chen, X. Facile preparation of layered melamine-phytate flame retardant via supramolecular self-assembly technology. *Journal of Colloid and Interface Science* **2019**, *553*, 364–371.

104. Li, W.-X.; Zhang, H.-J.; Hu, X.-P.; Yang, W.-X.; Cheng, Z.; Xie, C.-Q. Highly efficient replacement of traditional intumescent flame retardants in polypropylene by manganese ions doped melamine phytate nanosheets. *Journal of Hazardous Materials* **2020**, *398*, 123001.

105. Horrocks, A.; Price, D. *Fire Retardant Materials*; Woodhead Publishing: Cornwall, U.K., 2001.

106. Fan, Y.; Fowler, G. D.; Zhao, M. The past, present and future of carbon black as a rubber reinforcing filler–A review. *Journal of Cleaner Production* **2020**, *247*, 119115.

107. Dittrich, B.; Wartig, K. A.; Hofmann, D.; Mülhaupt, R.; Schartel, B. Carbon black, multiwall carbon nanotubes, expanded graphite and functionalized graphene flame retarded polypropylene nanocomposites. *Polymers for Advanced Technologies* **2013**, *24*, 916–926.

108. Yang, H.; Gong, J.; Wen, X.; Xue, J.; Chen, Q.; Jiang, Z.; Tian, N.; Tang, T. Effect of carbon black on improving thermal stability, flame retardancy and electrical conductivity of polypropylene/carbon fiber composites. *Composites Science and Technology* **2015**, *113*, 31–37.

109. Liu, L.; Zhao, X.; Ma, C.; Chen, X.; Li, S.; Jiao, C. Smoke suppression properties of carbon black on flame retardant thermoplastic polyurethane based on ammonium polyphosphate. *J. Therm. Anal. Calorim.* **2016**, *126*, 1821–1830.

110. Yan, L.; Xu, Z.; Deng, N.; Chu, Z. Synergistic effects of mono-component intumescent flame retardant grafted with carbon black on flame retardancy and smoke suppression properties of epoxy resins. *J. Therm. Anal. Calorim.* **2019**, *138*, 915–927.

111. Higginbotham, A. L.; Lomeda, J. R.; Morgan, A. B.; Tour, J. M. Graphite Oxide Flame-Retardant Polymer Nanocomposites. *ACS Applied Materials & Interfaces* **2009**, *1*, 2256–2261.

112. Qiu, M.; Wang, D.; Zhang, L.; Li, M.; Liu, M.; Fu, S. Electrochemical exfoliation of water-dispersible graphene from graphite towards reinforcing the mechanical and flame-retardant properties of poly (vinyl alcohol) composites. *Materials Chemistry and Physics* **2020**, *254*, 123430.

113. Laachachi, A.; Burger, N.; Apaydin, K.; Sonnier, R.; Ferriol, M. Is expanded graphite acting as flame retardant in epoxy resin? *Polymer Degradation and Stability* **2015**, *117*, 22–29.

114. Wang, Y.; Tang, G.; Zhao, J.; Han, Y. Effect of flaky graphite with different particle sizes on flame resistance of intumescent flame retardant coating. *Results in Materials* **2020**, *5*, 100061.

115. Guler, T.; Tayfun, U.; Bayramli, E.; Dogan, M. Effect of expandable graphite on flame retardant, thermal and mechanical properties of thermoplastic polyurethane composites filled with huntite&hydromagnesite mineral. *Thermochimica Acta* **2017**, *647*, 70–80.

116. Moradkhani, G.; Fasihi, M.; Parpaite, T.; Brison, L.; Laoutid, F.; Vahabi, H.; Saeb, M. R. Phosphorization of exfoliated graphite for developing flame retardant ethylene vinyl acetate composites. *Journal of Materials Research and Technology* **2020**, *9*, 7341–7353.

117. Lee, S.; Kim, H. m.; Seong, D. G.; Lee, D. Synergistic improvement of flame retardant properties of expandable graphite and multi-walled carbon nanotube reinforced intumescent polyketone nanocomposites. *Carbon* **2019**, *143*, 650–659.

118. Das, O.; Bhattacharyya, D.; Hui, D.; Lau, K.-T. Mechanical and flammability characterisations of biochar/polypropylene biocomposites. *Composites Part B: Engineering* **2016**, *106*, 120–128.

119. Das, O.; Bhattacharyya, D.; Sarmah, A. K. Sustainable eco–composites obtained from waste derived biochar: a consideration in performance properties, production costs, and environmental impact. *Journal of Cleaner Production* **2016**, *129*, 159–168.

120. Das, O.; Kim, N. K.; Kalamkarov, A. L.; Sarmah, A. K.; Bhattacharyya, D. Biochar to the rescue: Balancing the fire performance and mechanical properties of polypropylene composites. *Polymer Degradation and Stability* **2017**, *144*, 485–496.

121. Das, O.; Kim, N. K.; Hedenqvist, M. S.; Lin, R. J.; Sarmah, A. K.; Bhattacharyya, D. An attempt to find a suitable biomass for biochar-based polypropylene biocomposites. *Environmental Management* **2018**, *62*, 403–413.

122. Matta, S.; Bartoli, M.; Frache, A.; Malucelli, G. Investigation of Different Types of Biochar on the Thermal Stability and Fire Retardance of Ethylene-Vinyl Acetate Copolymers. *Polymers* **2021**, *13*, 1256.

123. Zhang, Q.; Zhang, D.; Xu, H.; Lu, W.; Ren, X.; Cai, H.; Lei, H.; Huo, E.; Zhao, Y.; Qian, M. Biochar filled high-density polyethylene composites with excellent properties: Towards

maximizing the utilization of agricultural wastes. *Industrial Crops and Products* **2020**, *146*, 112185.

124. Barbalini, M.; Bartoli, M.; Tagliaferro, A.; Malucelli, G. Phytic Acid and Biochar: An Effective All Bio-Sourced Flame Retardant Formulation for Cotton Fabrics. *Polymers* **2020**, *12*, 811.

125. Zheng, Z.; Liu, Y.; Dai, B.; Meng, C.; Guo, Z. Fabrication of cellulose-based halogen-free flame retardant and its synergistic effect with expandable graphite in polypropylene. *Carbohydrate Polymers* **2019**, *213*, 257–265.

126. Tong, C.; Zhang, S.; Zhong, T.; Fang, Z.; Liu, H. Highly fibrillated and intrinsically flame-retardant nanofibrillated cellulose for transparent mineral filler-free fire-protective coatings. *Chemical Engineering Journal* **2021**, *419*, 129440.

127. De Chirico, A.; Armanini, M.; Chini, P.; Cioccolo, G.; Provasoli, F.; Audisio, G. Flame retardants for polypropylene based on lignin. *Polymer Degradation and Stability* **2003**, *79*, 139–145.

128. Liang, D.; Zhu, X.; Dai, P.; Lu, X.; Guo, H.; Que, H.; Wang, D.; He, T.; Xu, C.; Robin, H. M.; Luo, Z.; Gu, X. Preparation of a novel lignin-based flame retardant for epoxy resin. *Materials Chemistry and Physics* **2021**, *259*, 124101.

129. Wu, K.; Xu, S.; Tian, X.-Y.; Zeng, H.-Y.; Hu, J.; Guo, Y.-H.; Jian, J. Renewable lignin-based surfactant modified layered double hydroxide and its application in polypropylene as flame retardant and smoke suppression. *International Journal of Biological Macromolecules* **2021**, *178*, 580–590.

130. Malucelli, G. Flame-Retardant Systems Based on Chitosan and Its Derivatives: State of the Art and Perspectives. *Molecules* **2020**, *25*, 4046.

131. Chen, R.; Luo, Z.; Yu, X.; Tang, H.; Zhou, Y.; Zhou, H. Synthesis of chitosan-based flame retardant and its fire resistance in epoxy resin. *Carbohydrate Polymers* **2020**, *245*, 116530.

132. Kundu, C. K.; Wang, X.; Hou, Y.; Hu, Y. Construction of flame retardant coating on polyamide 6.6 via UV grafting of phosphorylated chitosan and sol–gel process of organo-silane. *Carbohydrate Polymers* **2018**, *181*, 833–840.

133. Yang, Z.; Li, H.; Niu, G.; Wang, J.; Zhu, D. Poly(vinylalcohol)/chitosan-based high-strength, fire-retardant and smoke-suppressant composite aerogels incorporating aluminum species via freeze drying. *Composites Part B: Engineering* **2021**, *219*, 108919.

134. Alongi, J.; Carosio, F.; Malucelli, G. Layer by layer complex architectures based on ammonium polyphosphate, chitosan and silica on polyester-cotton blends: flammability and combustion behaviour. *Cellulose* **2012**, *19*, 1041–1050.

135. Carosio, F.; Alongi, J.; Malucelli, G. Layer by layer ammonium polyphosphate-based coatings for flame retardancy of polyester–cotton blends. *Carbohydrate Polymers* **2012**, *88*, 1460–1469.

136. Rajczak, E.; Arrigo, R.; Malucelli, G. Thermal stability and flame retardance of EVA containing DNA-modified clays. *Thermochimica Acta* **2020**, *686*, 178546.

137. Alongi, J.; Carletto, R. A.; Di Blasio, A.; Cuttica, F.; Carosio, F.; Bosco, F.; Malucelli, G. Intrinsic intumescent-like flame retardant properties of DNA-treated cotton fabrics. *Carbohydrate Polymers* **2013**, *96*, 296–304.

138. Cheng, X.-W.; Guan, J.-P.; Tang, R.-C.; Liu, K.-Q. Phytic acid as a bio-based phosphorus flame retardant for poly(lactic acid) nonwoven fabric. *Journal of Cleaner Production* **2016**, *124*, 114–119.

139. Costes, L.; Laoutid, F.; Brohez, S.; Delvosalle, C.; Dubois, P. Phytic acid–lignin combination: A simple and efficient route for enhancing thermal and flame retardant properties of polylactide. *European Polymer Journal* **2017**, *94*, 270–285.
140. He, S.; Gao, Y.-Y.; Zhao, Z.-Y.; Huang, S.-C.; Chen, Z.-X.; Deng, C.; Wang, Y.-Z. Fully Bio-Based Phytic Acid–Basic Amino Acid Salt for Flame-Retardant Polypropylene. *ACS Applied Polymer Materials* **2021**, *3*, 1488–1498.
141. Kim, Y. N.; Ha, Y.-M.; Park, J. E.; Kim, Y.-O.; Jo, J. Y.; Han, H.; Lee, D. C.; Kim, J.; Jung, Y. C. Flame retardant, antimicrobial, and mechanical properties of multifunctional polyurethane nanofibers containing tannic acid-coated reduced graphene oxide. *Polymer Testing* **2021**, *93*, 107006.
142. Liu, X.; Zhang, Q.; Peng, B.; Ren, Y.; Cheng, B.; Ding, C.; Su, X.; He, J.; Lin, S. Flame retardant cellulosic fabrics via layer-by-layer self-assembly double coating with egg white protein and phytic acid. *Journal of Cleaner Production* **2020**, *243*, 118641.
143. Liu, Z.; Shang, S.; Chiu, K.-l.; Jiang, S.; Dai, F. Fabrication of conductive and flame-retardant bifunctional cotton fabric by polymerizing pyrrole and doping phytic acid. *Polymer Degradation and Stability* **2019**, *167*, 277–282.
144. Zhang, W.; Yang, Z.-Y.; Tang, R.-C.; Guan, J.-P.; Qiao, Y.-F. Application of tannic acid and ferrous ion complex as eco-friendly flame retardant and antibacterial agents for silk. *Journal of Cleaner Production* **2020**, *250*, 119545.

Chapter 4

Self-Extinguishing Polyurethanes

Tuhin Ghosh[1] and Niranjan Karak[*,1]

[1]Advanced Polymer and Nanomaterial Laboratory, Department of Chemical Sciences, Tezpur University, Napaam, Tezpur 784028, India
*Emails: karakniranjan@gmail.com, nkarak@tezu.ernet.in

There is skyrocketing demand for polyurethane (PU)-based products in fields including coating, automobile, aerospace, and furniture industry due to their versatile and favorable properties. PU foams are extensively used as a cushioning material in cars, furniture, and mattresses. However, the shortcomings associated with these PU-based products include their high flammability, rapid-fire catching, and fire spreading tendencies. Once they come in contact with fire, a catastrophic series of reactions start rapidly igniting the combustible things that are present in their surroundings. As a result, versatile polymers should be flame retardant (FR), which makes them useful without any fear of fire-related issues. In this chapter, the authors highlight the necessity of self-extinguishing PU. In subsequent sections of the chapter, a comprehensive discussion is presented on the general approaches (incorporation of FR additives, covalent linking with reactive FRs, and coating with FR materials) to produce a FR-PU. For better insight, numerous literature reports are also discussed including the processes and methods involved in generating a FR-PU. Finally, the chapter provides future directions in this field with a concise conclusion and summary.

Introduction

In this modern era of polymer studies, polyurethane (PU) is the most important synthetic polymer. It is well known for its light weight, inherent water and chemical resistance, tunable thermomechanical properties, and excellent abrasion resistance properties along with low viscosity and thermal conductivity. Most interestingly, due to availability of a wide range of monomers, these properties can be tuned depending on the polymer's targeted applications ranging from the preparation of foams, coatings, packagings, adhesives, sealants, sporting goods, furniture, insulating materials, seat cushions in the automotive industry, building blocks, and aerospace components (1–3). From a survey based on the PU market, it is evident that approximately $59 billion was invested globally for PU production with that number increasing 6% annually. Moreover, the survey also mentioned that the amount will expand to more than $93 billion by the end of 2026 (4). In this

© 2021 American Chemical Society

current scenario, PU production ranked as the fifth most produced synthetic polymer (5). To better understand the production of all synthetic polymers, a pie chart is shown in Figure 1.

Production (Millions of tons)

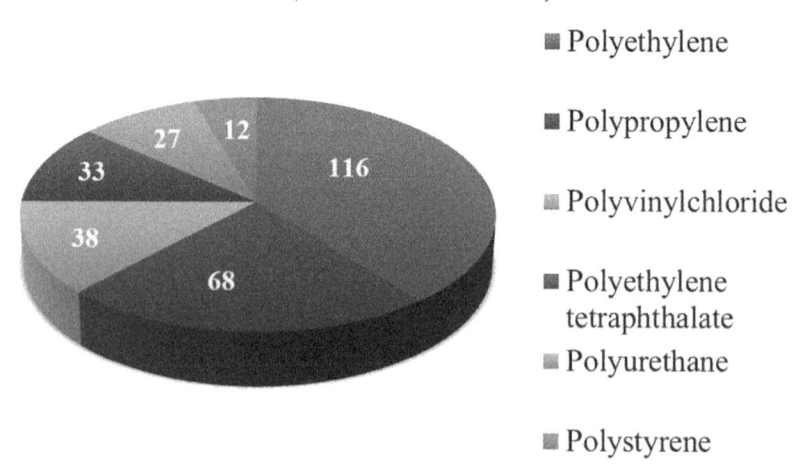

- Polyethylene
- Polypropylene
- Polyvinylchloride
- Polyethylene tetraphthalate
- Polyurethane
- Polystyrene

Figure 1. Production of synthetic polymers in millions of tons. Adapted with permission from reference (5). Copyright Danso, D.; Chow, J.; Streit, W. R., some rights reserved; exclusive licensee American Society for Microbiology. Distributed under a Creative Commons Attribution License 4.0 (CC BY) https://creativecommons.org/licenses/by/4.0/.

As these PU-based materials are extensively used in modern society, their protection and safety are major issues. Such materials are highly flammable and produce toxic gas or smoke and volatiles upon burning. High thermal decomposition behavior resides in the raw materials from which the PUs are commercially prepared. Most of the commercially available PUs are prepared from petroleum-based isocyanates, polyols, and chain extenders. However, petroleum-based precursors are not the only ones used. Bioderived compounds like vegetable oils (sunflower oil, castor oil, linseed oil, olive oil, and soybean oil), agricultural by-products, polysaccharides (starch, cellulose, chitin, sucrose, and isosorbide) are also used for synthesizing PU and preferred due to the rapid exhaustion of fossil fuels and growing environmental concerns. Whatever the monomers (bio- or petroleum-based), the problem of rapid flammability is the most serious issue for PU due to its elemental compositions with high flammability nature (mainly oxygen and hydrogen). As a result, incredible research is being conducted to produce PUs with more resistance toward fire.

Currently, several techniques are used to increase the flame retardant (FR) behavior of PU including incorporation of FR additives during processing, covalent functionalization of reactive FR compounds with the main polymeric chains, and coating with FR compounds (6). Generally, compounds containing chlorine, bromine, phosphorus, nitrogen, silicon, zinc, boron, iron, and aluminum are used to increase the FR behavior of the PUs. However, some of these compounds have severe drawbacks, such as environment and health problems. For example, halogenated FRs have great potential to increase the fire resistance behavior of PUs. However, these halogenated compounds have adverse effects on the environment and cause serious human health complications including disruption in neurological and thyroid development, endocrine disruption, and cancer (7). Brominated FRs like polybrominated diphenyl ethers, were mainly used in the automotive industry and are now banned completely due to their high toxicity and persistent behavior in environment. Such problems are resolved by the use of other organohalogens, but these kinds of PUs increase the production of toxins (furans, dioxins, and carcinogenic compounds) and corrosive gases during combustion. Henceforth, there is a huge demand for halogen-free FR-PUs.

Phosphorus-containing FRs are a promising alternative and have huge potential to reduce the rapid combustion behavior of PU. During the combustion process, phosphorus-based radicals (PO·, PO$_2$· etc.) form in the gaseous phase, which captures the radical responsible for combustion and inhibits the combustion process. Moreover, they also maximize the fire resistant behavior through catalyzing carbonization reaction. Furthermore, these compounds are less toxic than halogenated FR-PUs. Even though they have excellent properties, they are also responsible for discoloring PU foams (PUFs). In particular, aliphatic phosphate-containing PUs are more prone toward discoloration than the aromatic phosphate-containing PUs. Several metal oxides like aluminum trihydroxide and magnesium dihydroxide are also used as eco-friendly FR additives. In some cases, the migrating nature of FR additives during processing or during their service life leads to a decrease in their mechanical and thermal properties.

In order to address these issues and create a high-performing FR-PU, nanomaterials are amalgamated with PU. Graphitic carbon nitrides, graphene-based and graphene-modified nanomaterials, polyhedral oligomeric silsesquioxane, carbon nanotubes, carbon quantum dots, and organomodified montmorillonite are also used to make PU fire resistant. Even though these nanomaterials impart desired properties, there are still some problems as the use of different additives leads to a significant increase in the viscosity of the PU, which creates difficulty in industrial scale preparation of FR-PUs. The viscosity issue is not their only issue as their high cost, poor dispersion behavior, and leaching tendency from the PU matrix limits their industrial scale production.

To overcome the previously mentioned shortcomings, material scientists designed an innovative technique in which reactive functional FR oligomers and monomers are used as building blocks of the PU chain. This processing technique avoids leaching and migration of the FR components and enhances the compatibility with the PU matrix. As a result, the FR property of the PU increases without any adverse effect on its mechanical and thermal properties. Research also discusses the use of the layer-by-layer (LBL) deposition technique to enhance the FR behavior of the PUs. This is an effective and powerful surface modification technique in which various polyelectrolytes (bio-based and phosphorus-containing polyelectrolytes) are used to prolong the FR behavior of the PU. However, this approach is also not free from certain shortcomings, as a new design means investment of time, labor, and cost along with optimization of process.

In this chapter, the authors try to provide a comprehensive and critical discussion on self-extinguishing PUs with a discussion of prior research. This chapter also provides a clear picture of the mechanisms involved during the combustion of PUs with and without the addition of FR components. The chapter also focuses on the chronological development of FR-PUs along with their positive and negative aspects. Special attention is given to the current trends in developing FR-PUs and finally a concise conclusion discussing the future direction of the field of FR-PUs.

Basics of Flame-Retardant (FR) Behavior

In general, FR behavior in polymers that include PUs depends on several parameters like elemental composition, crystallinity, burning rate, flame spreading rate, ignition features, smoke generation, and total heat release. All of these parameters are strongly affected by the first parameter (elemental composition) (8). To understand the FR mechanism of the PUs, it is first necessary to understand the actual thermal decomposition process. The decomposition mainly occurs through a combination of chain scission, unzipping, and cross linking reactions (9). At the initial stage of thermal excitation, the entrapped volatiles present in the matrix are released. In the next step, various thermolabile covalent bonds present in the matrix undergo a complex type of vibration in their local free space. Upon further excitation, chain scission and depolymerization occur resulting in the generation of smaller fragments, radicals, and volatiles (CO_2, CO, and HCN). The radicals form in this step recombined or undergo a cross linking reaction with the existing functionalities. Finally, the decomposition ends with the formation of char, which is nondegradable above that temperature. The amount of char residue increases with the increase in nanomaterials and inorganic counterparts. Henceforth, in order to suppress, inhibit, or retard the rapid thermal degradation behavior, FR compounds are used as they generally act as the physicochemical barrier against the combustion process and enhance the thermal stability. Some studies cited the use of three different pathways including: (1) the use of suitable additives; (2) covalent modification; and (3) coating with FR materials. These pathways are elaborated on in the following sections.

Generation of FR-PUs Using Suitable Additives

Using an FR additive is the most common and widely accepted technique for making PU fire resistant. Most of the additives are used in a powdered form during the synthesis of PU. As a result, there is no covalent bonding between the additive and the PU (4). The additives are chosen based on their mode of action during combustion. Some examples include: (1) additives that decompose through highly endothermic reactions create a heat sink and reduce the burning rate; (2) additives that generates non-combustible volatiles and gases upon burning create a blanketing action against the rapid spreading of fire; (3) additives that have flame poisoning effects form reactive chemical species upon burning that rapidly stop the thermo-oxidation process of burning through scavenging OH and H radicals; and (4) additives that produce char layer and limit heat and mass transfer (10). The following sections provide a comprehensive discussion based on the aforementioned principles for better understanding.

Heat Sinking through Endothermic Reaction

From a chemistry point of view, these are the additives that undergo endothermic reaction during combustion as the reaction absorbs energy from the burning compounds. As a consequence of this heat transferring phenomenon, the temperature of the material falls below its combustion temperature and inhibits the burning process. For example, aluminum hydroxide or $Al(OH)_3$ (also known as aluminum trihydrate (ATH)) is used as a FR mineral in many PUs. During the combustion process, ATH converts into Al_2O_3 and water through an endothermic reaction and absorbs the heat from the PU surface. In that time, the released water vaporizes and creates a diluting effect by capturing the flame-forming radicals. Moreover, the aluminum produced during decomposition creates a protective char layer and prevents the degradation process. Like $Al(OH)_3$, several other compounds like magnesium hydroxide, calcium hydroxide, and layered double hydroxides (LDHs) are also used as FR additives.

Following the same chemistry, Peng et al. fabricated different compositions of PU by varying the amount (20, 15, 10, 5, or 0 wt%) of $Al(OH)_3$ and $Mg(OH)_2$ (11). PU nanocomposites with 20% of $Al(OH)_3$ showed an enhanced limited oxygen index (LOI) value compared to $Mg(OH)_2$-incorporated nanocomposites. The summation of the endothermic reaction, flame dilution effect, and carbonation phenomenon of $Al(OH)_3$ is responsible for this enhanced property. It is important to define the LOI. The LOI is a parameter that represents the FR efficiency of any polymeric materials. Actually, the LOI defines the minimum percentage of oxygen required for combustion of polymeric compounds in a mixture of nitrogen and oxygen atmosphere (10). The higher the LOI value (greater than 26%), the higher the FR behavior of the corresponding polymer. Chai et al. reported the fabrication of a series of FR-PUFs by combining both Al(OH)$_3$ and Mg(OH)$_2$ (12). This report mentions the endothermic degradation reactions of Al(OH)$_3$ and Mg(OH)$_2$, as shown in the equations (1) and (2).

$$2Al(OH)_3 \rightarrow Al_2O_3 + 3H_2O \quad (1)$$

$$Mg(OH)_2 \rightarrow MgO + H_2O \quad (2)$$

The results clearly show that flammability parameters are significantly influenced by the amount and nature of FR additives that are used (Table 1). Wang et al. reported the fabrication of FR polyisocyanurate-PU foam using the synergistic effect of ATH and expanded graphite (EG) (13). They mentioned that the LOI value for bare foam is 26.5%, whereas the LOI value increased drastically to 92.8% after incorporating ATG-EG. Wang et al. studied the enhancement of FR behavior of PU after incorporating EG-encapsulated Mg(OH)$_2$ (14). In this study, the LOI value of PU was enhanced from 22.2% to 32.6% after incorporating that particular nano-inorganic hybrid. Dike et al. fabricated a series of thermoplastic PU (TPU) containing zinc borate, huntite-hydromagnesite (HH, which is mainly a mixed mineral containing ATH and hydrated magnesium carbonates), or a combination of both (15). In this report, they mentioned that after incorporating equal amounts of zinc borate and HH, the LOI value increased from 21.2% to 26.8%. Similarly, Mohammadi et al. reported the fabrication of bio-based PU nanocomposites using two types of sulfonate-containing calix[4]arenes intercalated LDHs (16). They reported that the pristine PU exhibited an LOI value of 18.9%, whereas after nanocomposite formation, the LOI value was increased by more than 20%. Moreover, the cone calorimetric test revealed that smoke generated during burning nanocomposites was suppressed significantly after modification with the modified LDHs.

Table 1. Variations of Flammability Parameters like Ignition Time, Flameout Time, Peak Value of the Heat Release Rate, and Total Heat Release of the PUF With and Without FR Additives[a]

FR Additives[b]	Amount of FR (wt%)	Ignition Time (s)	Flameout Time (s)	Peak Heat Release Rate (kWm^{-2})	Total Heat Release (MJm^{-2})
Nil	0	5.67	252.33	185.54	24.29
ATH	2.5	15.33	232.00	191.47	29.14
ATH	5	12.00	206.00	200.43	20.42
ATH	7.5	20.50	180.00	219.59	24.37
ATH	10	17.67	216.00	238.05	20.94
MDH	2.5	10.67	310.00	200.00	18.17
MDH	5	16.33	181.00	250.26	29.39
MDH	7.5	12.00	184.00	246.32	31.73
MDH	10	11.67	206.00	277.41	31.65
ATH + MDH	2.5 + 2.5	10.00	140.00	128.51	13.07

Table 1. (Continued). Variations of Flammability Parameters like Ignition Time, Flameout Time, Peak Value of the Heat Release Rate, and Total Heat Release of the PUF With and Without FR Additives[a]

FR Additives[b]	Amount of FR (wt%)	Ignition Time (s)	Flameout Tim (s)	Peak Heat Release Rate (kWm^{-2})	Total Heat Release (MJm^{-2})
ATH + MDH	5 + 2.5	11.00	143.00	129.54	14.64
ATH + MDH	7.5 + 2.5	20.00	258.00	134.75	21.69
ATH + MDH	10 + 2.5	15.00	235.00	145.05	23.26
ATH + MDH	2.5 + 5	12.50	199.00	199.97	25.80
ATH + MDH	5 + 5	18.33	161.00	205.29	25.58
ATH + MDH	7.5 + 5	13.50	200.00	207.08	30.02
ATH + MDH	10 + 5	15.00	180.00	218.05	32.34
ATH + MDH	2.5 + 7.5	11.67	205.00	196.49	30.84
ATH + MDH	5 + 7.5	16.67	241.00	175.33	29.32
ATH + MDH	7.5 + 7.5	15.00	294.00	199.07	33.70
ATH + MDH	10 + 7.5	17.00	258.00	214.94	33.41
ATH + MDH	2.5 + 10	15.67	222.00	262.57	33.57
ATH + MDH	5 + 10	16.00	298.00	239.29	39.17
ATH + MDH	7.5 + 10	16.00	248.00	242.22	38.20
ATH + MDH	10 + 10	15.67	259.00	208.73	38.80

[a] Reproduced with permission from reference (12). Copyright 2019 Springer.   [b] ATH: AL(OH)$_3$ and MDH: Mg(OH)$_2$.

FR Behavior through Generation of Noncombustible and Inert Gases

In these cases, the chosen additives can release inert gases upon combustion and dilute the oxygen concentration near the burning spot both in solid and liquid states. As a result, the reaction reduces the flammability of the PU and increases the FR behavior. For example, Savas et al. reported that the fabrication of TPU composites contain microencapsulated red phosphorus and HH-hydromagnesite (17). Savas et al. studied the FR behavior of each additive separately as well as through the mixing both additives. In this report, the bare TPU showed a LOI value of 21.2%, whereas the LOI value increased to 25.7% after incorporating 50 wt% of HH. When the combination of both HH and microencapsulated red phosphorus was used, the LOI value jumped to 32.5%. The combination of endothermic degradation of minerals, dilution of the flame by the released CO_2 and H_2O in the condensed phase, and the generation of char residue in the solid phase are the main factors in increased FR behavior in the TPU. Similarly, Guler et al. reported the fabrication of a series of TPU composites containing EG and HH-hydromagnesite (18). The LOI value of pristine TPU was found to enhance from 21.2% to 31.2% after incorporating 60 wt% of HH. The LOI value of the TPU increased to 32.6% after incorporating EG and HH-hydromagnesite at equal amounts. Even though good FR behavior was achieved, mechanical properties decreased drastically with the incorporation of the additive amount. In this report, the TPU exhibits a tensile strength of approximately 24.8 MPa and an elongation value of 424%, whereas in the composite (equal amount of EG and HH-hydromagnesite), tensile strength is 9.6 MPa and the elongation value is 44%.

Fernández et al. demonstrated the preparation of carbonate-intercalated LDH of Mg and Al and further incorporated PU through in-situ polymerization technique (19). During the combustion process, LDH produces a lot of flame diluting components like various noncombustible gases (mainly CO_2 and H_2O) and HO^- ions. Moreover, the degradation of the mineral keeps the surrounding of the polymer cool. As a result, thermal stability of the modified LDH-based PUs increases significantly.

FR Behavior through Generation of Reactive Chemical Species

These reactive chemical species are additives that generate free radicals upon thermal degradation and their mode of action is mainly observed in gas phase. The radicals responsible for burning are scavenged by the radicals generated from the additives. As a result, the combustion process is suppressed and FR tendency enhances. Research studies advocate the presence of numerous PU composites where this technique was used to improve FR behavior. For example, Tang et al. studied the change in FR behavior of rigid PUF (RPUF) after incorporating different amounts of aluminum diethylphosphinate (ADP) (20). In this report, the LOI value of bare RPUF is 18.8%, whereas the LOI value increased to 23% after incorporation of 30 wt% of ADP. In this report, it is clearly mentioned that at the preliminary step of thermal excitation, ADP decreases and produces diethylphosphinic acid and oligomeric products of phosphinates that further transform into $PO\cdot$ and $P\cdot$ radicals. These radicals rapidly terminate the gas phase chain reaction though quenching $HO\cdot$ and $H\cdot$ radicals. Liu et al. tried to utilize the synergistic effect of dimethyl methylphosphonate (DMMP), ATH, EG and ammonium polyphosphate to increase the FR behavior of the PU (21).

FR Behavior of Additives through the Formation of Char Layer

This section discusses additives that upon burning produce an inert char layer above the matrix. This layer mainly exhibits very poor thermal conduction behavior. It reduces the transfer of heat and oxygen from the outer burning sight to the inner surface and it reduces the fuel flow for combustion. As a result of this phenomenon, thermal degradation kinetics reduce significantly (8). Yuan et al. reported the preparation of PU foam containing 9,10-dihydro-9-oxa-10-phosphaphenanthrene-10-oxide and melamine derivatives (22). In this report, it is clearly mentioned that the bare PU foam is highly flammable with a LOI value of 19%. However, after incorporating the previously mentioned additives in the PU, the LOI value was changed to 27%. Such changes in the LOI value correspond to the generation of a noncombustible shield of char.

From the detailed morphological analysis (Figure 2) of the char residue, it is clear that bare PU (Figure 2a) formed char with microvoids and cracks that influence the transfer of heat and mass during combustion. However, after incorporating FR additives, the morphology (Figure 2b–f) of char completely changes to a continuous layered structure that is chemically inert in nature and inhibits the transfer of required heat and mass to the inner surface for degradation. Shi et al. demonstrated the fabrication of EG-embedded PU foam (23). During the combustion process, PU foam covers its surface by producing a graphitic wormlike char residue that significantly minimizes the transfer of heat and mass from the burning surface to the polymer interface. In this case, the LOI value of bare PU was improved from 22.5% to 39.5% after incorporating 20 wt% of EG. Shi et al. further studied the effect of different particle sizes of EG on the FR behavior of the PU. They found that the EG with bigger particle sizes completely cover the burning surface and effectively prevent the combustion process through inhibiting the transfer of combustible gases to the inner surface of the PU foam.

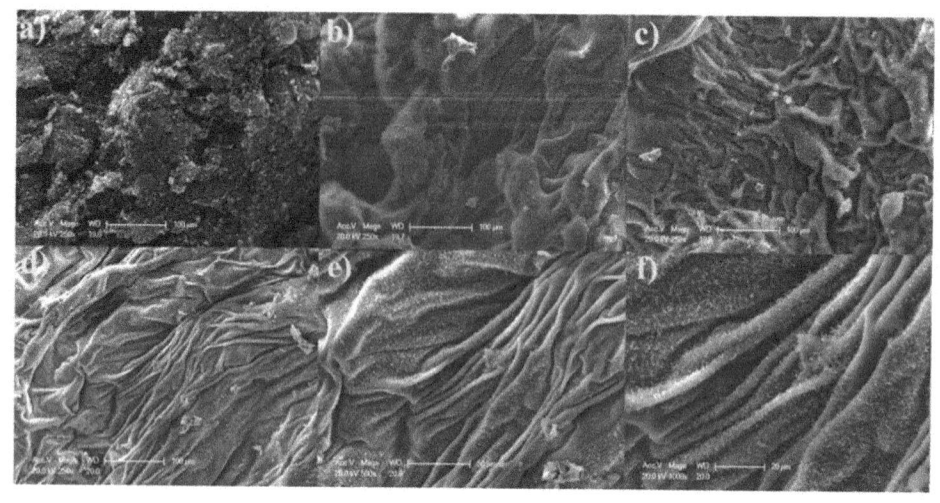

Figure 2. Scanning electron micrographs of char residue of (a) bare PUF, (b)PUF-9,10-dihydro-9-oxa-10-phosphaphenanthrene-10-oxide, (c) PUF-melamine-derived polyol, and (d–f) PUF-9,10-dihydro-9-oxa-10-phosphaphenanthrene-10-oxide -melamine-derived polyol (at different magnifications) after LOI tests. Reproduced with permission from reference (22). Copyright 2018 Elsevier.

A similar experiment was also reported by Thirumal et al. where EG with two different particle sizes (180 and 300 μm) was used to enhance the FR behavior of the RPUF (24). They observed that EG with a particle size of 300 μm showed better FR behavior compared to other EG with lower particle sizes. Moreover, the LOI value increases gradually with increased loading of the EG. In this report, it is mentioned that during the combustion process the intercalated species (sulphuric acids) between the graphitic layers produced CO_2, H_2O, and SO_2 gases, which create a dilution effect through reducing oxygen concentration. It not only creates a dilution effect but also expands the interlayer distance of EG in the char. The greater the amount and size of EG, the higher the production of the carbon-rich char layer. On the other hand, the amount of char produced in bare PU is comparatively lower than the EG-incorporated PUs. Due to its insulating nature, the produced char residue inhibits the conduction of thermal energy from the outer surface to the inner surface thus increasing the FR behavior.

Gao et al. used the synergistic effect of EG, modified LDH, and diethyl ethylphosphonate (DEEP) to enhance the FR behavior of the polyisocyanurate-PU foam (25). The LOI value of the pristine foam was enhanced from 22.1% to 31.2% after incorporating 45 php of EG and DEEP. This value was further increased to 32.5% after incorporating 3 php of modified LDH along with 45 php of EG and DEEP. The evolution of inert gases (NH_3, CO_2, H_2O, and SO_2) and the formation of the protective char layer during combustion are responsible for this improved FR behavior.

Wang et al. demonstrated the preparation of a series of FR-PU using ammonium polyphosphate and zinc hydroxystannate (26). After incorporating 13.5 wt% of ammonium polyphosphate and 1.5 wt% of zinc hydroxystannate, the LOI value of bare PU increased from 21.5% to 28.5%. During the combustion process of this PU, thermostable char residue was formed through the salt bridge formation between zinc hydroxystannate and ammonium polyphosphate. Furthermore, the presence of zinc hydroxystannate suppressed the generation of toxic smoke during the combustion process.

Ramanujam et al. reported the preparation of a bio-based polyol containing PUF with a variable amount of DMMP (27). In this case, the neat PUF showed a weight loss of 38% with a burning time of 115 s. However, after loading DMMP, both weight loss and burning time decreased to 5.5% and 3.5 s. Digital images (Figure 3) of different compositions of PUF before and after horizontal tests clearly support the enhanced FR behavior after incorporating DMMP. Reported PUF enhanced the self-extinguishing behavior in both the condensed (through formation of thermostable char) and gas phase (through formation of flame diluting reactive radicals) during decomposition as previously discussed.

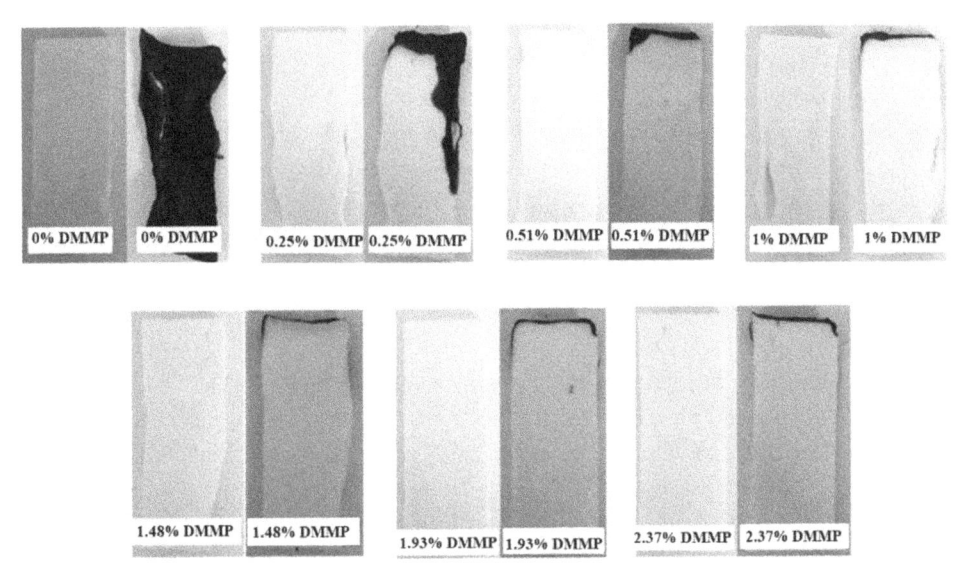

Figure 3. Digital images of PUF before and after horizontal burning tests. Adapted with permission from reference (27). Copyright Ramanujam, S.; Zequine, C.; Bhoyate, S.; Neria, B.; Kahol, P.; Gupta, R., some rights reserved; exclusive licensee MDPI. Distributed under a Creative Commons Attribution License 4.0 (CC BY) https://creativecommons.org/licenses/by/4.0/.

Agrawal et al. reviewed the FR behavior of a flexible PUF (FPUF) containing variable amounts of ceramic additives (aluminum and zirconia powder) (28). In this report, the RPUF was prepared from the transesterified product of castor oil. The ceramic additives incorporated (up to 6 wt%) in the bio-based RPUF showed better FR property compared to the pristine PUF. Generally, ceramic additives are highly stable and thermally insulating in nature. As a result, during combustion, the ceramic-containing additives produced high temperature resistant, thermo-insulating char residue around the burning site. As a result of this protective insulating char, the heat generated during combustion is unable to transmit to the inner surface of the RPUF, and due to lack of fuel flow, the FR behavior increases significantly. The same group further studied the effect of kaolinite clay and feldspar on the FR properties of bio-based RPUF (29). From this study, it is determined that the PU containing kaolinite showed enhanced FR behavior compared to feldspar-based PUF. They demonstrated that both additives are highly thermostable, but feldspar started to decompose above 1150 °C whereas kaolinite clay underwent dehydration at around 450 °C to produce metakaolinite, which converts into more thermostable silica and mullite upon further heating above 1000 °C. In addition, the high surface aspect ratio of kaolinite produces an enhanced barrier for heat conduction.

Hence, the char produced in the case of kaolinite-incorporated PUF provides better fire protection than feldspar-incorporated PUF. Even though incorporation of these additives, increases the FR behavior significantly, the mechanical property of the PU decreases drastically, which further limits its application.

Covalently Modified FR-PUs

To address the drawbacks of the FR additives, material scientists designed a new pathway in which reactive FRs are incorporated during formulation of the PU. These reactive FR compounds exhibited a reactive functional group, which covalently linked with polyisocyanate (mainly diisocyanate) and subsequently entered into the PU backbone. This technique inherently improves the FR behavior of the PU and resolves the drawbacks encountered in an additive-based PU, such as leaching problems, incompatibility with the PU, migration tendency, and high cost of the additives. Studies cited in this section help to create a better understanding for the reader.

Bhoyate et al. reported the synthesis of reactive FR polyols through the reaction of phenylphosphonic acid and propylene oxide (30). Synthesized polyol was further used in variable amounts to optimize the optimum FR performance of the resultant PU. From the reported data, it is evident that the self-extinguishing time of the pristine PU was reduced drastically from 81 s to 11.2 s after incorporating approximately 30 wt% of polyols. The self-extinguishing time and also the weight loss during combustion were reduced from 28% to 9.8%. Formation of a graphitic char layer during thermal decomposition is responsible for this enhanced FR behavior.

Cui et al. demonstrated the preparation of a triazine-based chain extender, which is encoded as 4,6-dimethoxy-1,3,5-triazine)-2-methyl-propane-1,3-diol (TMDP) and further incorporated into PU to improve the FR behavior (31). Ammonium polyphosphate (APP) was also used in combination with TMDP to produce the synergistic effect of both the components. ,It is determined that the LOI value of pure PU increases from 18% to 27% after incorporating 6 wt% of TMDP and 5 wt% of APP. Moreover, the heat release rate (HRR) and total heat release rate during decomposition also decreased from 686 KW.m^{-2} to 236 KW.m^{-2} and 54.5 MJ.m^{-2} to 32.8 MJ.m^{-2}, respectively. This phenomenon corresponds to the formation of a protective and compact char layer during the decomposition of TMDP and APP, which reduces the heat flow to the underlying section and decreases the combustion rate. A detailed surface morphology (Figure 4) of the obtained char residue was carried out for better insight into this phenomenon. However, the obtained micrographs are quite similar to the micrographs of Yung et al. as shown in Figure 2.

Zhang et al. reported the preparation of a FR-PU using a phosphorus-nitrogen-silicon-based chain extender (32). The LOI value of the PU was increased from 18.4% to 27.7% after incorporating this chain extender. The use of this chain extender also reduces the smoke production rate during the burning process. From the scanning electron microscopy images of the produced char, it is observed that the char layer of PU contains a large number of microvoids, whereas the surface of chain extender containing PU is highly dense and compact. A highly folded and wrinkled type of morphology was observed in place of microvoids or cracks. Presence of such wrinkled morphology significantly hinders the transfer of oxygen to the inner surface. As a result, HRR and THR values decrease significantly compared to the pristine PU.

Rao et al. reported the synthesis of a phosphorus-containing polyol through the reaction of phenylphosphonic dichloride and ethylene glycol (33). Several compositions with variable amounts of phosphorus-containing polyols were also synthesized to understand the change in FR behavior. They mentioned that the LOI value of flexible PU gradually increases with the increasing amount of the polyols. Digital images at different ignition times are also represented in Figure 5 to better understand the improvement in FR behavior. They further studied the migration behavior of the synthesized polyols. For that purpose, they compared the synthesized polyols containing PU with another PU containing the same amount of DMMP. In a thermal aging test, the LOI value of DMMP-containing PU decreases from 22.5% to 20.5% after 16 h at 140 °C. However, the same PU also loses its self-extinguishing behavior after the test due to the leaching and migration tendency of DMMP, whereas the PU based on phosphorus-containing polyols can retain the LOI value even after a 64-h thermal aging test at 140 °C. This phenomenon corresponds to the covalent functionalization of the synthesized polyol into the PU backbone. As a result, the leaching and migration behavior stopped completely.

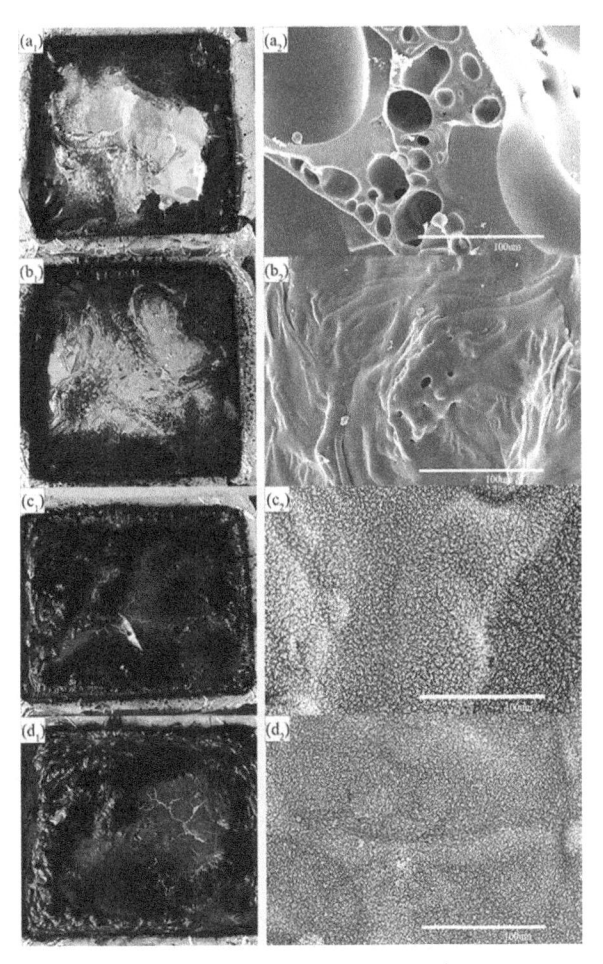

Figure 4. Digital images (a_1, b_1, c_1, and d_1) of char and scanning electron micrographs (a_2, b_2, c_2 & d_2) from char residue of bare PUF and covalently modified PUFs after LOI tests. Reproduced with permission from reference (31). Copyright 2021 Elsevier.

Similarly, Arora et al. reported the synthesis of FR moiety through the reaction of o-phenylenediamine and phenylphosphonic dichloride (34). The product is further reacted with 3-monochloro-1,2-propanediol to obtain the FR polyol. After that, the polyol was incorporated separately into the epoxy and PU systems. In this report, it is found that the PU system showed enhanced FR behavior compared to the epoxy system. The FR polyol not only increases the FR behavior but also imparts a self-extinguishing (within 10 s) nature on all the compositions. Moreover, the LOI value of the PU system increases from 21% to 32% after incorporation of these specially designed polyols.

Zeng et al. reported the synthesis of PUF using β-cyclodextrin (CD) and H_3PO_4-modified CD as a chain extender (35). They mentioned that the LOI value of PUF was increased from 17% to 18% after incorporating CD and to 24.5% after incorporating H_3PO_4-modified CD. Such an increase in the LOI value corresponds to the formation of a dense carbonaceous char layer, which significantly restricts the transfer of heat and oxygen to the underlying PU material. In this report, it is clearly mentioned that the amount of char formed in bare PUF is 1.51%. However, after the incorporation of CD and H_3PO_4-modified CD, the amount of char yields was enhanced to 6.15% and 20.58%, respectively. Yuan et al. investigated the synergistic effect of phosphorus- and nitrogen-containing polyols on the FR behavior of EG-PU composites (36). A phosphorus-containing polyol was synthesized through a reaction of benzene phosphorus oxidichloride and 1,4-butanediol, whereas the nitrogen-containing polyol was synthesized by reacting melamine, paraformaldehyde, and diethanol amines. In this study, they clearly mention that the synergistic effect of the polyols increased the LOI value of PU from 20% to 33.5%. Formation of a more compact carbonaceous char layer is responsible for this good FR behavior. However, the research advocates for the presence of numerous reports in which PUs are modified inherently to enhance the FR behavior. In some of these reports, complicated modification and synthetic procedures limit their practical applications in FR materials.

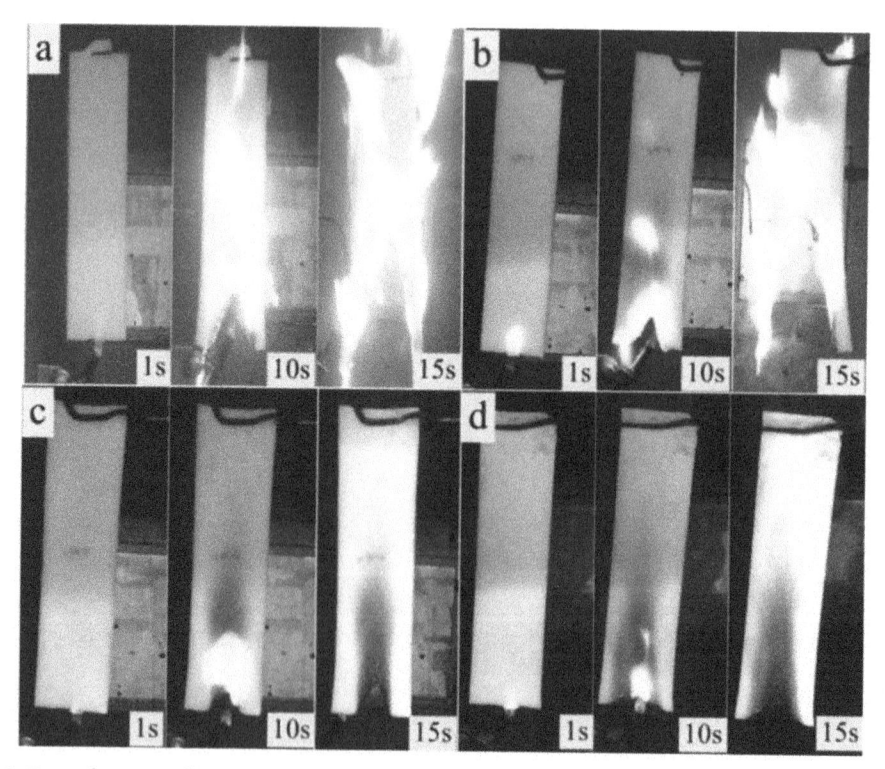

Figure 5. Digital images of (a) PU and (b–d) different compositions of phosphorus-containing polyol-based PUs at different ignition times. Reproduced with permission from reference (33). Copyright 2018 Elsevier.

PU Surface Coated with FR Materials

This is the most easy, efficient, cost effective, and widely acceptable technique for producing FR-PUs. This technique improves the FR behavior without affecting the inherent properties of the PU. Moreover, the mechanical and thermal properties of PU are also unaffected after surface coating. Recently, several surface coating techniques like the sol-gel process, LBL assembly, and plasma technique have been sued for this purpose. These techniques are discussed in the following section.

Sol-Gel Process

The sol-gel process is a wet chemical process involving several steps such as hydrolysis and condensation reaction of the metal alkoxides on the surface of the PU followed by gelation, drying, aging, and crystallization. There are several reports in which sol-gel process was used to develop a FR-PU. For example, Bellayer et al. used sol-gel formulation with different stoichiometric mixtures of methyl triethoxysilane, 3-aminopropyltriethoxysilane, diethylphosphatoethyltriethoxysilane, and diethyl phosphite to prepare a series of flexible PUs (37). In this report, it is clearly mentioned that the pristine PU rapidly catches fire when a flame is applied, whereas after modification, the self-extinguishing behavior enhances significantly. This modification not only increases the PU's self-extinguishing behavior but also suppresses smoke and CO formation during the combustion process. Furthermore, the same group in a subsequent study demonstrated the mechanism behind the FR behavior. In this study, the FR behavior occurs due to the formation of an insulating char layer of Si-O-P and SiO_2 on the surface of the PU foam and the formation of a radical scavenger during burning (7). Hosgor et al. developed a FR-PU-silica hybrid coating using sol-gel formulation (38).

Layer-by-Layer (LBL) Assembly

LBL assembly is the most common and environmentally friendly approach to make PU FR. In this process, the PU sample is dipped in oppositely charged polyelectrolyte solutions. The used polyelectrolytes are attached on the surface of PUF through electrostatic interaction, H-bonding interaction, donor and acceptor mechanism, and covalent interaction (2). In this technique, a highly FR-PU can be prepared at a low cost without hampering its inherent properties. Moreover, in the LBL technique, multilayer FR coatings applied to the PU surface interfere during the combustion process and minimize the flame. As a result, underlying PU is completely safe from the fire.

Hai et al. reported the preparation of a series of FR-PUs using the LBL technique. Herein, positively charged silica aerogel, negatively charged sodium alginate, and positively charged polyethyleneimine were deposited on the skeletal framework of PUF using electrostatic force (39). For better insight, a schematic diagram is presented in Figure 6 showing the LBL technique used in this article. The role of each component are clearly mentioned in this study. For example, silica aerogel was used to enhance the barrier property of the coating layer and sodium alginate was used to enhance the deposition of the silica aerogel on the surface of the PUF. As a synergistic effect of both the organic and inorganic components, the upper layer of PUF forms an insulating char layer and protects the inner PUF during combustion. In this case, the pristine PU burned rapidly (within 8 s) and produced lot of smoke, whereas after coating with four-trilayers of the previously mentioned materials, the self-extinguishing time increased to 36 s.

Liu et al. demonstrated the preparation of FR PUF using LDH suspension, alginate, and chitosan solution (40). Moreover, several compositions were also prepared by simply increasing the number of trilayers. In this case, two types of LDH such as nickel-aluminum LDH and magnesium-aluminum LDH were used to choose the optimum FR performance. It is also reported that nickel-aluminum LDH-based PUF showed better FR behavior than magnesium-aluminum LDH. From cone calorimetric data, it is evident that there are huge changes in HRR and THR values after modification. The pristine PUF showed HRR and THR values of 801 kW.m^{-2} and 22.1 MJ.m^{-2}, respectively, whereas after modification with 12 trilayers containing nickel-aluminum, the LDH showed HRR and THR values of 197 kW.m^{-2} and 16.2 MJ.m^{-2}, respectively. This good FR property of the PUF was ascribed to the enhanced thermal barrier property of the applied coating.

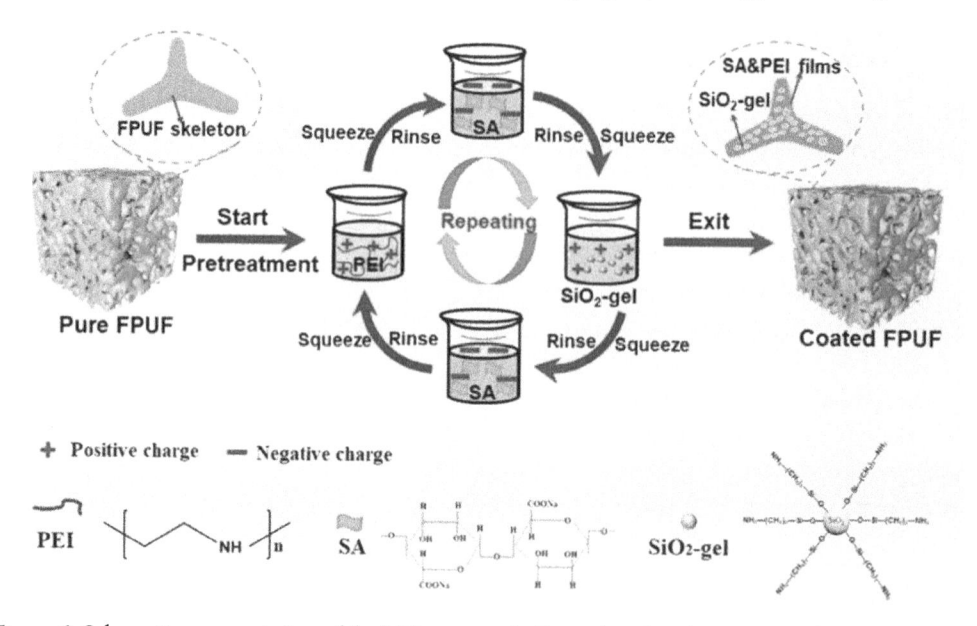

Figure 6. Schematic representation of the LBL approach. Reproduced with permission from reference (39). Copyright 2020 American Chemical Society.

Similarly, Pan et al. developed an effective trilayer approach to make FR FPUF (*41*). The trilayer mainly contains positively charged chitosan solution, negatively charged titanate nanotubes in suspension, and alginate solution. Details of the fabrication procedure are shown in Figure 7. Titanate nanotubes are used due to their excellent protective effects. Moreover, the high aspect ratio, thermoinsulating nature, and good adsorption effect are the added advantages of titanate nanotubes. Because of this, after deposition of eight trilayers, the HRR and THR values of FPUF decreased from 714 kW.m^{-2} to 212 kW.m^{-2} and from 17.8 MJ.m^{-2} to 15.5 MJ.m^{-2}, respectively. Even after the modification, the smoke production rate value of FPUF with eight trilayers was reduced from 0.0875 m^{2}s^{-1} to 0.0325 m^{2}s^{-1}. The high-quality FR and smoke suppressing capacity of FPUF are due to the compact protective coating of the nanotubes on the exterior of PUF, which significantly slows down the transfer of heat, smoke-forming particles, and volatiles. As a result, FR behavior of the PUF increases gradually with increasing trilayers.

Shi et al. constructed a trilayer of polyethyleneimine, graphene oxide (GO), and melamine nanoparticles (*42*). In this study, GO was used to generate enhanced barrier property of the coating, and melamine nanoparticles were used because of their preeminent radical scavenging activity and high thermostability. During combustion process of PUF, the chain scission reaction mainly takes place through a radical initiated mechanism. Therefore, the radicals responsible for burning are easily scavenged by the melamine nanoparticles. At the same time, GO initiates the formation of a dense char layer on the surface of the PUF that significantly hinders the transfer of heat to the underlying

PU and also inhibits the release of smoke from the inner surface to the surrounding environment. In this case, the neat PUF showed HRR and THR values of 388 kW.m^{-2} and 26 MJ.m^{-2}, respectively, whereas after two-trilayer deposition, the HRR and THR values decreased to 237 kW.m^{-2} and 23 MJ.m^{-2}, respectively.

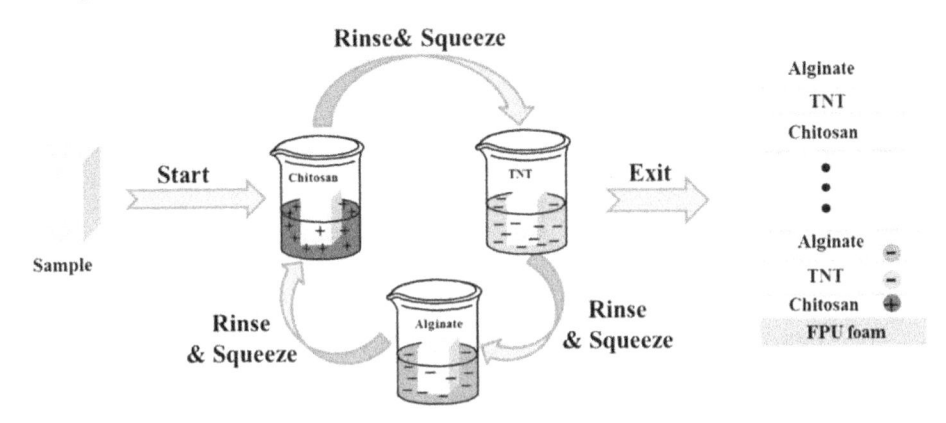

Figure 7. Schematic representation of the LBL approach used in Pan et al. Reproduced with permission from reference (41). Copyright 2015 American Chemical Society.

Current research standards advocate for the presence of numerous reports, where the LBL technique is successfully employed to develop self-extinguishing PU. For example, Kim et al. used a cationic layer containing carbon nanofibers and polyethyleneimine and an anionic layer containing poly(acrylic acid) in order to prepare PUF (43). Maddalena et al. employed a bilayer of GO and chitosan on the surface of PUF to improve the FR behavior of the PUF (44). Carosio and Alongi reported the fabrication of a FR PUF using an ultrafast LBL approach containing chitosan and poly(phosphoric acid) (45). Wang et al. used a trilayer of polyethyleneimine, MnO$_2$, and alginate to initiate the self-extinguishing behavior in the PUF (46).

Plasma Technology

Plasma technology is a promising environmentally friendly approach in which FR coating is deposited on the surface of PU through plasma treatment. Not only that, but the surface group of PU is also modified using FR material through plasma etching. In this approach, the FR efficiency is guided by some parameters like power types and atmospheric pressure. For example, Jimenez et al. modified the surface of the FPUF using diethylvinylphosphonate in the presence or absence of a cross linker through plasma treatment. It was determined that the surface modification in the presence of a cross linker significantly inhibits melt dripping and spreading of the flame over the PU (47).

Applications of FR-PUs

According to fire statistics, one household is engulfed in fire every two minutes, and half of the those involved in these incidents died or badly injured. In most of these cases, the source of the fire is a faulty electrical system or carelessness (1). Initially, the flame growth rate is slow, but upon contact with regular plastic counterparts, wooden or plastic furniture, mattresses, or cloths, the flame rapidly flashes over throughout the house. Because of this, a general arrangement should be taken to stop the fire initially by using FR-containing compounds. As discussed in earlier sections, most household items are now made by PUs. Most importantly, due to the desired flexibility, light weight, good heat, and strong chemical and insulating properties, the outer layer of the electrical wear and protective covers of electronic gadgets are made from FR-PUs. A modern household demands mechanically durable and long-lasting coatings to protect products from the natural aging process. In this field, using FR-PU coatings resolves both aging and fire safety issues. In the aerospace industry, the use of FR-PUs is significant. High-performing RPUFs are now widely used in skyscrapers as construction materials. However, to make these skyscrapers fireproof, FR RPUFs are used extensively. Moreover, FR PUFs are extensively used in furniture, bedding, and the underlying cushions of every vehicle to make them fire resistant (2).

Future Directions

In the last few decades, extensive research has been carried out in the field of self-extinguishing PU. Even though the field is well standardized, there have been numerous recent challenges and drawbacks. Generally, in most of the cases, the desired FR property was achieved through the loose inherent properties of the PU. As a result, the field demands greater modification through which a desired FR property can be achieved without losing its mechanical or other desired properties. A simple and easy technique should be developed in place of complicated and time-consuming processes. For example, the LBL technique involves several deposition cycles and subsequent drying, which reduces their industrial demands. Henceforth, a cost-effective facile technique should be developed to maximize its applications in every possible field. More research should be focused on the use of suitable nanomaterials to enhance the self-extinguishing behavior of PU. From the sustainable point of view, greener modification techniques should be employed in place of hazardous techniques. Emphasis should be put on the use of bio-based FRs like modified vegetable oils, chitosan, lignin (kraft lignin and sulfonated lignin), xylan, cellulose, phosphorylated phloroglucinol, levulinic acid, cardanol, and phytic acid. The char-forming ability of the bio-based FRs is significantly higher than in petro-based FRs. These bio-derived FRs can also be procured or prepared at a low cost using simple techniques. Structural manipulation using suitable reactive compounds should be carried out on a molecular level to make intrinsic FR-PU.

Summary and Conclusion

In this chapter, the authors highlighted the fundamentals related to self-extinguishing PUs. The introduction section provided a clear presentation of the current market trends in PU production and their increasing use in various fields, including available raw materials used for PU preparation and their drawbacks, the necessity of FR-PUs, and the chronological development of FR-PU. In the subsequent sections, a comprehensive discussion was carried out on the basics of generating FR-PU along with the chemistry behind this FR behavior. Three general approaches and their subcategories (sol-gel process, LBL assembly, and the plasma technique) used to make FR-PU were presented. Numerous studies were cited throughout the discussion for better understanding. The chapter also provided a clear indication that the future direction of PU research resides in the field of self-extinguishing PU. Furthermore, the present chapter tried to provide required information to the readers on the structure-property relationship (mainly FR property) of PU, especially upon incorporation of additives, chemical modification, and coating formation.

References

1. Visakh, P. M.; Semkin, A. O.; Rezaev, I. A.; Fateev, A. V. Review on Soft Polyurethane Flame Retardan. *Constr. Build. Mater.* **2019**, *227*, 116673.
2. Yang, H.; Yu, B.; Song, P.; Maluk, C.; Wang, H. Surface-Coating Engineering for Flame Retardant Flexible Polyurethane Foams: A Critical Review. *Compos. Part B Eng.* **2019**, *176*, 107185.
3. Levchik, S. V.; Weil, E. D. Thermal Decomposition, Combustion and Fire-Retardancy of Polyurethanes - A Review of the Recent Literature. *Polym. Int.* **2004**, *53* (11), 1585–1610.
4. Vahabi, H.; Rastin, H.; Movahedifar, E.; Antoun, K.; Brosse, N.; Saeb, M. R. Flame Retardancy of Bio-Based Polyurethanes: Opportunities and Challenges. *Polymers (Basel)* **2020**, *12* (6), 1234.
5. Danso, D.; Chow, J.; Streit, W. R. Plastics: Environmental and Biotechnological Perspectives on Microbial Degradation. *Appl. Environ. Microbiol.* **2019**, *85* (19), 1–27.
6. Singh, H.; Jain, A. K. Ignition, Combustion, Toxicity, and Fire Retardancy of Polyurethane Foams: A Comprehensive Review. *J. Appl. Polym. Sci.* **2008**, *116* (5), 1115–1143.
7. Bellayer, S.; Jimenez, M.; Prieur, B.; Dewailly, B.; Ramgobin, A.; Sarazin, J.; Revel, B.; Tricot, G.; Bourbigot, S. Fire Retardant Sol-Gel Coated Polyurethane Foam: Mechanism of Action. *Polym. Degrad. Stab.* **2018**, *147*, 159–167.
8. Bourbigot, S.; Duquesne, S. Fire Retardant Polymers: Recent Developments and Opportunities. *J. Mater. Chem.* **2007**, *17* (22), 2283–2300.
9. Chattopadhyay, D. K.; Webster, D. C. Thermal Stability and Flame Retardancy of Polyurethanes. *Prog. Polym. Sci.* **2009**, *34* (10), 1068–1133.
10. Camino, G.; Costa, L.; Luda di Cortemiglia, M. P. Overview of Fire Retardant Mechanisms. *Polym. Degrad. Stab.* **1991**, *33* (2), 131–154.
11. Peng, H. K.; Wang, X. X.; Li, T. T.; Lou, C. W.; Wang, Y. T.; Lin, J. H. Mechanical Properties, Thermal Stability, Sound Absorption, and Flame Retardancy of Rigid PU Foam Composites Containing a Fire-Retarding Agent: Effect of Magnesium Hydroxide and Aluminum Hydroxide. *Polym. Adv. Technol.* **2019**, *30* (8), 2045–2055.
12. Chai, H.; Duan, Q.; Jiang, L.; Sun, J. Effect of Inorganic Additive Flame Retardant on Fire Hazard of Polyurethane Exterior Insulation Material. *J. Therm. Anal. Calorim.* **2019**, *135* (5), 2857–2868.
13. Wang, W.; He, K.; Dong, Q.; Zhu, N.; Fan, Y.; Wang, F.; Xia, Y.; Li, H.; Wang, J.; Yuan, Z.; Wang, E.; Lai, Z.; Kong, T.; Wang, X.; Ma, H.; Yang, M. Synergistic Effect of Aluminum Hydroxide and Expandable Graphite on the Flame Retardancy of Polyisocyanurate-Polyurethane Foams. *J. Appl. Polym. Sci.* **2014**, *131* (4), 1–7.
14. Wang, Y.; Wang, F.; Dong, Q.; Yuan, W.; Liu, P.; Ding, Y.; Zhang, S.; Yang, M.; Zheng, G. Expandable Graphite Encapsulated by Magnesium Hydroxide Nanosheets as an Intumescent Flame Retardant for Rigid Polyurethane Foams. *J. Appl. Polym. Sci.* **2018**, *135* (39), 1–9.
15. Dike, A. S.; Tayfun, U.; Dogan, M. Influence of Zinc Borate on Flame Retardant and Thermal Properties of Polyurethane Elastomer Composites Containing Huntite-Hydromagnesite Mineral. *Fire Mater.* **2017**, *41* (7), 890–897.

16. Mohammadi, A.; Wang, D. Y.; Hosseini, A. S.; De La Vega, J. Effect of Intercalation of Layered Double Hydroxides with Sulfonate-Containing Calix[4]Arenes on the Flame Retardancy of Castor Oil-Based Flexible Polyurethane Foams. *Polym. Test.* **2019**, *79* (August), 106055.

17. Savas, L. A.; Deniz, T. K.; Tayfun, U.; Dogan, M. Effect of Microcapsulated Red Phosphorus on Flame Retardant, Thermal and Mechanical Properties of Thermoplastic Polyurethane Composites Filled with Huntite & Hydromagnesite Mineral. *Polym. Degrad. Stab.* **2017**, *135*, 121–129.

18. Guler, T.; Tayfun, U.; Bayramli, E.; Dogan, M. Effect of Expandable Graphite on Flame Retardant, Thermal and Mechanical Properties of Thermoplastic Polyurethane Composites Filled with Huntite and Hydromagnesite Mineral. *Thermochim. Acta* **2017**, *647*, 70–80.

19. Gómez-Fernández, S.; Ugarte, L.; Peña-Rodriguez, C.; Zubitur, M.; Corcuera, M. Á.; Eceiza, A. Flexible Polyurethane Foam Nanocomposites with Modified Layered Double Hydroxides. *Appl. Clay Sci.* **2016**, *123*, 109–120.

20. Tang, G.; Zhou, L.; Zhang, P.; Han, Z.; Chen, D.; Liu, X.; Zhou, Z. Effect of Aluminum Diethylphosphinate on Flame Retardant and Thermal Properties of Rigid Polyurethane Foam Composites. *J. Therm. Anal. Calorim.* **2020**, *140* (2), 625–636.

21. Liu, Y.; He, J.; Yang, R. Effects of Dimethyl Methylphosphonate, Aluminum Hydroxide, Ammonium Polyphosphate, and Expandable Graphite on the Flame Retardancy and Thermal Properties of Polyisocyanurate-Polyurethane Foams. *Ind. Eng. Chem. Res.* **2015**, *54* (22), 5876–5884.

22. Yuan, Y.; Ma, C.; Shi, Y.; Song, L.; Hu, Y.; Hu, W. Highly-Efficient Reinforcement and Flame Retardancy of Rigid Polyurethane Foam with Phosphorus-Containing Additive and Nitrogen-Containing Compound. *Mater. Chem. Phys.* **2018**, *211*, 42–53.

23. Shi, L.; Li, Z.-M.; Xie, B.-H.; Wang, J.-H.; Tian, C.-R.; Yang, M.-B. Flame Retardancy of Different-Sized Expandable Graphite Particles for High-Density Rigid Polyurethane Foams. *Polym. Int.* **2006**, *55* (8), 862–871.

24. Thirumal, M.; Khastgir, D.; Singha, N. K.; Manjunath, B. S.; Naik, Y. P. Effect of Expandable Graphite on the Properties of Intumescent Flame-Retardant Polyurethane Foam. *J. Appl. Polym. Sci.* **2008**, *110* (5), 2586–2594.

25. Gao, L.; Zheng, G.; Zhou, Y.; Hu, L.; Feng, G.; Zhang, M. Synergistic Effect of Expandable Graphite, Diethyl Ethylphosphonate and Organically-Modified Layered Double Hydroxide on Flame Retardancy and Fire Behavior of Polyisocyanurate-Polyurethane Foam Nanocomposite. *Polym. Degrad. Stab.* **2014**, *101* (1), 92–101.

26. Wang, B.; Sheng, H.; Shi, Y.; Song, L.; Zhang, Y.; Hu, Y.; Hu, W. The Influence of Zinc Hydroxystannate on Reducing Toxic Gases (CO, NOx and HCN) Generation and Fire Hazards of Thermoplastic Polyurethane Composites. *J. Hazard. Mater.* **2016**, *314*, 260–269.

27. Ramanujam, S.; Zequine, C.; Bhoyate, S.; Neria, B.; Kahol, P.; Gupta, R. Novel Biobased Polyol Using Corn Oil for Highly Flame-Retardant Polyurethane Foams. *J. Carbon Res.* **2019**, *5* (1), 13.

28. Agrawal, A.; Kaur, R.; Singh Walia, R. Flame Retardancy of Ceramic-Based Rigid Polyurethane Foam Composites. *J. Appl. Polym. Sci.* **2019**, *136* (48), 1–10.

29. Agrawal, A.; Kaur, R.; Walia, R. S. Investigation on Flammability of Rigid Polyurethane Foam-Mineral Fillers Composite. *Fire Mater.* **2019**, *43* (8), 917–927.

30. Bhoyate, S.; Ionescu, M.; Kahol, P. K.; Chen, J.; Mishra, S. R.; Gupta, R. K. Highly Flame-Retardant Polyurethane Foam Based on Reactive Phosphorus Polyol and Limonene-Based Polyol. *J. Appl. Polym. Sci.* **2018**, *135* (21), 16–19.
31. Cui, M.; Li, J.; Qin, D.; Sun, J.; Chen, Y.; Xiang, J.; Yan, J.; Fan, H. Intumescent Flame Retardant Behavior of Triazine Group and Ammonium Polyphosphate in Waterborne Polyurethane. *Polym. Degrad. Stab.* **2021**, *183*, 109439.
32. Zhang, P.; Fan, H.; Tian, S.; Chen, Y.; Yan, J. Synergistic Effect of Phosphorus-Nitrogen and Silicon-Containing Chain Extenders on the Mechanical Properties, Flame Retardancy and Thermal Degradation Behavior of Waterborne Polyurethane. *RSC Adv.* **2016**, *6* (76), 72409–72422.
33. Rao, W.-H.; Zhu, Z.-M.; Wang, S.-X.; Wang, T.; Tan, Y.; Liao, W.; Zhao, H.-B.; Wang, Y.-Z. A Reactive Phosphorus-Containing Polyol Incorporated into Flexible Polyurethane Foam: Self-Extinguishing Behavior and Mechanism. *Polym. Degrad. Stab.* **2018**, *153*, 192–200.
34. Arora, S.; Mestry, S.; Naik, D.; Mhaske, S. T. O-Phenylenediamine-Derived Phosphorus-Based Cyclic Flame Retardant for Epoxy and Polyurethane Systems. *Polym. Bull.* **2020**, *77* (6), 3185–3205.
35. Zeng, S.; Xing, C.; Chen, L.; Xu, L.; Li, B.; Zhang, S. Green Flame-Retardant Flexible Polyurethane Foam Based on Cyclodextrin. *Polym. Degrad. Stab.* **2020**, *178*, 109171.
36. Yuan, Y.; Yang, H.; Yu, B.; Shi, Y.; Wang, W.; Song, L.; Hu, Y.; Zhang, Y. Phosphorus and Nitrogen-Containing Polyols: Synergistic Effect on the Thermal Property and Flame Retardancy of Rigid Polyurethane Foam Composites. *Ind. Eng. Chem. Res.* **2016**, *55* (41), 10813–10822.
37. Bellayer, S.; Jimenez, M.; Barrau, S.; Bourbigot, S. Fire Retardant Sol-Gel Coatings for Flexible Polyurethane Foams. *RSC Adv.* **2016**, *6* (34), 28543–28554.
38. Hosgor, Z.; Kayaman-Apohan, N.; Karatas, S.; Gungor, A.; Menceloglu, Y. Nonisocyanate Polyurethane/Silica Hybrid Coatings via a Sol-Gel Route. *Adv. Polym. Technol.* **2012**, *31* (4), 390–400.
39. Hai, Y.; Wang, C.; Jiang, S.; Liu, X. Layer-by-Layer Assembly of Aerogel and Alginate toward Self-Extinguishing Flexible Polyurethane Foam. *Ind. Eng. Chem. Res.* **2020**, *59* (1), 475–483.
40. Liu, L.; Wang, W.; Hu, Y. Layered Double Hydroxide-Decorated Flexible Polyurethane Foam: Significantly Improved Toxic Effluent Elimination. *RSC Adv.* **2015**, *5* (118), 97458–97466.
41. Pan, H.; Wang, W.; Pan, Y.; Song, L.; Hu, Y.; Liew, K. M. Formation of Layer-by-Layer Assembled Titanate Nanotubes Filled Coating on Flexible Polyurethane Foam with Improved Flame Retardant and Smoke Suppression Properties. *ACS Appl. Mater. Interfaces* **2015**, *7* (1), 101–111.
42. Shi, X.; Yang, P.; Peng, X.; Huang, C.; Qian, Q.; Wang, B.; He, J.; Liu, X.; Li, Y.; Kuang, T. Bi-Phase Fire-Resistant Polyethylenimine/Graphene Oxide/Melanin Coatings Using Layer by Layer Assembly Technique: Smoke Suppression and Thermal Stability of Flexible Polyurethane Foams. *Polymer (Guildf)* **2019**, *170*, 65–75.
43. Kim, Y. S.; Davis, R.; Cain, A. A.; Grunlan, J. C. Development of Layer-by-Layer Assembled Carbon Nanofiber-Filled Coatings to Reduce Polyurethane Foam Flammability. *Polymer (Guildf)* **2011**, *52* (13), 2847–2855.

44. Maddalena, L.; Carosio, F.; Gomez, J.; Saracco, G.; Fina, A. Layer-by-Layer Assembly of Efficient Flame Retardant Coatings Based on High Aspect Ratio Graphene Oxide and Chitosan Capable of Preventing Ignition of PU Foam. *Polym. Degrad. Stab.* **2018**, *152*, 1–9.
45. Carosio, F.; Alongi, J. Ultra-Fast Layer-by-Layer Approach for Depositing Flame Retardant Coatings on Flexible PU Foams within Seconds. *ACS Appl. Mater. Interfaces* **2016**, *8* (10), 6315–6319.
46. , W.; Pan, Y.; Pan, H.; Yang, W.; Liew, K. M.; Song, L.; Hu, Y. Synthesis and Characterization of MnO2 Nanosheets Based Multilayer Coating and Applications as a Flame Retardant for Flexible Polyurethane Foam. *Compos. Sci. Technol.* **2016**, *123*, 212–221.
47. Jimenez, M.; Lesaffre, N.; Bellayer, S.; Dupretz, R.; Vandenbossche, M.; Duquesne, S.; Bourbigot, S. Novel Flame Retardant Flexible Polyurethane Foam: Plasma Induced Graft-Polymerization of Phosphonates. *RSC Adv.* **2015**, *5* (78), 63853–63865.

Chapter 5

Highly Flame-Retardant Polyurethane

Young Nam Kim,[1,2] Hyunsung Jeong,[1,3] Sooyeon Ryu,[1,4] and Yong Chae Jung[*,1]

[1]Institute of Advanced Composite Materials, Korea Institute of Science and Technology (KIST), 92, Chudong-ro, Bongdong-eup, Wanju-gun, Jeonbuk 55324, Republic of Korea
[2]Department of Chemical and Biomolecular Engineering, Yonsei University, 262 Seongsanno, Seodaemun-gu, Seoul 03722, Republic of Korea
[3]School of Chemical Engineering, Sungkyunkwan University (SKKU), Suwon 16419, Republic of Korea
[4]Department of Organic Materials and Textile Engineering, Jeonbuk National University, Baekje-daero, Deokjin-gu, Jeonju-shi, Jeonbuk 54896, Republic of Korea
*Email: ycjung@kist.re.kr

Polyurethane is a representative engineering plastic that is widely used for its excellent properties. Since its discovery in 1937, its application value has exceeded our imagination. This chapter introduces research results and trends of polyurethane's high flame retardancy.

Introduction

What Is Polyurethane?

Polyurethane (PU) is a polymer that is vital to many applications for manufacturing surface coating reagents, foams, composite materials, and adhesives because of its intrinsic characteristics such as high hardness, high wear resistance, low thermal conductivity, and low water absorption (*1–3*).

It was first discovered in 1937 by Otto Bayer and his colleagues in the Interessengemeinschaft Farbenindustrie AG in Leverkusen, Germany. In earlier studies, PU products obtained from aliphatic diisocyanate and diamine generated from polyurea were emphasized, and those studies continued until the characteristics of PU obtained from aliphatic diisocyanate and glycol materialized. Subsequently, in terms of polyisocyanate, since the end of World War II (1952), the commercial-scale production of PU from toluene diisocyanate and polyester polyol was enabled, and polyester polyol was gradually replaced by polyether polyol because of its advantages, including ease of handling and better hydrolysis stability. In 1956, DuPont was the first commercialized polyether polyol achieved by polymerizing tetrahydrofuran, and poly(tetramethylene ether) glycol was introduced, followed by Baden Aniline and Soda Factory and Dow Chemical produced polyalkylene glycol in 1957 (*4*).

© 2021 American Chemical Society

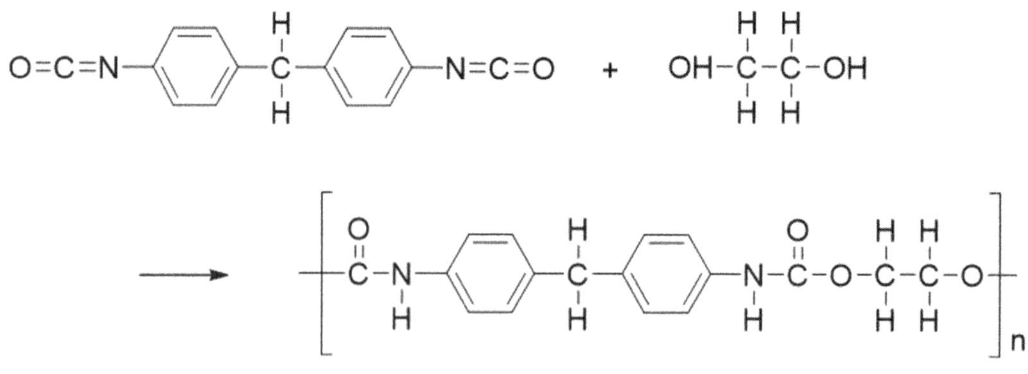

Scheme 1. Molecular structure of PU. Adapted with permission from reference (2). Copyright 2007 Elsevier.

For a few decades, PU has enabled the use of polymeric isocyanates—for example, foaming agents such as polyether polyols and polymethylene diphenyl diisocyanate (PMDI). Hence, the rigid PU foam (RPUF), rather than the flexible PU foam (FPUF), was utilized. Such PMDI-based PU foam (PUF) demonstrates high heat resistance and flame retardancy, as shown in Scheme 1 (2, 5).

However, PU is easily ignited and has a fast fire propagation speed because of its high inflammability, low thermal conductivity, and porous structure. It emits a significant volume of poisonous gases during combustion. Hence, flame-retardant (FR) PU is being actively researched and developed (6–9).

To enhance the flame retardancy of PU, FRs can be directly added or the raw material with high flame-retarding performance can be used by chemically combining polyol or isocyanate with FR components such as phosphorus, nitrogen, or halogen. Until now, flame retardancy has been typically enhanced through physical methods such as adding or coating FRs. However, a few requirements apply—for example, the high mixability of raw materials and additives, lack of effect on the final mechanical properties, and low generation of carcinogen and poisonous gases during combustion.

The most widely used additive FRs (AFRs) include halogen-based, phosphorus-based, nitrogen-based, and inorganic FRs. The most representative halogen-containing phosphorus-based FRs (i.e., tris 2-chloropropyl phosphate, tris 2-chloroethyl phosphate, and phosphinyl alkyl phosphate ester) show a flame retardancy effect because of the halogen and phosphorus inside the FR. Halogen becomes a molecule or atom in gaseous form because of combustion and shows the flame retardancy effect by stabilizing the free radical (OH·, H·). However, because halogen corrodes metals with gases generated during combustion and produces gases that are harmful to the human body, it is regarded as an environmentally regulated substance, and its usage is limited. Its flame retardancy reduces over time.

To overcome these challenges, studies for enhancing the flame retardancy of PU are actively being conducted using various methods such as reforming and adding an inorganic FR, adding a functional group with flame retardancy and heat resistance (such as imide and isocyanurate) to the PU chain, using an organic and inorganic hybrid, adding expandable graphite or nanoclay, and encapsulating an FR—selectively imparting an FR effect (10–16).

Phosphorus-based FRs have the most potential to replace halogen-based ones (17), and the major flame-retarding mechanism of phosphorus applies in both gas and solid phases. The dehydration and carbonization reactions by phosphorus generated via pyrolysis, in addition to the radical catabolism reaction of hydrogen and hydroxyl radical–containing phosphorus, are attributed to its flame retardancy. In particular, PU contains oxygen and benzene rings in the main chain; hence, char forms easily during combustion, which results in a beneficial flame retardancy effect (*18*).

The performance of phosphorus-based FRs depends significantly on their phosphorus content, and it is generally known that the flame-retarding performance is better with higher phosphorous content. In particular, phosphorus and nitrogen increase the formation of char and, consequently, demonstrate a synergistic effect in the flame-retarding performance; hence, studies on FRs with high phosphorus content and ones containing both phosphorus and nitrogen are being actively performed (*19*).

In this section, we report the present research on FR-PU. Specifically, we describe the flame retardancy evaluation method and flame retardancy of PU based on nonhalogen FRs. Finally, the research trend of the flame retardancy of PU using nano- or microsized organic–inorganic additives is described.

Methods for Analyzing FR Properties

Limiting Oxygen Index

The limiting oxygen index (LOI) test method is a standardized fire testing method that reproduces an actual fire environment and is a representative method to measure the flame retardancy of polymers by limiting the oxygen index (according to ISO 4589, ASTM D 2863, and NFT 51-07), as shown in Figure 1.

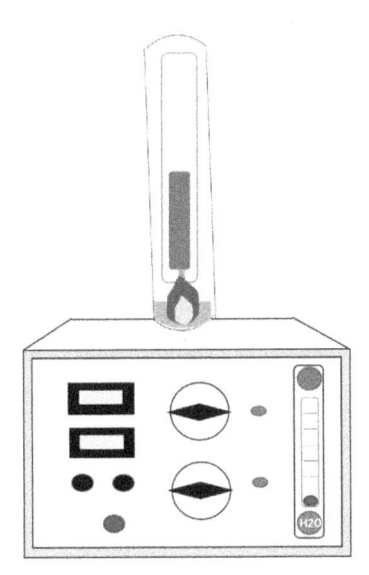

Figure 1. Limiting oxygen index (LOI) test method. Reproduced with permission from reference (20). Copyright 2021 Wiley-VCH GmbH.

In this testing method, the measurement is conducted after fixing the specimen vertical to the tube under the conditions that the relative concentrations of nitrogen and oxygen required for the measurement specimen to burn for 3 min after the beginning of fire can change. During this process, flame retardancy is evaluated by measuring the minimum oxygen volumetric percentage in mixed air. The LOI is expressed as follows:

$$\text{LOI} = [O_2] / [(O_2) + (N_2)] \times 100 \quad (1)$$

where O_2 is the oxygen flow rate (L/min), and N_2 is the nitrogen flow rate (L/min).

An LOI value of less than 20.9 implies an effective combustion, even in air, whereas an LOI value of between 20.9 and 27 implies a relatively slow burn in air. An LOI value exceeding 27 implies a self-extinguishing specimen that is difficult to burn in air.

Cone Calorimeter

Cone calorimetry is a testing method that measures the release rate of heat generated from the specimen under the heat supplied by the conical heater, smoke generation rate, ignition time, amount of oxygen consumed, amounts of carbon monoxide and carbon dioxide generated, and mass reduction rate, as shown in Figure 2. This testing method is based on the basic principle that 13.1 MJ/kg of calorie is generated when 1 kg of oxygen is consumed, which is based on the fact that the net combustion calorie is proportional to the amount of oxygen required for combustion. The heat release rate (HRR) generated from the specimen, ignition time, amount of oxygen consumed, amount of carbon monoxide and carbon dioxide generated, and flow rate of the combusted gases are measured. During this process, it is proven that the standard deviation for the released calorie of each material is within $\pm 5\%$. The constant, 13.1 MJ/kg of O_2, is the same for all hydrocarbon materials (materials composed of carbon and hydrogen), except for a few materials such as acetylene.

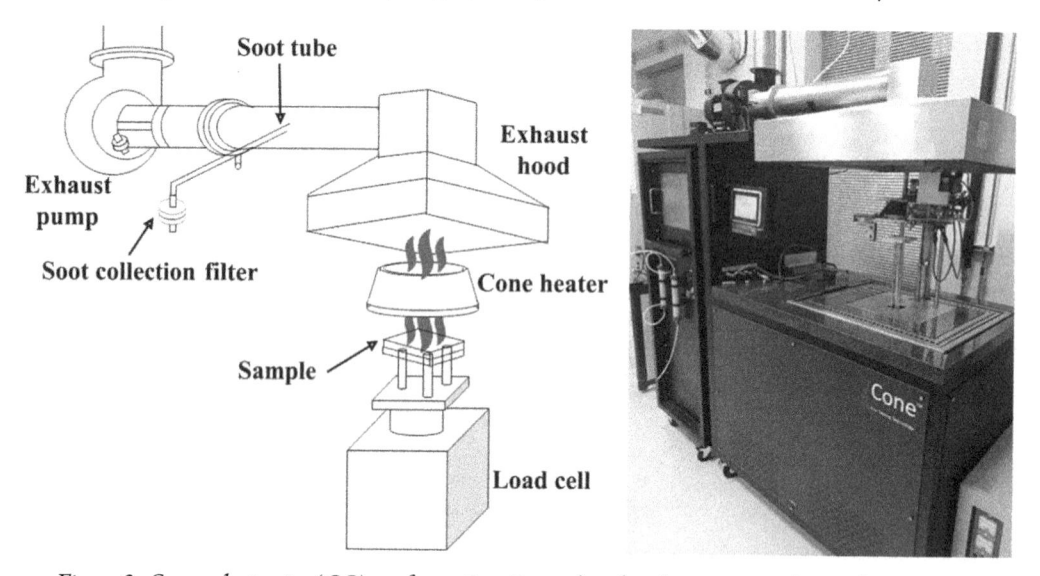

Figure 2. Cone calorimeter (CC) configuration. Reproduced with permission from reference (20). Copyright 2021 Wiley-VCH GmbH.

The testing method for HRR conforms to the typical ISO 5660-1 (cone calorimeter, or CC) specifications and is based on the fact that the HRR is proportional to the amount of oxygen required for combustion (in the case of combustion). It evaluates the heat release properties of the specimen exposed to the radiant heat by measuring the concentration of oxygen in the combustion product flow and the amount of oxygen consumed, which is induced from the flow rate.

The heat of combustion is generally based on the fact that it is proportional to the amount of oxygen required for combustion. While receiving the regulated external radiant heat, the specimen burns under the surrounding air condition. During this process, the HRR is estimated by measuring the oxygen concentration and exhaust gas flow rate. A CC evaluates the contribution of a material or product to the HRR while it is being exposed to fire.

Underwriters Laboratories 94

Underwriters Laboratories (UL) is an organization that evaluates stability by establishing standards to assess the safety grade of electric appliances or electronics. Representative UL regulation types include UL-94 (assesses flame retardancy), UL-746A (assesses electrical properties), UL-746B (assesses the long-term heat resistance temperature of plastics), UL-1446 (assesses electrical insulation systems), and UL-746D (certification for molder). UL-94 is the most representative method to assess the flame retardancy of plastics, and the gradation is based on the burning degree, thereby allowing the flame retardancy grade to be rated.

The UL-94 test method is categorized into the horizontal combustion test (applied to noncombustible resins), vertical combustion test (applied to FR resins), and flat combustion test, as shown in Figure 3.

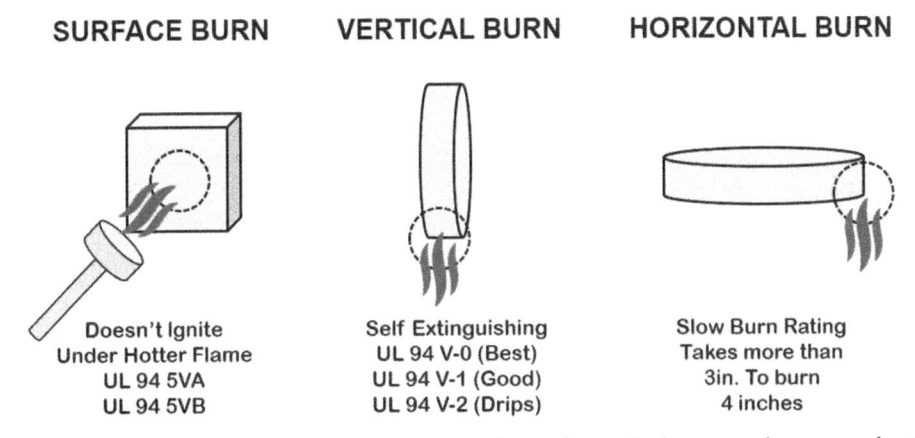

Figure 3. UL-94 method. Reproduced with permission from reference (20). Copyright 2021 Wiley-VCH GmbH.

Highly FR-PU

Organic Materials

The development of PU using natural oils is being actively investigated. Among natural oils, castor oil is an important renewable resource that is widely used as feed material in many industrial products (21–23). However, PU obtained from castor oil presents a few disadvantages, including low attributability because of its low hydroxyl content, slow curing speed because of a secondary hydroxyl group, and low flame retardancy (24). Hence, Zhang et al. (25) developed a PUF using glycerolysis castor oil (GCO), and castor oil phosphate FR polyol (COFPL), which is a castor oil with flame retardancy by reacting GCO with diethyl phosphate to develop an FR PUF. Expanded graphite (EG)—for example, a flame-retarding additive with a high flame-retarding efficiency that does not pollute the environment—or reformed EG is a general flame-retarding material for various polymers, including PUF. Because EG is advantageous in terms of its cost effectiveness, high specific surface area, and distinct structure (wormlike), its flame retardancy effect is prominent.

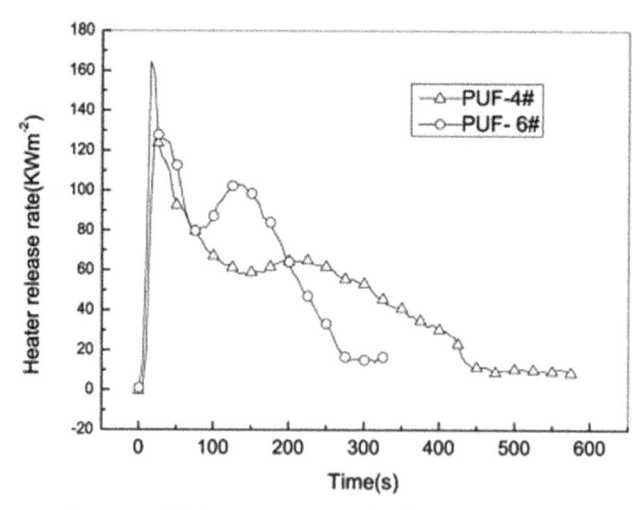

Figure 4. HRR curves for GCO-filled PUF, COFPL-filled PUF. Reproduced with permission from reference (25). Copyright 2013 Elsevier.

Figure 4 shows that the PUF functionalized with phosphorus (PUF-6#, only using COFPL) has a lower peak heat release rate (PHRR) value (127.78 kW/m^2) compared with that of the PUF obtained from castor oil (PUF-4#, only using GCO).

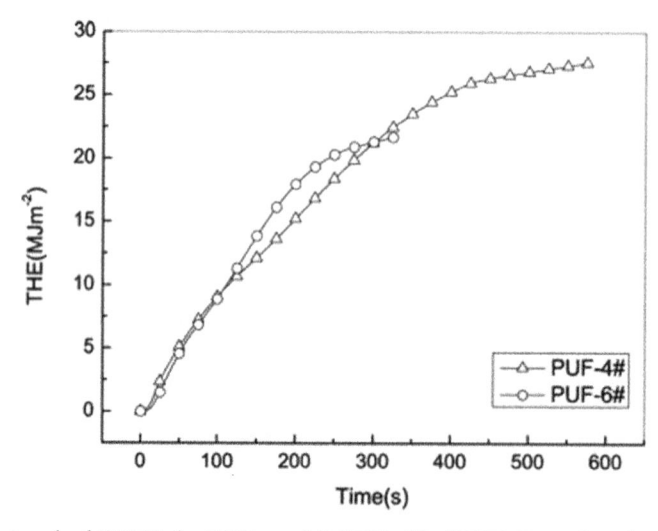

Figure 5. Total heat evolved (THE) for GCO- and COFPL-filled PUF. Reproduced with permission from reference (25). Copyright 2013 Elsevier.

As shown in Figure 5, which presents the total heat evolved (THE), the THE of PUF-6# reduced to 21.80 MJ/m^2 compared with that of PUF-4#. Such results indicate that the PUF sample was not completely burned but was protected. This is because the EG formed a protective film with a distinct structure (wormlike) on the PUF surface when the thermal expansion and adsorption of the system occurred at 180–300 °C. The corresponding protective film served as a physical protection barrier that prevented heat from transferring into the inside of the sample and, hence, reduced thermal emission, demonstrating a cumulative effect between EG and COFPL on flame inhibition. Such a synergistic effect occurred because diethyl phosphate decomposed and generated phosphinic acid (which demonstrates strong dehydration during heating), phosphinic acid accelerated the formation of char, and char and EG formed a compact carbonaceous layer in a stable structure.

Zhang et al. (26) reformed glycerol-1-allyl ether obtained from orange peel oil and manufactured PUF. Although they assigned flame retardancy using dimethyl methyl phosphonate (AFR), the PUF using phosphorus indicated deteriorated mechanical strength. Hence, they manufactured a flame-retarding PUF by assigning flame retardancy to a PUF utilizing 6-tribromophe-nol (reactive FR [RFR]) and achieved a prominent flame retardancy effect, in addition to high mechanical strength. Because the bromine included in their PUF can inhibit OH or H radicals generated during combustion, a combustion chain reaction was avoided because of these radicals; hence, the flame retardancy effect was demonstrated (27).

Cho et al. (28) assigned flame retardancy to PUF using polydopamine (PDA), which is a nontoxic, environmentally friendly FR. PDA has the ability to scavenge free radicals because of its hydroquinone structure with OH groups attached, and because the particle is nanosized, the surface-to-volume ratio is increased to amplify its flame retardancy because more reactive sites are created (29). In their study, PDA was coated on the PUF, and the catechol structure of the PDA can remove O_2 radicals generated during combustion. In terms of PDA3D, where PDA was coated three times, the lowest HRR value was indicated, as shown in Figure 6 (239 kW/m^2).

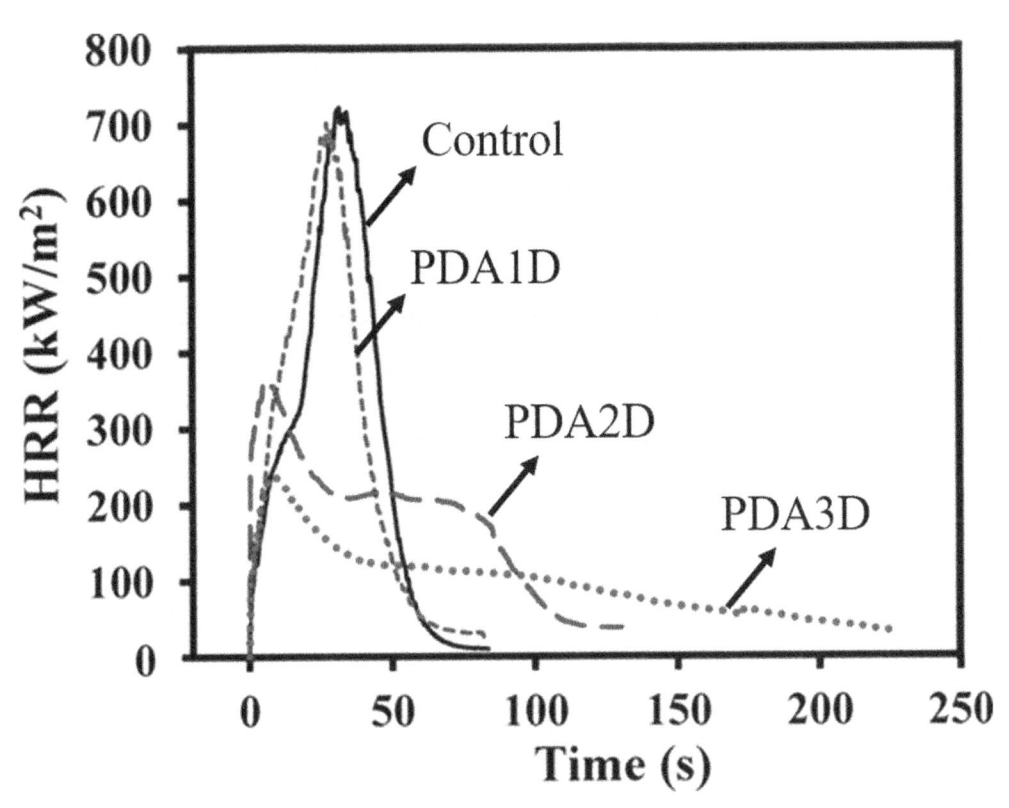

Figure 6. HRR of control (solid line), PDA1D (short-dashed line), PDA2D (long-dashed line), and PDA3D (dotted line) as a function of time during CC testing. Reproduced with permission from reference (28). Copyright 2015 American Chemical Society.

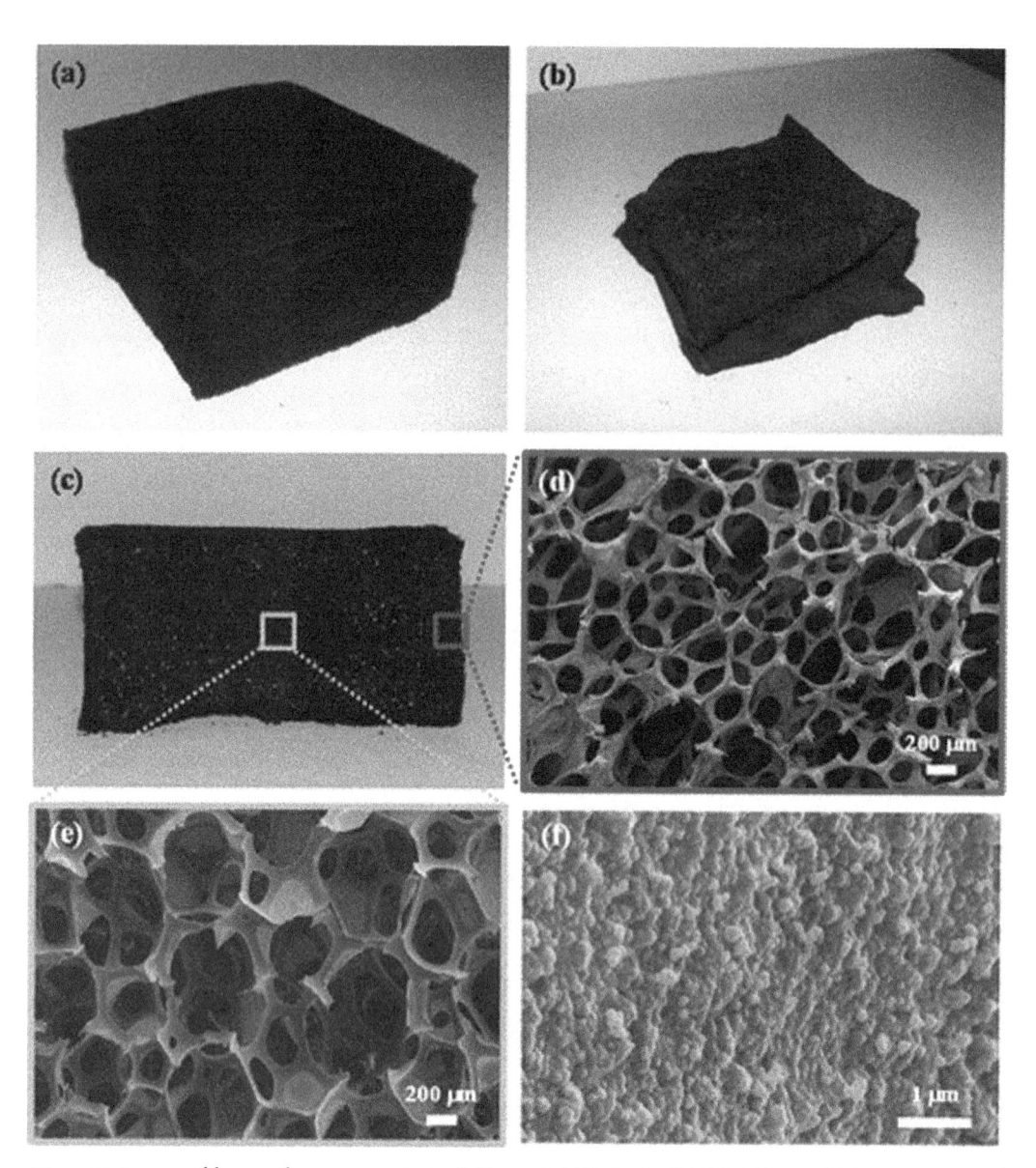

Figure 7. Images of foams after exposure to torch flame. (a) PDA3D, (b) PDA1D, and (c) cross-section of PDA3D foam. Scanning electronic microsopy images of cross-section of PDA3D specimen (d) at the surface and (e) in center of specimen, showing retention in cellular structure. (f) Scanning electronic microsopy image of char surface of post-burn PDA3D foam. Reproduced with permission from reference (28). Copyright 2015 American Chemical Society.

As shown in Figure 7, although the PDA3D foam was exposed to flame and combusted, it not only maintained its stable structure but also inhibited flame spread after a certain duration and then extinguished itself, demonstrating its high flame retardancy. The process by which it first combusted, gradually inhibited flame spread, and then extinguished itself is because of the radical removal ability of the catechol. Although the surface directly burned by the flame showed a partially contracted form char structure (Figure 7e), the overall PU structure was well preserved (Figure 7d). The char form generated on the foam surface after combustion is presented in Figure 7f, which shows a similar surface as the PDA coating before combustion. These properties indicate that the PDA coating structure was preserved during combustion.

In contrast, the PDA1D foam did not readily extinguish itself during combustion, and the flame spread rapidly to the entire sample with significant heat generation after ignition, and considerable thermal contraction was observed (Figure 7b). However, the PDA thin film coating on the surface of the sample left char residue on the foam surface after combustion; consequently, the original cellular structure of the PU was preserved. Therefore, it was concluded that a significant decrease in the PHRR because of the PDA coating can effectively inhibit combustion and factors contributing to rapid flame spread.

Liu et al. (30) assigned flame retardancy to thermoplastic PU (TPU) using chitosan (CS), as shown in Figures 8 and 9. First, to introduce additional thermal stability and flame retardancy to CS, benzaldehyde, salicylaldehyde (SCS), hydroxybenzaldehyde, and ammonium polyphosphate were utilized. After mixing the reformed CS and ammonium polyphosphate with TPU separately, the TPU composite was manufactured, and flame retardancy was then assessed.

As shown in Figure 8, the HRR and THR of the composite incorporated with an FR were significantly lower compared with those of TPU (reference sample). In particular, the TPU–SCS sample showed the most superior flame retardancy effect because SCS contained the O-hydroxyl, which can be transformed into ortho-quinone methide and resorcinol during decomposition. This compound is vital for promoting a stable cross-link orientation network formation, and it can enhance char formation (Figure 9).

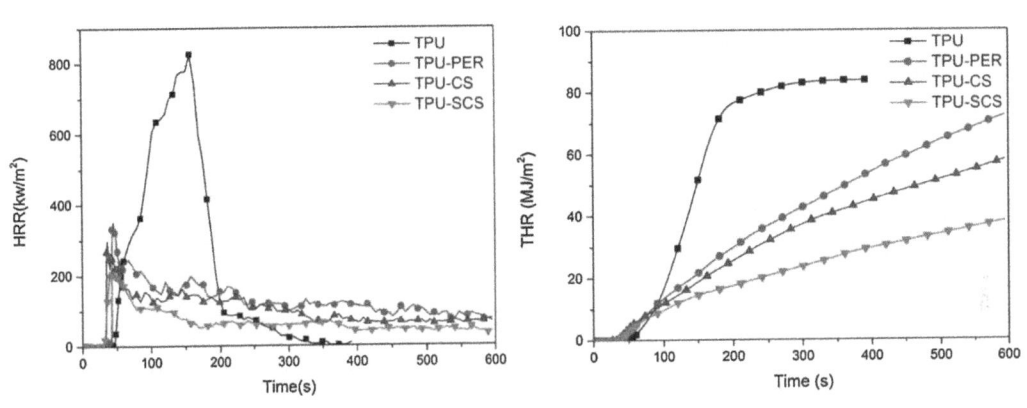

Figure 8. HRR and THR curves of TPU composite samples. Reproduced with permission from reference (30). Copyright 2017 Elsevier.

Nanocarbon Molybdenum Disulfide and EG

Previously, a separate FR was typically added to endow widely used polymers with flame retardancy properties. Nanocomposites have been manufactured using carbon-based nanofillings, such as carbon nanotube, graphene, and nanocarbon molybdenum disulfide (MoS_2), and results pertaining to the correlation between their structural control and flame retardancy have been reported.

Nanocomposites are manufactured by uniformly dispersing nanofillings into the organic polymer; they exhibit superior mechanical and thermal properties compared to simple polymers or existing composite materials. MoS_2, which is widely utilized in various fields because of its superior structural and chemical properties, is a representative nanofilling.

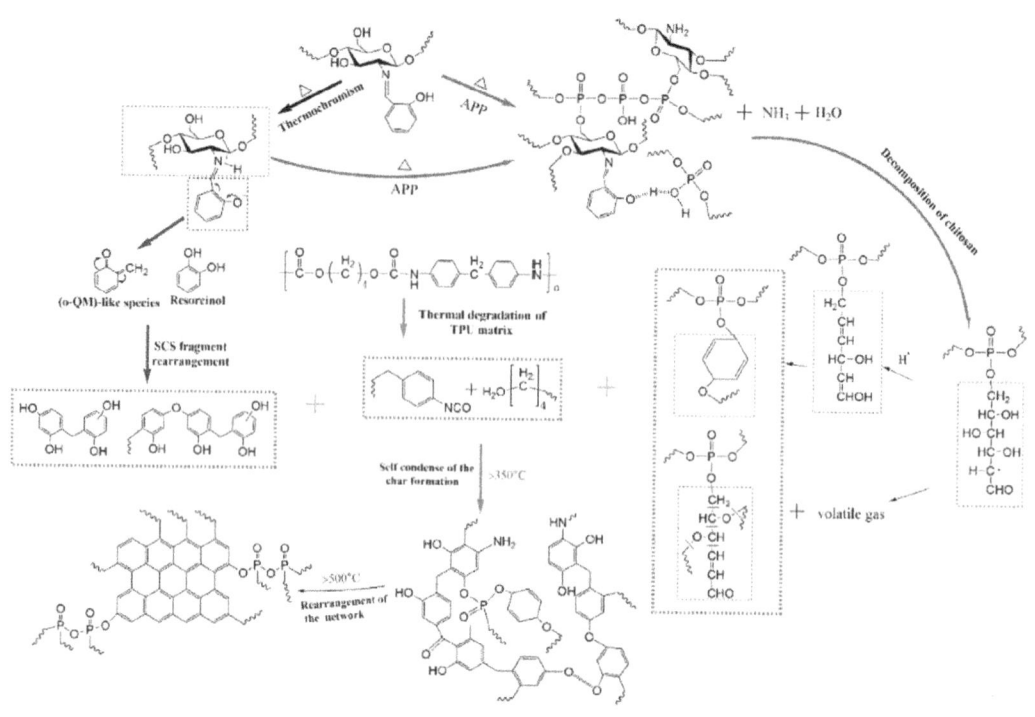

Figure 9. Proposed pyrolysis mechanism of TPU–SCS composite. Reproduced with permission from reference (30). Copyright 2017 Elsevier.

MoS_2 is composed of three S–Mo–S layers bound by a weak van der Waals force and is widely utilized in many fields, including nanoelectronic engineering, sensors, and catalysts. Based on results reported, MoS_2 exhibits favorable mechanical properties such as high elastic modulus, yield stress, and flexural modulus of elasticity; it particularly exhibits high structural and thermal stability. Its electronic properties are superior because of its layered structure, which is similar to graphite, and it has a high microwave attenuation potential because of dielectric and magnetic losses attributed to polymorphism. In addition, MoS_2 is garnering attention as an FR because of its high thermal stability and ability to inhibit gas generation during combustion.

Cai et al. (31) manufactured flame-retarding PU by adding MoS_2 to TPU. They functionalized 9,10-dihydro-9-oxa-10-phosphaphenanthrene-10-oxide (DOPO) and polyethyleneimine to MoS_2 and enhanced the interfacial strength between the filler and matrix. Additionally, they developed a nanocomposite with improved flame retardancy. Figure 10 shows a comparison of the flame retardancy of flame-retarding urethane manufactured under different conditions and pure TPU. Compared with pure TPU, TP-1.0 (which contained functionalized MoS_2 [f-MoS_2]m) indicated a better flame retardancy effect. TPU-2.0 with 2 wt % f-MoS_2 added showed a better flame retardancy effect compared with TPU-1.0 with 1 wt % f-MoS_2 added. As shown in Figure 10, the TPU-2.0 sample with f-MoS_2 added emitted fewer organic gases (e.g., isocyanate, ether, and CO_2 gases serving as organic gas fuels), suggesting its ability to enhance fire safety.

After analyzing the flame retardancy mechanism of TPU with f-MoS_2 added in Figure 11, it was discovered that because the f-MoS_2 easily spreads to basic materials and can establish a strong attractive force with them, f-MoS_2 can assign a flame retardancy effect through barrier action during combustion. Because of the structural properties of f-MoS$_2$, an enhanced flame retardancy effect is expected because of the physical and chemical adsorption of pyrolysis byproducts generated from the TPU basic materials during combustion. By forming a char layer on the surface of the basic materials, additional thermal contact was prevented—resulting in the enhanced flame retardancy effect.

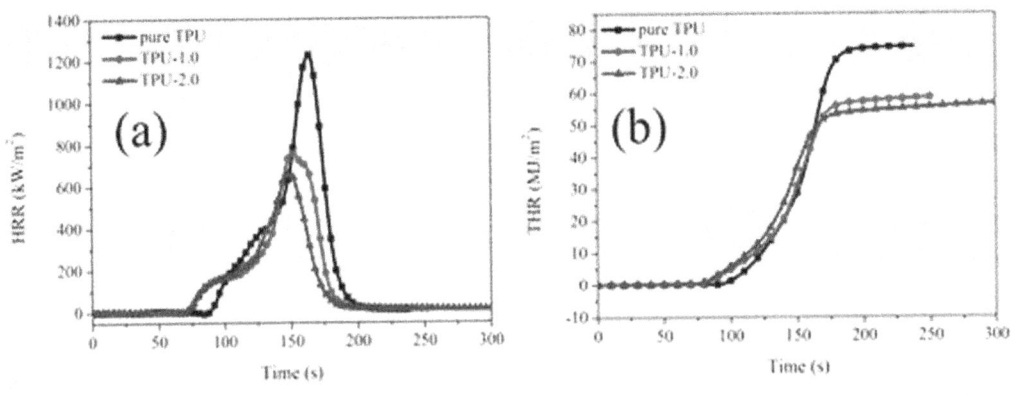

Figure 10. (a) HRR and (b) THR versus time curves of TPU and its composites. Reproduced with permission from reference (31). Copyright 2017 American Chemical Society.

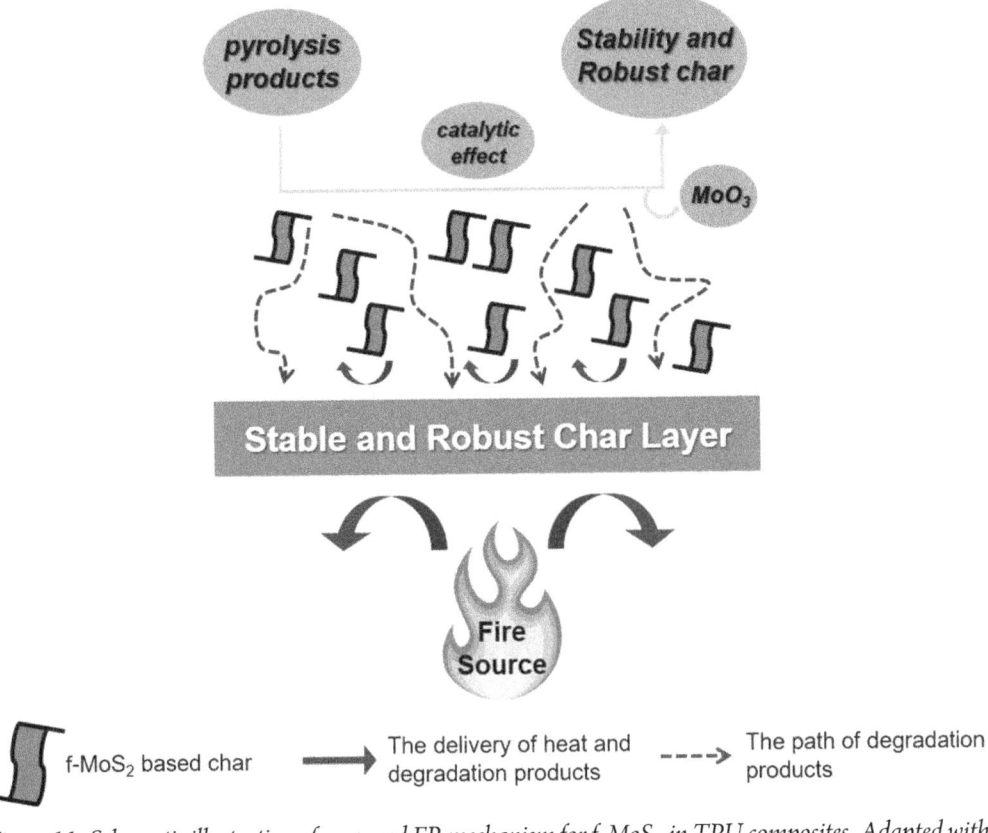

Figure 11. Schematic illustration of proposed FR mechanism for f-MoS$_2$ in TPU composites. Adapted with permission from reference (31). Copyright 2017 American Chemical Society.

Zhi et al. (32) manufactured an FPUF and investigated its flame retardancy. As FPUF exhibits low density, high resilience, and strong mechanical strength, it is widely used in various applications such as automobile materials, soundproof materials, and cushions. However, because of its large specific surface area and open-cell structure, it poses the risk of fire. Therefore, by directly functionalizing DOPO to MoS$_2$, flame retardancy was assigned to FPUF. To compare flame retardancy, each FR was combined with the FPUF, and a comparative analysis was subsequently performed.

As shown in Figure 12, the PHRR of the MoS$_2$-DOPO/FPUF sample decreased 41.3% compared with that of the reference sample, FPUF. Meanwhile, the THR of FPUF increased significantly within 40 s after ignition, suggesting rapid heat generation, whereas the THR of the DOPO- f-MoS$_2$ increased gradually, and the total heat decreased from 21.3 to 15.4 MJ/m^2. Different tendencies were observed depending on the type of FR.

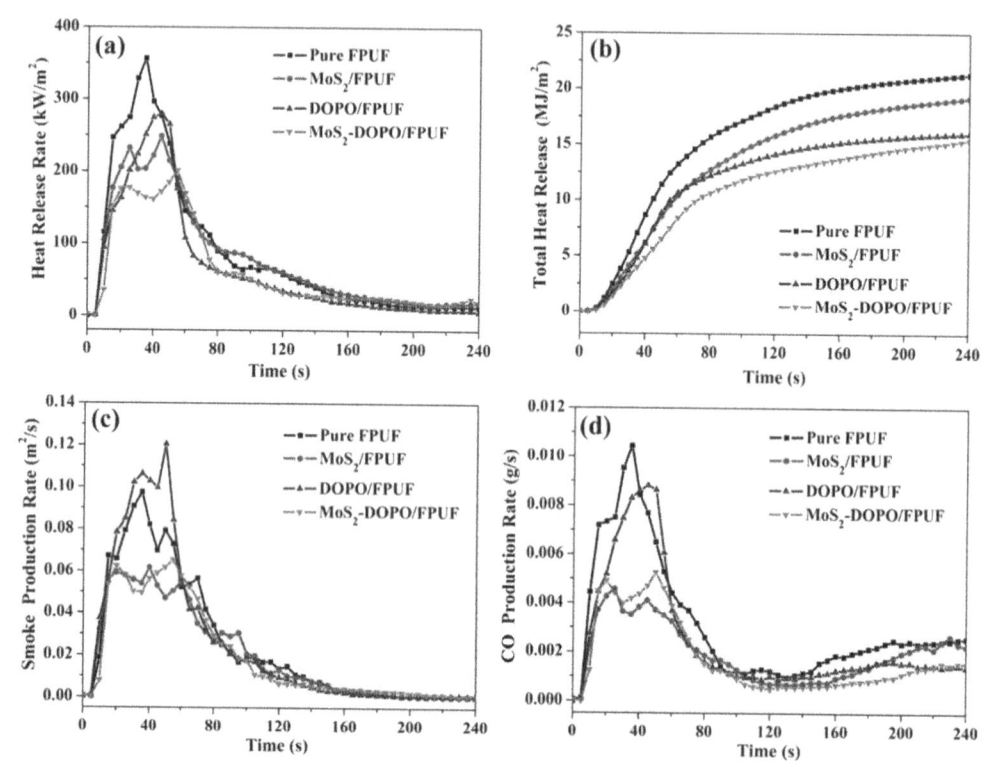

Figure 12. (a) HRR, (b) THR, (c) smoke production rate (SPR), and (d) CO production rate curves for pure and FR FPUFs. Reproduced with permission from reference (32). Copyright 2020 American Chemical Society.

Regarding the amount of gas generated during combustion, it was discovered that the composite sample including an FR emitted relatively fewer harmful gases, and the DOPO-f-MoS$_2$ produced the lowest amount of harmful gases (Figure 13). This is because the P=O group (observed at 254 °C) generated the PO radical, removed H and OH radicals formed during combustion, and then inhibited H and OH radical chain reactions.

The DOPO- f-MoS$_2$ exhibited superior dispersibility inside the basic material because of the strong surface interaction. The physical barrier effect afforded by the structural properties of the two-dimensional nanosheet enabled the release of flammable pyrolysis products and delayed the penetration of heat and oxygen (Figure 14).

Malkappa et al. (33) manufactured an FR filler by functionalizing melamine to MoS_2 and incorporated it into PU to achieve a high flame-retarding PU.

Melamine-based FRs can be a substitute for halogen-based ones, and their usage is increasing primarily in the western portions of Europe, in addition to phosphorus-based and inorganic FRs, as the demand for the development of new FRs increases. Melamine-based FRs are less toxic and easier to manage compared with their halogen-based counterparts. In particular, during the pyrolysis of FPUF products containing melamine, no toxic gases are generated, and they produce less smoke compared with other FRs.

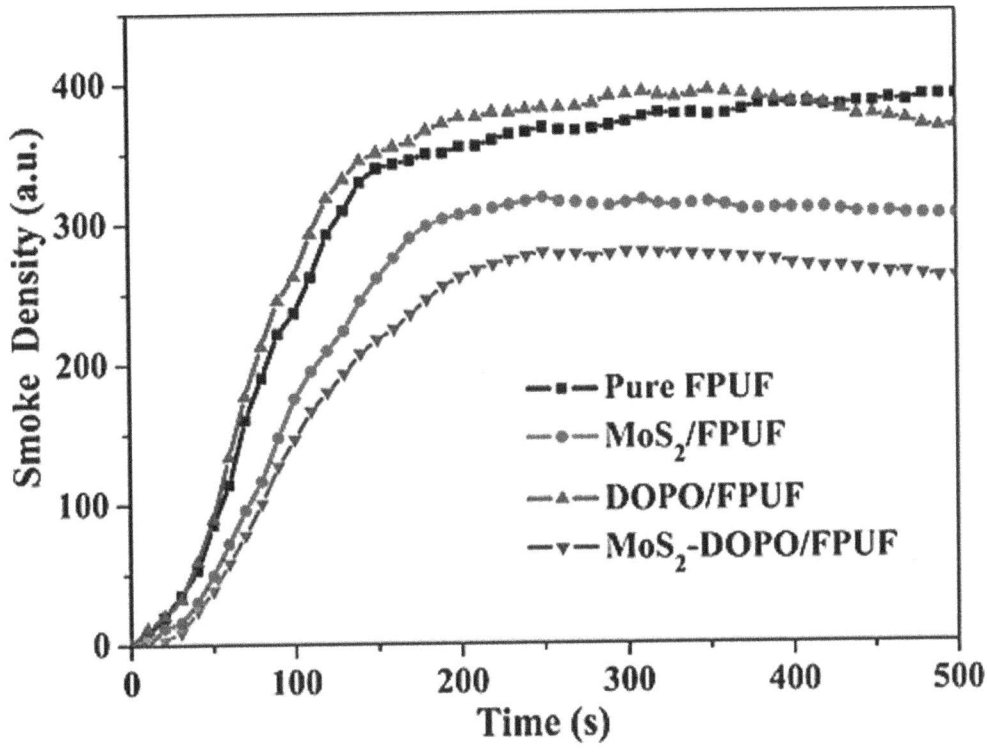

Figure 13. Smoke density curves of pure and FR FPUFs. Reproduced with permission from reference (32). Copyright 2020 American Chemical Society.

Looking at the results of the Malkappa et al. (33), in terms of the combustion gas of each sample, the sample with MoS_2 added (PU/MoS_2) emitted relatively fewer combustion gases such as CO, CO_2, and HCN compared with PU, and the PU($PU/M-MoS_2$) sample with melamine- f-MoS_2 added showed superior flame retardancy compared with the PU/MOS_2 sample. The $PU/M-MoS_2$ sample showed the highest flame retardancy based on the CC results. This was because of the high thermal stability afforded by the structural stability of MoS_2 and the effect of air being obstructed from the outside via the rapid generation of char by the surface functionalization of melamine during combustion. It was discovered that the increase in the FR content increased the prominence of this effect.

This effect can be identified through the flame-retarding mechanism (33). The char layer of the flame-retarding PU, in which PU and MoS_2 are combined, exhibited a prominent flame retardancy effect because MoS$_2$ demonstrated thermal stability at high temperatures, and the melamine- f- MoS$_2$ can yield a stronger char layer, as it can form hydrogen bonds with polymer. Hence, even if combustion gases are generated during combustion, the cross-linked char layer obstructs it, and the shielding effect from the surrounding heat is greater, thereby resulting in a better flame retardancy effect.

Xi et al. (34) developed an RPUF with flame retardancy using EG and phosphorus. EG is a promising inorganic FR and is used extensively in RPUF. When EG is rapidly heated, there is a flame- retarding mechanism in which the sulfuric acid between graphite layers is heated and gas is emitted, and exfoliation and expansion of graphite flakes occur; consequently, the surface of the matrix is covered. However, because the viscosity of the system increases when EG is added, the amount of EG must be maintained at a low level to ensure better fluidity. Reactive liquid compounds such as [bis(2-hydroxyethyl)amino]-methyl-phosphonic acid dimethyl ester (BH) contain more phosphorus and can form a PU matrix by reacting with isocyanate. Hence, it is used to synthesize a better FR. Using this AFR, the flame-retarding behavior and mechanism of the BH/EG/RPUF system were investigated.

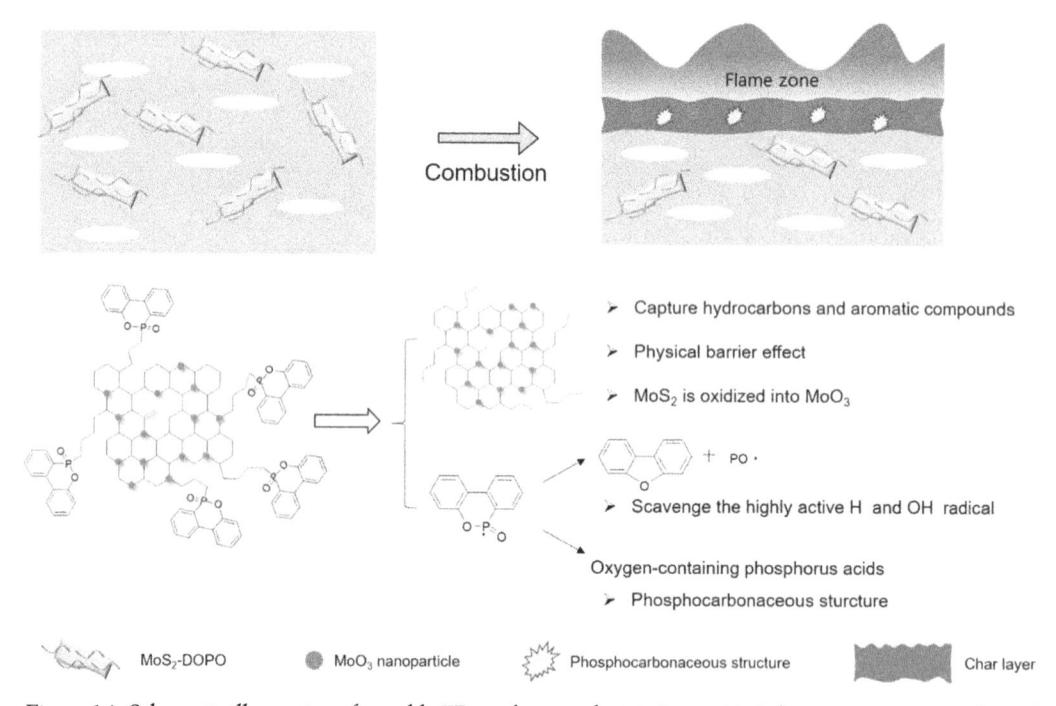

Figure 14. Schematic illustration of possible FR mechanism for MoS$_2$–DOPO/FPUF composite. Adapted with permission from reference (32). Copyright 2020 American Chemical Society.

Based on the LOI results of the manufactured composite, it was discovered that the LOI of the reference sample, RPUF, was 19.4%, whereas in the sample with FR-EG added, it increased to 24.3%. This is because the EG expanded instantaneously and formed a worm-structured char layer to inhibit heat transfer at high temperatures. The 18% BH/8% EG/RPUF composite using two FRs simultaneously demonstrated the most superior flame retardancy effect of 33.0%, compared with the 8% EG-added and 18% BH-added composites. In other words, an enhanced synergistic effect in physical properties can be achieved using more than two FRs simultaneously instead of only using one.

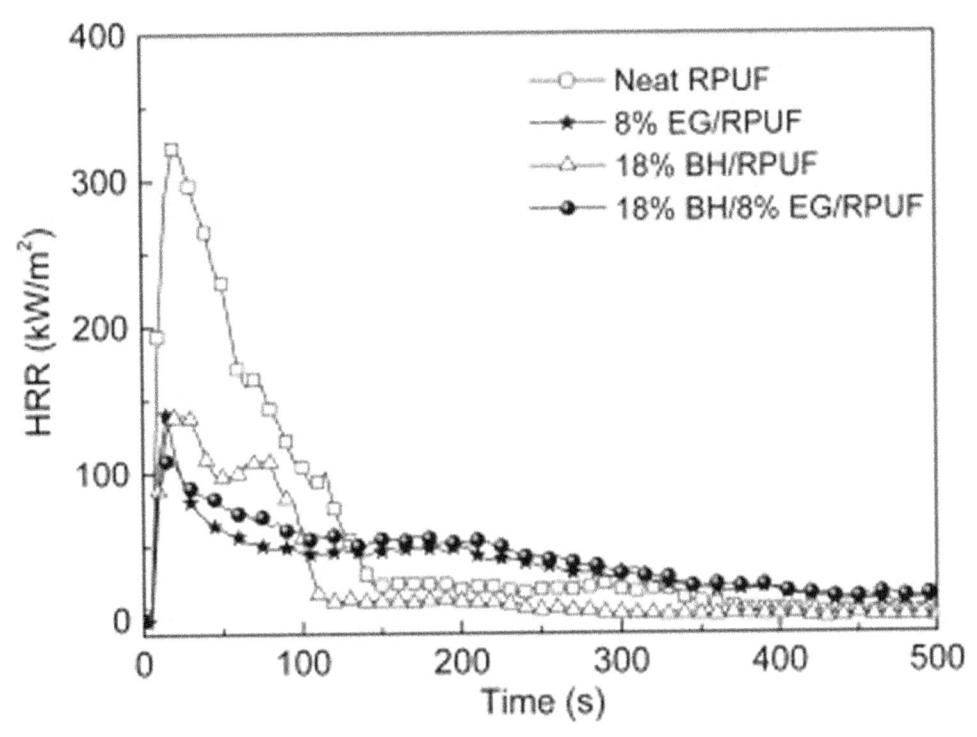

Figure 15. HRR curves of neat RPUF and FR–RPUF. Reproduced with permission from reference (34). Copyright 2015 Elsevier.

Based on the PHRR results of the manufactured composite (Figure 15), the PHRR of the RPUF increased rapidly at the maximum combustion strength after ignition. However, the PHRR value RPUF with EG and BH added decreased gradually. This suggests that the EG reduced the combustion strength of the matrix during combustion, EG and BH reacted simultaneously because of the quenching and carbonization effect of BH during combustion, and the combustion speed decreased significantly. The flame-retarding mechanism is illustrated in Figure 16.

BH first generates dimethyl methyl phosphonate gas; subsequently, its is pyrolyzed to free PO and PO_2 radicals during combustion, eliminate the flammable free radical inside the matrix, and accelerate the formation of firm remaining char containing phosphorus and nitrogen. Because this char combines with EG in the condensed phase and creates a firm char layer, it exhibits a prominent defense effect toward heat and fire. Based on these results, the flame retardancy effect of BH and EG is shown in both gas and condensed phases; the results suggest that more effective fire resistance was assigned to the RPUF.

Thirumal et al. (35) assigned flame retardancy to PUF using only EG, in which sulfuric acid was incorporated between carbon layers. When adding EG to PUF, graphite combines with the sulfuric acid inside the EG layers at high temperatures; emits CO_2, SO_2, and H_2O gases (which results in the expansion of the material); and inhibits and extinguishes flame while serving as a physical barrier to heat and mass transfer. It was discovered that because char was formed by sacrificing flammable gases, the amount of char increased with the EG content because of the relationship between the amount of char and fire resistance. The gas and char generated as a result of this process diluted the oxygen concentration near the flame, prevented further combustion, and consequently increased the LOI value.

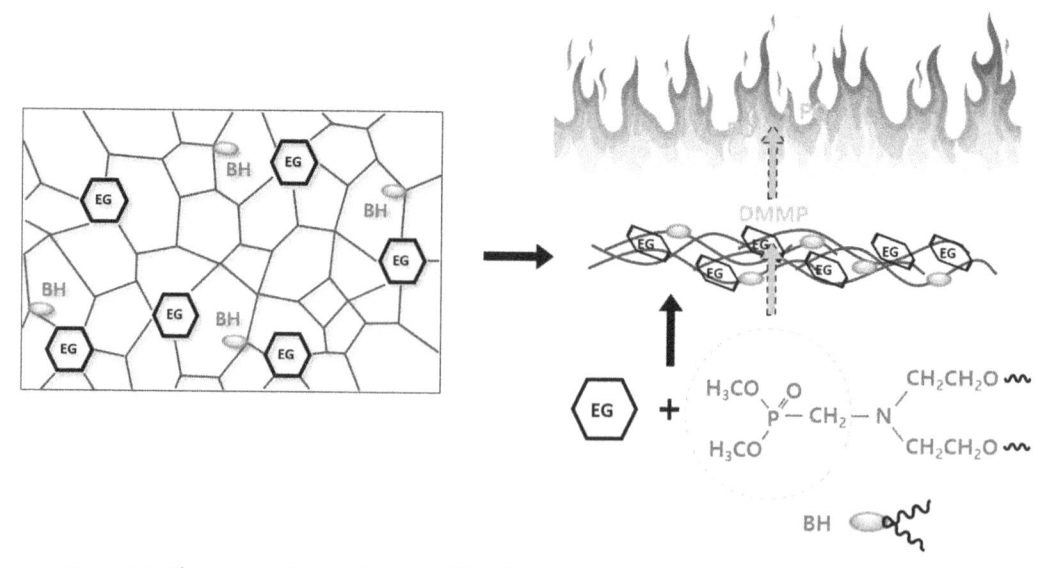

Figure 16. Flame-retarding mechanism of BH/EG. Adapted with permission from reference (34). Copyright 2015 Elsevier.

Others

RPUF is utilized for various purposes because of its high thermal performance, low density, high strength, high dimensional stability, high adhesion strength, and high resistance. However, when an RPUF is ignited, it is not easily self-extinguished and emits a significant volume of toxic gas, which is detrimental to human lives. Therefore, FRs are necessary to reduce combustibility and inhibit toxic steam and smoke generated by the RPUF after ignition (36).

Yang et al. (37) created a phosphate-based FR and synthesized a phosphate FR additive—hexa-(phosphite-hydorxyl-methyl-phenoxyl)-cyclotriphosphazene (HPHPCP)—as shown in Figure 17. HPHPCP is an RFR comprised of multifunctional groups that can improve the stable structure of the network by enhancing the cross-link range. The RPUF composite was manufactured by mixing the synthesized FR with an RPUF.

Through a comparison of RPUF and FR–PUF at a weight decomposition rate of 5%, it was discovered that the thermal stability of the FR–RPUF was superior. This is because of the rigid cross-link inside the PU, which suggests that at a higher cross-link density, the thermal stability exhibits a higher level of dependence than the molecular weight. =HPHPCP is an RFR comprised of multifunctional groups that improve the cross-link range and network structure.

Because PU with a high cross-link density requires more thermal energy to destroy additional bonding in the cross-link, it exhibits high thermal stability in oxidative and inert environments. The remaining amount of FR–RPUF was dependent on an increase in HPHPCP content, and the composite with 25 wt % HPHPCP added showed a significant weight reduction rate of 38.3 wt %. This can be explained by the fact that phosphorus is included in HPHPCP and accelerates the formation of the remaining char during pyrolysis.

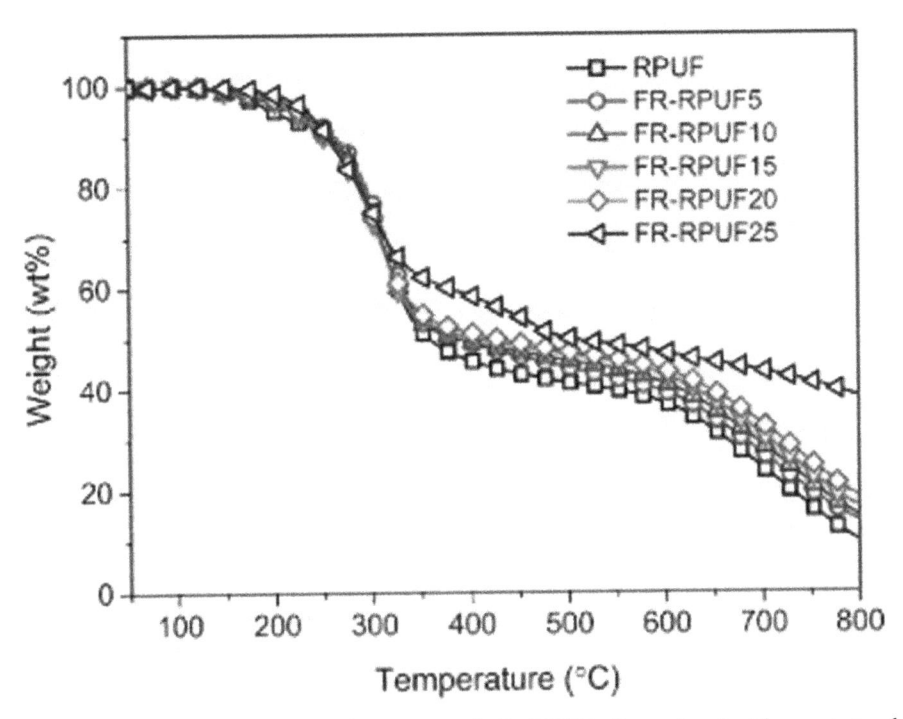

Figure 17. Thermogravimetric curves for RPUF and FR–RPUFs. Reproduced with permission from reference (37). Copyright 2015 Elsevier.

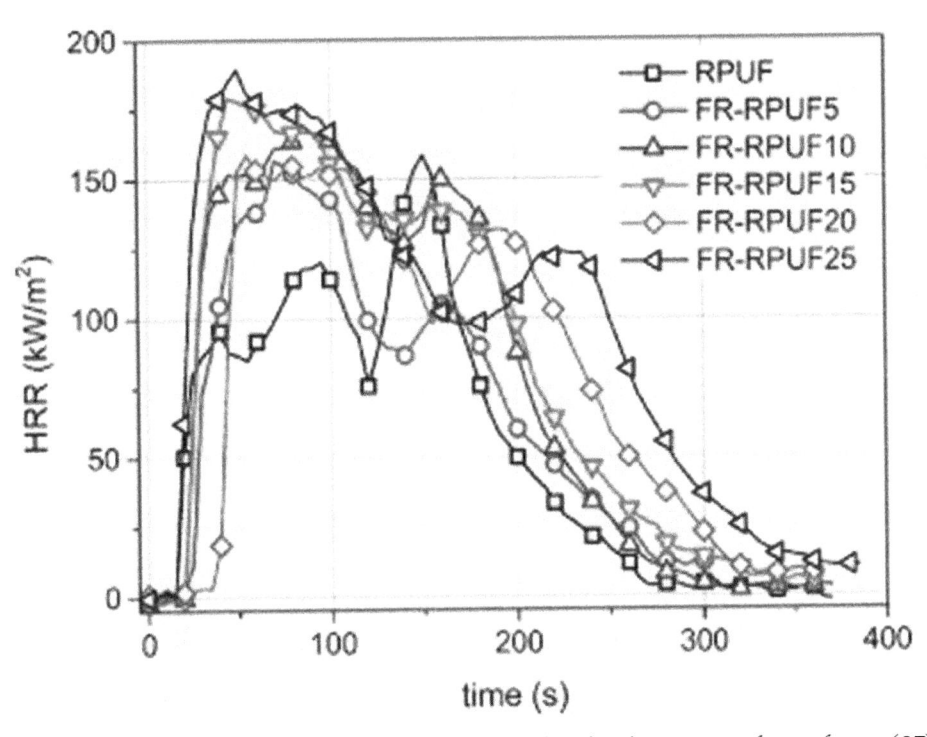

Figure 18. HRR curves of RPUF and FR–RPUFs. Reproduced with permission from reference (37). Copyright 2015 Elsevier.

As shown in Figure 18, the RPUF and FR–RPUF indicated two PHRR peaks. The first PHRR peak value of the FR–RPUF was higher than that of the RPUF because of increased HPHPCP content in the FR–RPUF and density. The second PHRR peak appeared at 150 s when the collapse of the structure was initiated. At this instant, the FR–RPUF showed a lower peak value than the RPUF. This is because HPHPCP formed a dense char layer that can serve as a stable barrier with the RPUF during combustion, resulting in a flame retardancy effect.

Yang et al. (38) developed a flame-retarding PU using a base material (i.e., Schiff) and phosphate. The structure of the Schiff base affords superior thermal stability and exhibits an azomethine (−CH=N−) structure. Hence, it can accelerate the cross-link of melting and, consequentially, is being utilized in the FR field because of the antidripping property and high flame retardancy of polymers. Therefore, the application of halogen-free and environmentally friendly FRs is a new trend in the fire safety industry (39, 40).

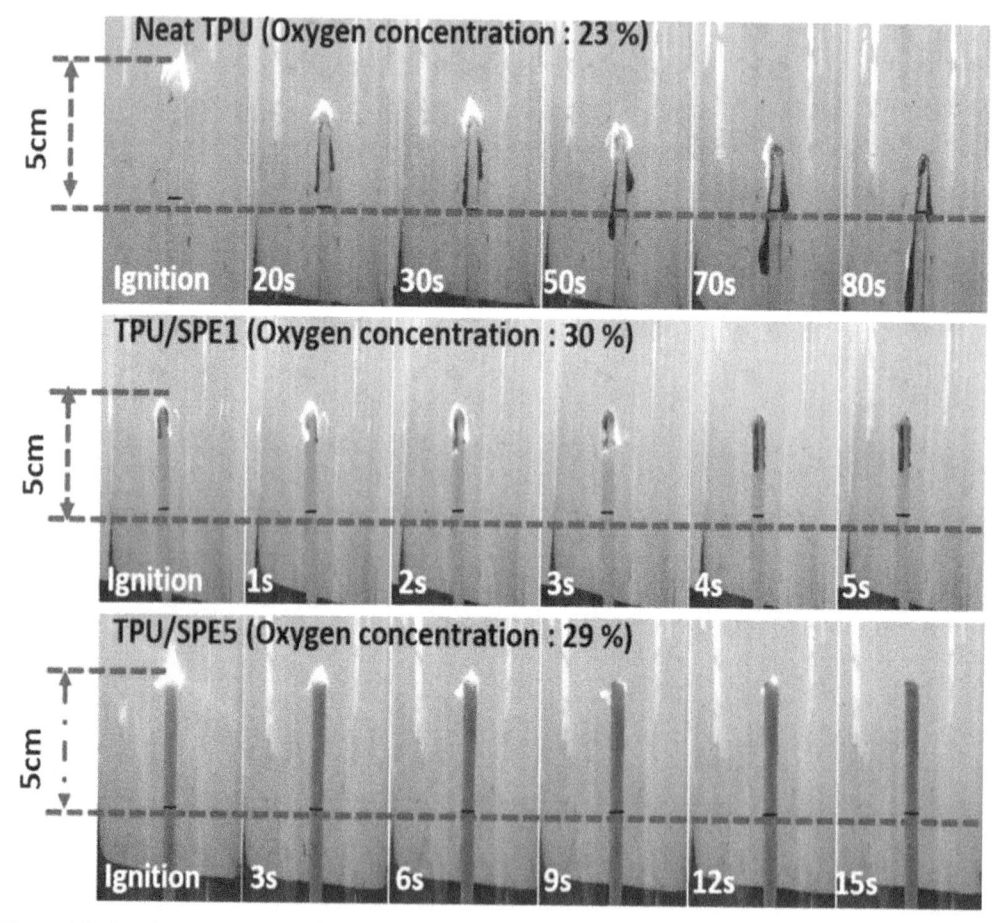

Figure 19. Combustion process of neat TPU, TPU/SPE1, and TPU/SPE5 at corresponding LOIs of 23, 30, and 29%, respectively. SPE indicates the Schiff base containing polyphosphate ester. Reproduced with permission from reference (38). Copyright 2017 Elsevier.

As shown in Figure 19, TPU/SPE5 (SPE indicates the Schiff base containing polyphosphate ester), which contained 5 wt % FR (i.e., SPE), did not produce a high amount of combustibles after combustion. Although dripping was observed, the amount of it was low, and the cotton did not catch fire during the test. In this case, TPU/SPE5 achieved a UL-94 V-0 grade.

Regarding TPU/SPE1, although dripping occurred at an LOI of 30%, it was rapidly self-extinguished. During this combustion process, heat was rapidly discharged via dripping. When the SPE content was low, the effect of SPE on the condensed and gas phases of TPU/SPE was extremely weak. When the SPE increased to 5 wt %, the dripping behavior of TPU/SPE improved significantly, and the sample rapidly self-extinguished. However, the combustion time was prolonged. When the SPE content in TPU was increased, more inflammable components were emitted because of the delay of self-emission, resulting in a decrease in the LOI.

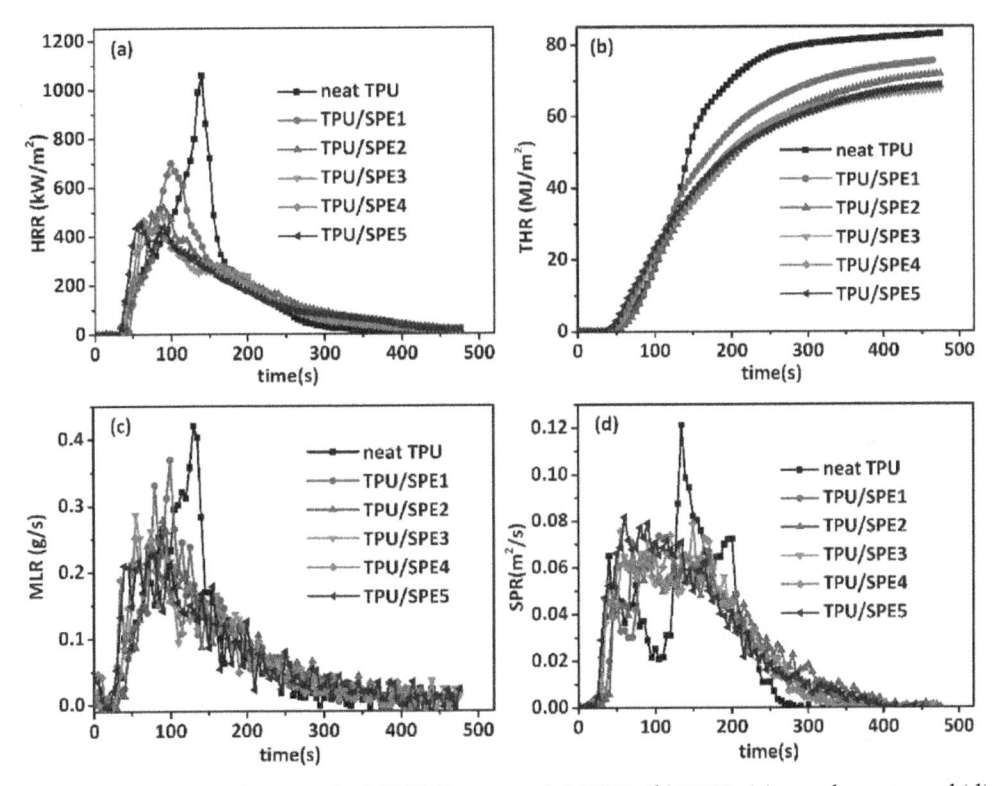

Figure 20. CC curves of TPU and TPU/SPE systems: (a) HRR, (b) THR, (c) mass loss rate, and (d) SPR. Reproduced with permission from reference (38). Copyright 2017 Elsevier.

After the complexation of SPE, the PHRR of TPU reduced significantly. At 5 wt % SPE, the PHRR of TPU/SPE decreased by 61.7% compared with that of the neat TPU, and the THR of TPU/SPE5 was lower than that of the TPU. The smoke production rate (SPR) peak of TPU/SPE was lower than that of the neat TPU. However, the total smoke rate of the TPU did not reduce after SPE was added because SPE functioned as an FR inside the TPU without any issues (Figure 20).

Conclusion

This chapter describes the flame retardancy of materials used in daily supplies. This topic is investigated to minimize fire risk and satisfy fire safety requirements. Flame retardancy is an essential function in widely used materials such as composite materials, adhesives, paints, insulators, rubber materials, and PUs. To express excellent flame retardancy, the flame retardancy improvement technology of PU and PUF (hard and soft foam) using an organic or nanocarbon-based additive was studied. In addition, several FR evaluation methods were also described.

To improve flame retardancy and reduce harmful gases generated during combustion, carbon-based fillers with excellent thermal stability and mechanical properties were used. Looking at the research results, the size of the filler gradually decreases from micro- to submicro- or nanosize, and EG and MoS_2 are typical examples. By using nanocarbon materials, various functionalities other than flame retardancy could be expected at the same time. In particular, the expression and improvement of new functionalities without degrading the matrix properties are noteworthy.

Although not mentioned in this chapter, a great deal of research on FR-PU based on renewable eco-friendly material or biomaterials has been reported. It can be said that FR-PU is no exception in the development of materials combining renewable systems from limited resources, and it is expected that its ripple effect will be great in the future. Through advancement of the processing method of these materials in the future, it is expected that this approach can be an applied to a composite (fiber-reinforced plastic) of retardant reinforcement. Finally, it is hoped that this study will benefit researchers who wish to investigate high-efficiency flame-retarding composites.

References

1. Malshe, V. C.; Sikch, M. *Basics of Paint Technology Part II*; Antar Prakash Centre for Yoga: Mumbai, India, 2008.
2. Chattopadhyay, D. K.; Raju, K. Structural Engineering of Polyurethane Coatings for High Performance Applications. *Prog. Polym. Sci.* **2007**, *32* (3), 352–418.
3. Qian, L.; Feng, F.; Tang, S. Bi-phase Flame-Retardant Effect of Hexa-phenoxy-cyclotriphosphazene on Rigid Polyurethane Foams Containing Expandable Graphite. *Polymer* **2014**, *55* (1), 95–101.
4. Wood, G. *The ICI Polyurethane Book*, 2nd ed.; John Wiley & Sons: New York, 1990.
5. Hepburn, C. *Polyurethane Elastomers*; Springer: Netherlands, 2012.
6. Drysdale, D. *Fire and Cellular Polymers*; Elsevier Applied Science: London, 1987.
7. Thomas, S.; Datta, J.; Haponiuk, J.; Reghunadhan, A. *Polyurethane Polymers: Composites and Nanocomposites*; Cambridge, MA, United States: Elsevier, 2017.
8. Tang, Z.; Maroto-Valer, M. M.; Andrésen, J. M.; Miller, J. W.; Listemann, M. L.; McDaniel, P. L.; Morita, D. K.; Furlan, W. R. Thermal Degradation Behavior of Rigid Polyurethane Foams Prepared with Different Fire Retardant Concentrations and Blowing Agents. *Polymer* **2002**, *43* (24), 6471–6479.
9. Levchik, S. V.; Weil, E. D. Thermal Decomposition, Combustion and Fire-Retardancy of Polyurethanes—A Review of the Recent Literature. *Polym. Int.* **2004**, *53* (11), 1585–1610.
10. Thirumal, M.; Singha, N. K.; Khastgir, D.; Manjunath, B.; Naik, Y. Halogen-Free Flame-retardant Rigid Polyurethane Foams: Effect of Alumina Trihydrate and Triphenylphosphate on the Properties of Polyurethane Foams. *J. Appl. Polym. Sci.* **2010**, *116* (4), 2260–2268.
11. Liu, J.; Ma, D. Study on Synthesis and Thermal Properties of Polyurethane–Imide Copolymers with Multiple Hard Segments. *J. Appl. Polym. Sci.* **2002**, *84* (12), 2206–2215.
12. Luchkina, L.; Askadskii, A.; Bychko, K. Composite Polymeric Materials Based on Polyisocyanurates and Polyurethanes. *Russ. J. Appl. Chem.* **2005**, *78* (8), 1337–1342.
13. Mahfuz, H.; Rangari, V. K.; Islam, M. S.; Jeelani, S. Fabrication, Synthesis and Mechanical Characterization of Nanoparticles Infused Polyurethane Foams. *Compos. Part A Appl. Sci. Manuf.* **2004**, *35* (4), 453–460.

14. Zatorski, W.; Brzozowski, Z. K.; Kolbrecki, A. New Developments in Chemical Modification of Fire-Safe Rigid Polyurethane Foams. *Polym. Degrad. Stab.* **2008**, *93* (11), 2071–2076.
15. Thirumal, M.; Khastgir, D.; Singha, N. K.; Manjunath, B.; Naik, Y. Effect of a Nanoclay on the Mechanical, Thermal and Flame Retardant Properties of Rigid Polyurethane Foam. *J. Macromol. Sci. Pure. Appl. Chem.* **2009**, *46* (7), 704–712.
16. Ni, J.; Tai, Q.; Lu, H.; Hu, Y.; Song, L. Microencapsulated Ammonium Polyphosphate with Polyurethane Shell: Preparation, Characterization, and Its Flame Retardance in Polyurethane. *Polym. Adv. Technol.* **2010**, *21* (6), 392–400.
17. Kim, J.; Lee, K.; Lee, K.; Bae, J.; Yang, J.; Hong, S. Studies on the Thermal Stabilization Enhancement of ABS; Synergistic Effect of Triphenyl Phosphate Nanocomposite, Epoxy Resin, and Silane Coupling Agent Mixtures. *Polym. Degrad. Stab.* **2003**, *79* (2), 201–207.
18. Jang, B.; Choi, J. Research Trends of Flame Retardant and Flame Retardant Resin. *Int. Polym. Sci. Technol.* **2009**, *20* (1), 8–15.
19. Modesti, M.; Zanella, L.; Lorenzetti, A.; Bertani, R.; Gleria, M. Thermally Stable Hybrid Foams Based on Cyclophosphazenes and Polyurethanes. *Polym. Degrad. Stab.* **2005**, *87* (2), 287–292.
20. Kim, Y.-O.; Jung, Y. C. Flame retardancy of bioepoxy polymers, their blends and composites. In *Bio-Based Epoxy Polymers, Blends and Composites : Synthesis, Properties, Characterization and Application*; Parameswaranpillai, J., Rangappa, S. M., Siengchin, S., Jose, S., Eds.; Wiley-VCH Verlag GmbH: Weinheim, Germany, 2021; Chapter 10.
21. Javni, I.; Petrović, Z. S.; Guo, A.; Fuller, R. Thermal Stability of Polyurethanes Based on Vegetable Oils. *J. Appl. Polym. Sci.* **2000**, *77* (8), 1723–1734.
22. Javni, I.; Zhang, W.; Petrović, Z. S. Effect of Different Isocyanates on the Properties of Soy-Based Polyurethanes. *J. Appl. Polym. Sci.* **2003**, *88* (13), 2912–2916.
23. Ogunniyi, D. S. Castor Oil: A Vital Industrial Raw Material. *Bioresour. Technol.* **2006**, *97* (9), 1086–1091.
24. Knaub, P.; Camberlin, Y. Castor Oil as a Way to Fast-Cured Polyurethane Ureas. *Eur. Polym. J.* **1986**, *22* (8), 633–635.
25. Zhang, L.; Zhang, M.; Zhou, Y.; Hu, L. The Study of Mechanical Behavior and Flame Retardancy of Castor Oil Phosphate-Based Rigid Polyurethane Foam Composites Containing Expanded Graphite and Triethyl Phosphate. *Polym. Degrad. Stab.* **2013**, *98* (12), 2784–2794.
26. Zhang, C.; Bhoyate, S.; Ionescu, M.; Kahol, P.; Gupta, R. K. Highly Flame Retardant and Bio-Based Rigid Polyurethane Foams Derived from Orange Peel Oil. *Polym. Eng. Sci.* **2018**, *58* (11), 2078–2087.
27. Pettigrew, A. Halogenated Flame Retardants. In *Kirk-Othmer Encyclopedia of Chemical Technology*, 4th ed.; John Wiley & Sons: New York, 1993; Vol. 10, pp 954–976.
28. Cho, J. H.; Vasagar, V.; Shanmuganathan, K.; Jones, A. R.; Nazarenko, S.; Ellison, C. J. Bioinspired Catecholic Flame Retardant Nanocoating for Flexible Polyurethane Foams. *Chem. Mater.* **2015**, *27* (19), 6784–6790.
29. Ju, K.-Y.; Lee, Y.; Lee, S.; Park, S. B.; Lee, J.-K. Bioinspired Polymerization of Dopamine To Generate Melanin-like Nanoparticles Having an Excellent Free-Radical-Scavenging Property. *Biomacromolecules* **2011**, *12* (3), 625–632.

30. Liu, X.; Gu, X.; Sun, J.; Zhang, S. Preparation and Characterization of Chitosan Derivatives and Their Application as Flame Retardants in Thermoplastic Polyurethane. *Carbohydr. Polym.* **2017**, *167*, 356–363.

31. Cai, W.; Zhan, J.; Feng, X.; Yuan, B.; Liu, J.; Hu, W.; Hu, Y. Facile Construction of Flame-Retardant-Wrapped Molybdenum Disulfide Nanosheets for Properties Enhancement of Thermoplastic Polyurethane. *Ind. Eng. Chem. Res.* **2017**, *56* (25), 7229–7238.

32. Zhi, M.; Liu, Q.; Zhao, Y.; Gao, S.; Zhang, Z.; He, Y. Novel MoS2–DOPO Hybrid for Effective Enhancements on Flame Retardancy and Smoke Suppression of Flexible Polyurethane Foams. *ACS Omega* **2020**, *5* (6), 2734–2746.

33. Malkappa, K.; Ray, S. S.; Kumar, N. Enhanced Thermo-Mechanical Stiffness, Thermal Stability, and Fire Retardant Performance of Surface-Modified 2D MoS2 Nanosheet-Reinforced Polyurethane Composites. *Macromol. Mater. Eng.* **2019**, *304* (1), 1800562.

34. Xi, W.; Qian, L.; Chen, Y.; Wang, J.; Liu, X. Addition Flame-Retardant Behaviors of Expandable Graphite and [Bis (2-hydroxyethyl) amino]-methyl-phosphonic acid dimethyl ester in Rigid Polyurethane Foams. *Polym. Degrad. Stab.* **2015**, *122*, 36–43.

35. Thirumal, M.; Khastgir, D.; Singha, N. K.; Manjunath, B.; Naik, Y. Effect of Expandable Graphite on the Properties of Intumescent Flame-Retardant Polyurethane Foam. *J. Appl. Polym. Sci.* **2008**, *110* (5), 2586–2594.

36. Lorenzetti, A.; Modesti, M.; Gallo, E.; Schartel, B.; Besco, S.; Roso, M. Synthesis of Phosphinated Polyurethane Foams with Improved Fire Behaviour. *Polym. Degrad. Stab.* **2012**, *97* (11), 2364–2369.

37. Yang, R.; Hu, W.; Xu, L.; Song, Y.; Li, J. Synthesis, Mechanical Properties and Fire Behaviors of Rigid Polyurethane Foam with a Reactive Flame Retardant Containing Phosphazene and Phosphate. *Polym. Degrad. Stab.* **2015**, *122*, 102–109.

38. Yang, A.-H.; Deng, C.; Chen, H.; Wei, Y.-X.; Wang, Y.-Z. A Novel Schiff-Base Polyphosphate Ester: Highly-Efficient Flame Retardant for Polyurethane Elastomer. *Polym. Degrad. Stab.* **2017**, *144*, 70–82.

39. Liao, S.-F.; Deng, C.; Huang, S.-C.; Cao, J.-Y.; Wang, Y.-Z. An Efficient Halogen-Free Flame Retardant for Polyethylene: Piperazinemodified Ammonium Polyphosphates with Different Structures. *Chin. J. Polym. Sci.* **2016**, *34* (11), 1339–1353.

40. Lu, S.-Y.; Hamerton, I. Recent Developments in the Chemistry of Halogen-Free Flame Retardant Polymers. *Prog. Polym. Sci.* **2002**, *27* (8), 1661–1712.

Chapter 6

The Role of Polyurethane Foam Indoors in the Fate of Flame Retardants and Other Semivolatile Organic Compounds

Mesut Genisoglu,[1] Sait C. Sofuoglu,[*,1] and Aysun Sofuoglu[2]

[1]Department of Environmental Engineering, Izmir Institute of Technology, 35430 Izmir, Turkey
[2]Department of Chemical Engineering, Izmir Institute of Technology, 35430 Izmir, Turkey
*Email: cemilsofuoglu@iyte.edu.tr

Flame retardant chemicals are added to polyurethane foams (PUFs) during production. These chemicals are released to the environment during the use of PUF containing furniture or building materials. In contrast, organic pollutants such as polychlorinated biphenyls, polycyclic aromatic hydrocarbons, synthetic musk compounds, and volatile organic compounds could be sorbed by PUF depending on the concentration gradient, ambient temperature, and the physicochemical properties. Most of these substances tend to accumulate by adhering to organic matter in dust, particles, and surfaces, as they do not degrade for long periods of time. Sorption-emission cycles of PUF-associated organic compounds prolong their presence in indoor environments, which could increase human exposure. Since these organic compounds might have carcinogenic or chronic-toxic health effects on living organisms, it is important to understand the role of PUF in exposure to these substances in indoor environments. This chapter reviews the literature on the relationship of organic substances with PUF in indoor environments.

Introduction

Persistent bioaccumulative toxic (PBT) pollutants are toxic chemicals to organisms and ecosystems, and are resistant to degradation. They accumulate along the food chain and can be transported over long distances by atmospheric processes (*1*). While most of these chemicals are manmade, some, such as polycyclic aromatic hydrocarbons (PAHs), are released to the environment as a by-product of incomplete combustion that may be natural (*2*). Most PBTs partition between gas and particle phases and surfaces. Indoor concentrations of persistent organic pollutants have been generally determined to be higher than those of outdoors because of the abundance of indoor sources. Flame retardants (FRs), polychlorinated biphenyls (PCBs), phthalate esters, PAHs, pesticides, and synthetic musk compounds (SMCs) are present in indoor environments (*3–5*).

© 2021 American Chemical Society

Commercial production of some semivolatile organic compounds (SVOCs), such as PCBs, polybrominated biphenyls (PBBs), polybrominated biphenyls (PBDEs), and some pesticides were banned because of their persistence and adverse effects on humans (6). These legacy SVOCs can still be determined in indoor and outdoor environments due to their persistence and partitioning to organic matter.

Polycyclic Aromatic Hydrocarbons

PAHs (Figure 1) are undesirable by-products of incomplete combustion of organic materials by humans for heating, cooking, smoking, residential heating, industry, and traffic, as well as in natural phenomena such as forest fires and volcanos. They can pose health risks such as haematological changes, genotoxicity, hepatoxicity, lung cancer, and neurological dysfunctions (2). Benz[a]anthracene, benzo[a]pyrene, benzo[b,k]fluoranthene, chrysene, dibenz(a,h)anthracene, and indeno(1, 2, 3-c, d)pyrene are classified as B2 carcinogen compounds by the Environmental Protection Agency (7).

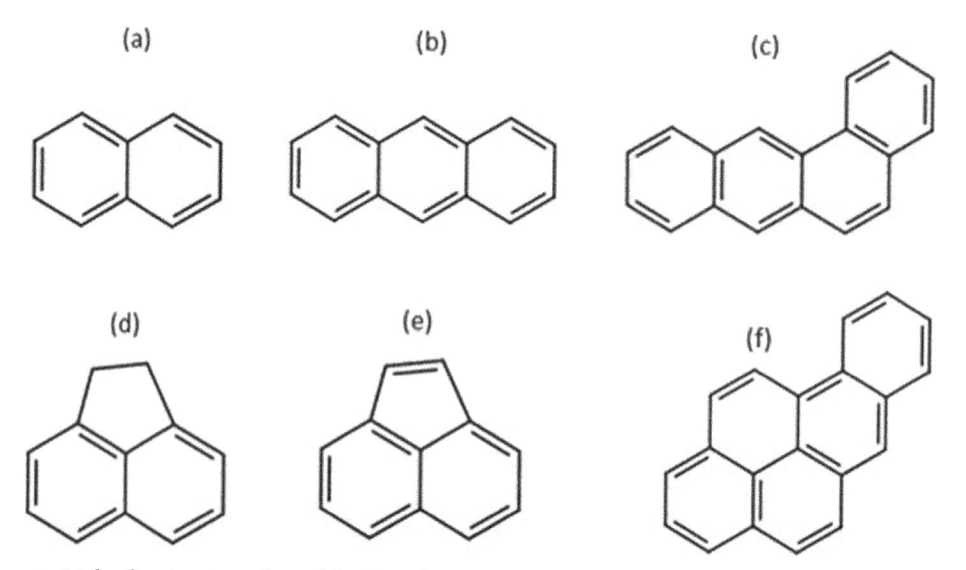

Figure 1. Molecular structure of six of the United State Environmental Protection Agency's 16 priority PAH compounds: (a) naphthalene, (b) anthracene, (c) benz[a]anthracene, (d) acenaphthene, (e) acenaphthylene, (f) benzo[a]pyrene.

Flame Retardants

FRs have been synthesized in numerous products to protect consumers from fire. These chemicals have been added to furniture foam, upholstery, bed foam, curtains, plastics, and electronic devices such as computers and televisions because of the requirements of fire regulations (8). Most FRs are characterized by their long lifespans because of their persistence, accumulation among the food chain, and toxicity (9). They can be found in all environmental media such as soil (10), vegetation (11), water sources (12), and air (13, 14). Historically, PCBs were used FR production because of their thermal stability, but after they were banned, bromine has been one of the most used chemicals in FR production. Brominated and organophosphate FRs are currently the most used commercial chemicals.

Brominated FRs (BFRs) (Figure 2) can be classified into the three subgroups based on their molecular structure: (1) aromatic (2) cycloaliphatic, and (3) aliphatic (15). Some examples include PBDEs and tetrabromobisphenol A (TBBP-A) for aromatic, hexabromocyclododecane (HBCDD) for cycloaliphatic, and dibromoneopentyl glycol (DBNPG) for aliphatic BFRs. Because of their stability in products, BFRs can be used as additive, reactive, and polymeric BFRs. Their stability is ranked as additive < reactive < polymeric BFRs, based on either their chemical bond or incorporation into the framework of the materials. Unfortunately, additive BFRs do not form a chemical bond with materials or products; therefore, the risk of exposure is higher because their release is easier. Consequently, because of their low physical stability, additive BFRs have a higher risk of environmental PBT contamination than the other kinds (16).

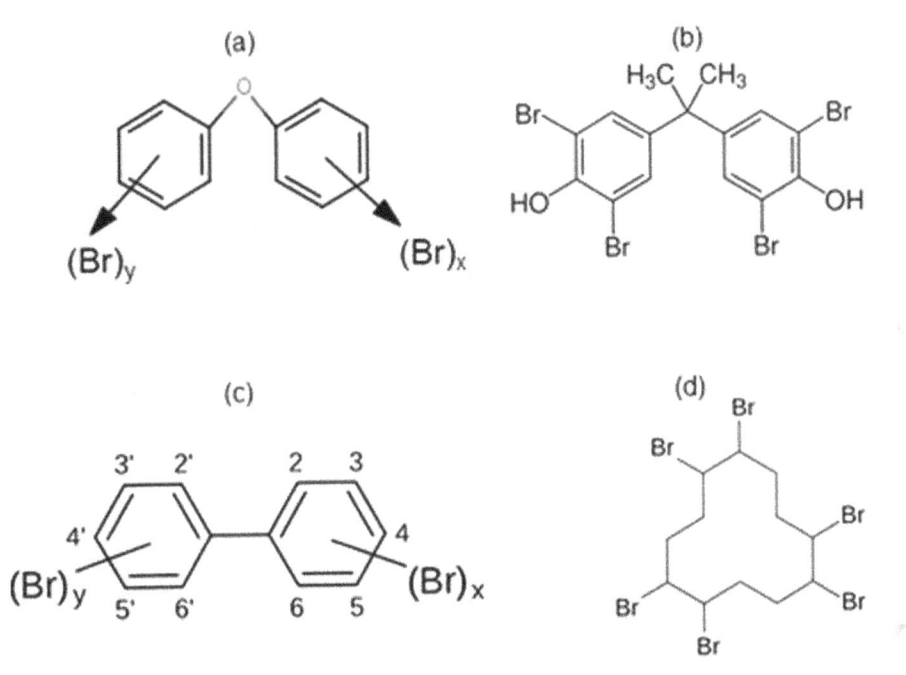

Figure 2. Molecular structure of some BFRs: (a) PBDEs, (b) TBBP-A, (c) PBBs, and (d) HBCDD.

PBDEs are a class of brominated hydrocarbons that have been used as FR chemicals for a long time (17). They can be released to the environment via volatilization, abrasion, and dissolution from PBDE-containing products (18, 19), recycling of electronic wastes, burning of residential waste dumps (20), and steel production processes (17, 21). Production of certain PBDEs has been banned in Europe and the US (22, 23). The commercial penta- (24) and octa-BDE mixtures (25) were added to the Stockholm Convention's list of persistent organic pollutants in 2008—commercial deca-BDE was added in 2016 (26). Their persistence and lipophilicity (27) result in bioaccumulation in the environment and the food chain; therefore, exposure to PBDEs is a serious matter. Animal and *in vitro* studies showed that thyroid disorders, reproductive health effects, and neurobehavioral and developmental anomalies are some of the associated health effects (28). PBDEs are classified as SVOCs, which can be more persistent indoors because of their partitioning to particle phase and lower degradation rates compared to that in ambient air (29). PBDEs have been found to be mainly associated with submicron particles (30), which present higher health risks because they have higher deposition rates in the respiratory airways and lungs compared to those of the coarse particles (31).

Production of non-PBDE brominated alternative novel FRs (NBFRs) has increased since the ban on PBDEs. Some of widely used NBFRs are 2,4,6-tribromophenyl allyl ether (TBP-AE), 2-bromoallyl-2,4,6-tribromophenyl ether (BATE), isomers of 1,2-dibromo-4-(1,2-dibromoethyl) cyclohexane (α-, β-, γ- and δ-DBE-DBCH), bis(2-ethylhexyl)-3,4,5,6-tetrabromophthalate (BEH-TEBP), 1,2-bis(2,4,6-tribromophenoxy)ethane (BTBPE), isomers of dechlorane plus (syn- and anti- DP), 2,4,6-tribromophenyl 2,3-dibromopropyl ether (TBP-DBPE), 2-ethylhexyl-2,3,4,5-tetrabromobenzoate (EH-TBB), hexabromobenzene (HBB), isomers of (α- and γ) HBCDD, octabromo-1,3,3-trimethyl-1-phenylindane (OBTMPI), and 1,2,5,6-tetrabromocyclooctane (TBCO). They have been found at higher levels indoors than PBDEs, reaching a median Σ_{12}NBFR dust concentration of 9.1 μg/g in offices in Istanbul, Turkey, and a maximum of 94 μg/g measured in a computer room, while those values for Σ_{12}PBDEs were 1.9 μg/g and 32 μg/g, respectively (22).

An increase in the use of organophosphate FRs (OPFRs) has also been observed after the ban on PBDEs (31). The production amount of commercial Deca-BDE mixtures decreased from 2300 tons in 2012, to around 11 tons in 2015 because of the regulations. Similar to additive BFRs, OPFRs also do not form a chemical bond with the products in which they are used; therefore, they can easily spread into the environment and pose health risks to those who are exposed. Because they have higher vapor pressures than PBDEs, inhaling them in gas phase is a serious risk. However, organic chemicals on submicron particles can reach deep into the lungs, to the alveoli, therefore posing higher health risks for residents. Studies on OPFRs, whose toxicity is less studied than BFRs, have focused on their toxic effects (32). Chlorinated OPFRs, tris (2-chloroethyl) phosphate (TCEP) and tris- (2-chloropropyl) phosphate (TCPP), are suspected to be carcinogenic, and TCEP is known to be an endocrine disruptor. Triphenyl phosphate (TPhP) and tricresyl phosphate (TCP) cause reproductive disorders and adversely affect neural and immune systems. One study reported that the highest cytotoxicity was observed in BDE-47 exposure, followed by TPhP, TCP, TCPP, and TCEP (33).

Polychlorinated Biphenyls

PCBs (Figure 3) are stable compounds with lipophilic structures, and they were widely used in various industrial applications such as in capacitors, transformers, and paint between 1930 and 1975. Production of PCBs have been halted and prohibited in many countries (34) because of their association with diseases and anomalies related to cancer, reproductive disorders, neurological disorders, immunological disorders, and endocrine systems. Additionally, they are classified as carcinogens for humans in Group 1 by the International Agency for Research on Cancer. Therefore, production and use of PCBs are limited in accordance with the Stockholm Convention on persistent organic pollutants (35). They are structurally very stable, have very long half-lives, and accumulate in the food chain because of their lipophilicity. The atmosphere plays an important role in transporting these pollutants to water and terrestrial environments. PCBs have been detected in the North Pole and other remote areas of the world, even where they were not being used. PCBs were manufactured since 1929 and were intensively used in industrial processes after the 1950s until their ban in 1979. Although many studies have shown that PCB sources are decreasing, they are still being detected in the environment 30 years after their ban (36). Despite their prohibition, they are still in nature after being released from PCB-containing waste sites, leakage from old electrical transformers, and burning of PCB-containing wastes in incinerators. The current primary sources of PCBs are industrial by-products of thermal processes, such as metal smelting and refining processes, thermal energy generation, cement kilns, wood burning and other biomass fuels used in combustion, and chlorination of wastewater treatment plants (37).

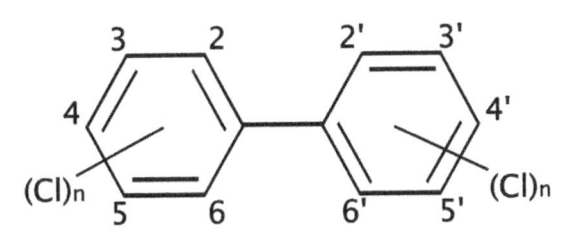

Figure 3. Chemical structure of PCBs.

PCBs were first banned in the production of building materials and later in the production of electric transformers and capacitors (38). However, PCB levels in the blood of people who lived in contaminated buildings (e.g., from sealing materials) were determined to be 30 times higher than in the control group (who lived in uncontaminated buildings) (39). In addition to the use of PCB-containing building materials and electrical equipment, cleaning routines are also important factors in PCB exposure. Frederiksen et al. (38) determined that vacuuming, dusting, floor washing, and airing frequencies significantly affected the indoor air concentrations in PCB-contaminated buildings.

To understand the circumstances that cause human PBT exposure, it is necessary to examine the media in which these compounds are found in indoor environments. While higher molecular weight PBTs dominate particle-phase concentrations, lower molecular weight compounds are found at higher percentages in the gaseous phase (>90%) (40, 41). In addition to being in gas and particle phases, these substances adhere to the organic substances on indoor surfaces. Polyurethane (PU) foams (PUFs) could be an important media for the fate of organic compounds in indoor environments because of their high surface area relative to their volume and adsorption affinity. In fact, it has been successfully used as a passive sampler for SVOCs for a long time. Furthermore, PUFs are present in high amounts in indoor environments with incorporated FR content that varies among countries with different regulations. Therefore, they play an in important role in the fate of SVOCs such as PCBs, PBDEs, and OPFRs indoors. Specifically, furniture with PUF can act as both a sink and source, having a determining effect on the exposure levels of these toxic and carcinogenic organic compounds, especially at home, where people spend most of their daily lives. Therefore, this chapter compiles the literature on the relationships between SVOCs and PUFs.

PUFs: A Sink and Source of SVOCs in Indoor Environments

PUFs are a class of PU productions; they correspond to 50% of the PU global consumption because of their low density and light weight. They can be flexible or rigid. Therefore, they can be used in different domestic and industrial products. In indoor environments, PUFs act as a source. Sofa and bed PUFs contain higher amounts of FR chemicals as an extra safety measure. FR chemicals are abundant in PUFs to give upholstered furniture flame resistance. A commercial penta-BDE mixture was the most commonly used FR in PUFs before they were banned (42). After the ban of commercial penta-BDE mixtures, deca-BDE mixtures had been abundantly used before they were added to the Stockholm Convention list in 2016 (26). Nowadays, alternative halogenated FRs are used to provide flame resistance of upholstered furniture (5). A positive correlation was determined for BDE-47 between FR-impregnated PUF and indoor settled dust and human serum levels. Release of particles from a PUF sofa into indoor air makes it a major settled-dust associated FR source (42).

In addition to being primary emission sources, they may act as sinks; FRs and other SVOCs could be sorbed by interior PUF media from indoor air emitted from other indoor SVOC sources and transported from outdoors to indoors with air exchange by natural or mechanical ventilation and infiltration. The sorbed SVOCs could then be re-emitted depending upon environmental and physicochemical variables, making the PUF a secondary source. The sorption and re-emission cycle of the SVOCs in indoor environment prolongs their indoor prevalence and increases the risk of human exposure (43).

Equilibrium Partitioning between PUFs and SVOCs

The equilibrium concentration ratio between two different media is referred as the partition coefficient (44). Equation 1 is used to experimentally determine the partition coefficients of compounds. However, low volatility compounds need a long time (they sometimes need a few years) to reach equilibrium. Thus, Shoeib and Harner (45) suggested the octanol-air partitioning coefficient-based empirical model to estimate the K_{PUF-A} values of SVOCs (eq 2, r^2=0.87). The estimation of K_{PUF-A} (45) was limited to the passive PUF disc sampling media to determine ambient air concentrations. High K_{PUF-A} levels of organic compounds show the sorption affinity of PUF and indicates that they are not only a useable sampling media but also could be an ideal sink for indoor air organic pollutants.

$$K_{PUF-A} = \frac{C_{i-PUF}}{C_{i-Air}} \tag{1}$$

where K_{PUF-A} is the PUF-air partition coefficient, C_{i-PUF} is the concentration of compound i on the PUF (g VOC/m³ PUF or g VOC/g PUF), and C_{i-A} is the concentration of compound i in air (g VOC/m³ air).

$$\log K_{PUF-A} = 0.6366 \log K_{OA} - 3.1774 \tag{2}$$

where K_{OA} is the octanol-air partition coefficient of compound i.

Indoor PUFs could also be a sink for volatile organic compounds (VOCs) because of their high sorption capacities. Like SVOCs, VOCs could be directly emitted from PUFs because of the residuals from manufacturing or re-emission of those that were sorbed from indoor air. Zhao et al. investigated the dynamic sink effect of the indoor PUF with an unstable VOC concentration (46). New PUFs were found to be a source for production-residual VOCs. The sorption–desorption (emission) cycle increases the indoor presence and exposure levels of VOCs like the SVOCs. The concentration gradient is the main mechanism that determines the source–sink behavior of PUFs. The amount of sorbed VOCs increases with increasing indoor VOC concentrations, whereas decreasing indoor VOC levels reverses the sorption phenomenon to desorption (emission). The sorption–emission phenomenon of PUFs on VOCs significantly affects their indoor air levels (46). The K_{PUF-A} of seven VOCs and one SVOC were experimentally determined using eq 1. Table 1 shows the K_{PUF-A}, PUF phase diffusion coefficient, and time to equilibrium values, as reported by Zhao et al. (46).

Table 1. Experimental K, D, t_e, P_V, and MW Values of Organic Compounds. Reproduced with permission from (46). Copyright (2004) American Society of Civil Engineers.

Compound	K^a	D^b (m²/s)	t_e^c (h)	P_V^d (Pa)	MW^e	b^f (L/mole)	S_M^g (Å²)
Naphthalene	6400	0.0066×10⁻¹³	39	23.6	128	-	190
1,2,4-Trimethylbenzene	440	1.0×10⁻¹³	3.4	304	120	-	153
Styrene	310	1.9×10⁻¹³	2.7	927	104	-	143
p-Xylene	130	2.7×10⁻¹³	1.5	1005	106	0.1809	137
Ethylbenzene	110	3.7×10⁻¹³	1.2	1173	106	0.1667	137
Chlorobenzene	140	3.3×10⁻¹³	1.4	14115	113	0.1453	129
Toluene	58	4.2×10⁻¹³	0.85	3582	92	0.1463	128
Benzene	19	7.0×10⁻¹³	0.67	11790	78	0.1554	107

[a]K: PUF-Air partition coefficients
[b]D: Diffusion coefficient on PUF phase
[c]t_e: time to equilibrium
[d]P_V: Vapor pressure
[e]MW: Molecular weight
[f]b: van der Waals molar volume
[g]S_M: molecular free surface area

K_{PUF-A} and P_V values of VOCs were significantly correlated (r^2=0.98) (46). The regression equation of K_{PUF-A} and P_V values of organic compounds in Table 1 were estimated using eq 3. Diffusivities of organic compounds were correlated with the Van der Waals molar volume(b) and molar free surface area(S_M). The regression equations of diffusivity versus Van der Waals molar volume (r^2=0.71) and diffusivity versus molar free surface (r^2=0.94) are expressed in eqs 4 and 5. While the K_{PUF-A} values of organic compounds are significantly correlated with the compound specific vapor pressures, diffusion coefficients of organic compounds of PU was moderately correlated to Van der Waals molar volume and significantly correlated to molar free surface area. In terms of the association between K_{PUF-A} and vapor pressure of organic compounds, increasing PUF partitioning with decreasing vapor pressure indicates the accumulation potentials of SVOCs on PUFs.

$$\log(K_{PUF-A}) = 5.03(\pm 0.19) - 0.93(\pm 0.06) P_V \tag{3}$$

$$\log(D_{PUF}) = -17.8(\pm 2.74) - 6.31(\pm 3.38) \log(b) \tag{4}$$

$$\log(D_{PUF}) = -9.14(\pm 0.38) - 0.026(\pm 0.003) S_M \tag{5}$$

Fugacity-Based Modelling of the Fate of SVOCs

The literature on fate of SVOCs in indoor environments is relatively limited. Bennett and Furtaw modeled indoor pesticide concentrations in air, carpet, vinyl flooring, and walls and ceilings using fugacities (47). Although the model resulted in overestimation in general, its performance was better for chlorpyrifos. Zhang et al. (43), modified the fugacity-based model (47) for modelling the fate of PBDEs by adding PUF furniture in the model. Equations 6–9 show the fugacity capacity calculation equations in PUF compartments.

$$K_{PUF-A} = \left(\frac{C_{PUF}}{C_{Air}}\right)_{Equilibrium} = \left(\frac{Z_{PUF} \times f_{PUF}}{Z_{Air} \times f_{Air}}\right)_{Equilibrium} \tag{6}$$

At equilibrium: $f_{PUF} = f_{Air}$

$$Z_{PUF} = K_{PUF-A} \times Z_{Air} \tag{7}$$

$$Z_{Q,i} = K_{PUF-A,i} \times Z_{Air} \times \rho_{Q,i} \times 10^{-9} \times Z_A \tag{8}$$

$$Z_{B,PUF} = \frac{Z_{PUF} \times \delta_{PUF} + \sum_{i=1}^{6}\left(Z_{Q,i} \times \frac{PL_{PUF,i}}{\rho_{Q,i}}\right)}{\delta_{PUF} + \sum_{i=1}^{6}\left(\frac{PL_{PUF,i}}{\rho_{Q,i}}\right)} \tag{9}$$

where Z is the fugacity capacities $\left(\frac{mol}{m^3 \times Pa}\right)$ of PUF material (Z_{PUF}), air (Z_A), particle in the PUF compartment ($Z_{Q,i}$), bulk PUF ($Z_{B,PUF}$), $\rho_{Q,i}$ is the particle density $\left(\frac{kg}{m^3}\right)$, δ_{PUF} is the thickness of PUF (m), and $PL_{PUF,i}$ is the mass of i particles loaded on PUF $\left(\frac{kg}{m^2}\right)$.

Diffusive Transport of SVOCs between Air and PUFs

Zhang et al. modeled diffusive transport of PBDEs (43) by using eqs 10–15. While eqs 10 and 11 are for the gaseous diffusive transport from air to the PUF and from the PUF to air, respectively, eqs 12 and 13 are for particle deposition from air to the PUF and resuspension from the PUF to air, respectively. Advective transport from air to the PUF and the PUF to air through bouncing were included in eqs 14 and 15, respectively.

$$D_{G,A-PUF} = 1 \bigg/ \left(\frac{\delta_{bl-PUF}}{B_A A_{A-PUF} Z_A} + \frac{h_{PUF} ln2}{B_A V_{A-PUF}^{4/3} A_{A-PUF} Z_A} \right) \quad (10)$$

$$D_{G,PUF-A} = D_{G,A-PUF} \quad (11)$$

$$D_{Q,A-PUF} = \sum_{i=1}^{6} U_{P,i} A_{A-PUF} fr_{A,Q,i} Z_{Q,i} \quad (12)$$

$$D_{Q,PUF-A} = 0.1 \times D_{Q,A-PUF} \times \frac{v_c}{v_{c,0}} \quad (13)$$

$$D_{adv,A-PUF} = G_{A-PUF} Z_A = CR A_{A-PUF} h_{PUF} v_c Z_A \quad (14)$$

$$D_{adv,PUF-A} = G_{PUF-A} Z_P = CR A_{PUF-A} h_{PUF} v_c Z_{PUF} \quad (15)$$

where δ_{bl-PUF} is the boundary layer thickness (m) between air and PUF compartments, B_A is the molecular diffusivity in air $\left(\frac{m^2}{h}\right)$, A_{A-PUF} is the boundary area between air and PUF (m^2), h_{PUF} is the thickness of the PUF, $h_{PUF} ln2$ is the average diffusivity length in the PUF, V_{A-PUF} is the volumetric fraction of air in the PUF, $U_{P,i}$ is the particle deposition rate $\left(\frac{m}{h}\right)$ on the PUF, $fr_{A,Q,i}$ is the volumetric fraction of particle in air, i indicates the size fraction of particle, CR is the compression ratio of the PUF, and v_c is the frequency $\left(\frac{1}{h}\right)$ of compression.

Next, diffusive and advective transport terms are used to construct a mass balance for the PUF compartment (eq 16).

$$\frac{dM_{PUF}}{dt} = (D_{G,A-PUF} + D_{Q,A-PUF} + D_{adv,A-PUF}) f_A - (D_{G,PUF-A} + D_{Q,PUF-A} + D_{adv,PUF-A}) f_{PUF} \quad (16)$$

where M_{PUF} is the molecular amount (mol) of PBDE in the PUF, t is the time (h), and f_A and f_{PUF} are the fugacity of air and the PUF.

Fate of Indoor SVOCs

Indoor SVOCs were assumed to be sourced by outdoor-to-indoor advection of bulk air and particles, emission from SVOC impregnated PUFs, carpets, and electronic equipment. At the same time, SVOC loss from indoor environments through air advection, dust removal, and chemical transformation were also considered in the model. PUFs were predicted to be the sink of 20% of emissions. Because of the indoor concentrations being dominated by highly brominated PBDEs, the dominant movement mechanism was associated to the particle phase (dust settling). Bouncing on PUF furniture (Figure 4) "accelerated gas and particle exchange between the bulk phase PUF and air due to the compression caused by adults sitting and (some) children bouncing on furniture" was also considered in the model by Zhang et al. (43), who estimated that low SVOC concentrations in PUFs increase the mass transport from air to PUF furniture, which might be caused by the decreased boundary layer thickness and increased resuspension of particles. However, higher concentrations in PUFs result in increasing reemission while bouncing, especially for lighter compounds. Increased air exchange rate was determined to be a reason for the decrease in the air–PUF deposition flux caused by the increasing concentration gradient ($C_{PUF} \gg C_{air}$) between indoor air, dilution with outdoor air, and PUF media.

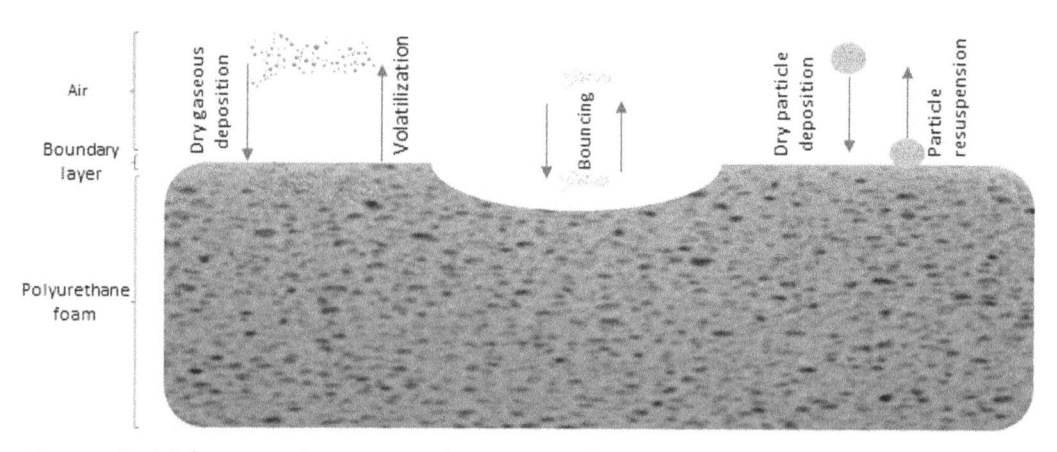

Figure 4. SVOC fate process between air and PUF in an indoor environment. Reproduced with permission from reference (43). Copyright (2009) American Chemical Society.

pp-LFER-Based Sorption Model of Organic Compounds on PUFs

Kamprad and Goss (48) used the polyparameter linear free energy (pp-LFER) relationship to describe the sorption of polar and nonpolar compounds to PUFs and to determine the enthalpies of the sorption (eq 17). In their equation, the lowercase letters are the descriptors of the sorption properties of the PUF, and capital letters are the compound specific parameters for different types of interactions. Equation 17 is the modification of the Abraham equation, which is used for dispersive interactions (E-parameters) instead of Van der Waals interactions and the cavity (V-parameter) between sorbent and compound.

$$K_{PUF-A} = l_{PUF}L_i + s_{PUF}S_i + a_{PUF}A_i + b_{PUF}B_i + v_{PUF}V_i + C_{PUF} \tag{17}$$

where L_i is the logarithmic value of the hexadecane-air partition coefficient (at 25 °C, $\frac{m^3_{air}}{m^3_{hexadecane}}$), V_i is the McGowan volume ($\left(\frac{cm^3}{mol}\right)/100$), A_i indicates the parameters of the hydrogen donor (electron acceptor) properties, B_i indicates the parameters of the hydrogen acceptor (electron donor) properties, and S_i is the descriptor of dipolarity or polarizability.

A calibrated pp-LFER model (48), shown as eq 18 (r^2=0.997, SE=0.09), was constructed using experimental data (15 °C, PU-ether LM2033, n=103). The b-descriptor was eliminated because the b-value of the foam was not significantly different from zero. As such, the foam was not affected by the hydrogen bond acceptor phenomenon. High fitted values of a and s indicated that the retention of hydrogen donors (high A_i), compounds (alcohols and acids), and strong sorption of high S value compounds (nitrogenous compounds and anilines). Sorbent descriptors for other PUFs are shown in Table 2. Sorption enthalpies could be determined using eq 19 (r^2=0.85).

$$\log K_{PUF-A} = (0.71 \pm 0.02)L_i + (1.69 \pm 0.05)S_i + (3.66 \pm 0.05)A_i + (0.36 \pm 0.09)V_i - (0.15 \pm 0.05) \tag{18}$$

$$\Delta H_i (kJ/mol) = (-4.3 \pm 1.0)L_i + (-17.6 \pm 2.3)S_i + (-46.6 \pm 4.0)A_i + (-12.8 \pm 4.0)V_i + (2.7 \pm 2.6) \tag{19}$$

Table 2. Sorbent Descriptors for PU Ether and Ester Foams. Reproduced with permission from reference (48). Copyright (2007) American Chemical Society.

PUF	$v_{PUF/air}$	$l_{PUF/air}$	$b_{PUF/air}$	$a_{PUF/air}$	$s_{PUF/air}$	$c_{PUF/air}$	n	r^2	SE
PU ether									
LM 2033	0.36±0.09	0.71±0.02	0	3.66±0.05	1.69±0.05	-0.15±0.05	103	0.99	0.09
TEX 2519	-0.10±0.31	0.88±0.07	0	3.52±0.21	1.61±0.15	-0.26±0.19	20	0.98	0.09
THE4030	0.23±0.18	0.84±0.05	0	3.76±0.15	1.72±0.10	-0.45±0.12	49	0.98	0.10
PU ester									
Ester CH	-0.23±0.30	0.79±0.08	0	3.63±0.25	2.43±0.17	-0.54±0.20	44	0.94	0.16
Ester R 90%rh	0.11±0.69	0.45±0.18	0	3.38±0.62	2.22±0.39	-0.25±0.52	21	0.85	0.26
R200 U	-1.03±0.41	1.11±0.11	0.79±0.21	3.79±0.27	2.33±0.24	-1.67±0.26	44	0.92	0.20
Willsorp	-0.21±0.31	0.90±0.09	0.35±0.16	3.91±0.20	2.55±0.17	-1.22±0.19	45	0.95	0.15
P 100 Z	-0.03±0.33	0.86±0.08	0	4.22±0.23	2.97±0.18	-1.51±0.21	49	0.95	0.19

PUF-Affected Indoor Levels of FRs

La Guardia and Hale determined the effects of FR-impregnated PUF blocks in gymnastic pits on indoor air and dust concentration levels (49). The targeted FRs were PBDEs (BDE-28, -47, -66, -85, -99, -100, -153, -154, -183, -206, and -209), alternative BFRs (α-, β-, and γ- isomers of HBCDD, TBBPA, Decabromodiphenyl ethane [DBDPE], BTBPE, Bis [2-ethylhexyl] 3, 4, 5, 6-tetrabromophthalate [TBPH], and EH-TBB), and chlorinated organophosphate FRs (TCEP, TCPP, and TDCPP). The total FR concentration of PUF blocks in gyms were determined to range from 12,100 to 25,800 μg/g, which corresponded to 1.2–2.6% of their weight. EH-TBB and TBPH were determined to be the major FRs in gym PUF blocks. Penta-BDE was found to contribute to 33.6% of the total FR concentrations in old PUF blocks. PUF additive FRs were determined to be at higher levels in gym air and dust samples than those in residences. Gym coaches were exposed to significantly higher levels of FRs than gymnasts. Occupational exposure to FRs emitted by impregnated PUFs in gyms might pose significant health risks.

Some of the most FR-abundant indoor environments are gymnastic studios because of the availability of the PUFs in pits and surfaces. The air concentrations of FRs near the PUF pit were five to six folds higher than in the other sides of the gym area (50). Ceballos et al. (51) determined FR compositions before and after the replacement of PUF cubes in gym pits and exposure levels of gymnasts. Penta-BDE exposure was decreased when foams were replaced with penta-BDE-free PUFs. While penta-BDE levels could be lowered with the replacement, it was not the case for EH-TBB and BEH-TEBP because both old and new PUFs had significant levels of the two NBFRs.

Gaylor et al. investigated the accumulation of PBDEs in house crickets by direct exposure to PBDE-impregnated PUFs (52). Their findings indicated that the insects might ingest PUFs of currently used or unclaimed furniture from waste disposal sites. The body burden of PBDE levels on the insects might increase with the ingestion of FR-loaded PUFs, and this could be an important factor for transporting foams along the food chain. Since PUFs accumulate organic chemicals in the environment, they also carry the accumulated chemicals along the food chain in their transport to distant areas.

Conclusion

PUFs contain high amounts of FR chemicals in order to meet fire regulations. While they are one of the primary sources of FRs, they can act as a sink for environmental organic pollutants such as PCBs, PAHs, SMCs, phthalate esters, pesticides, siloxanes, and VOCs in indoor environments. Sorbed organic compounds could be emitted back from PUFs depending on environmental and chemical factors such as ambient temperature and concentration gradients. The sorption–emission phenomenon results in the secondary emission of the organic compounds to indoor environments, which prolongs their presence indoors. High K_{PUF-A} values make indoor PUFs ideal sinks for organic compounds. In addition to the accumulation of gas phase organic compounds on PUFs, dry particle deposition is also an important transport mechanism between air and the foams. Indoor PUFs are responsible for the deposition of 20% of the emitted organic compounds. Linear correlation was determined between K_{PUF-A}-P_V and K_{PUF-A}–K_{O-A}. As a result, organic compounds tend to accumulate on the PUFs. Organic compounds accumulated on the PUFs, might be reemitted into indoor environment because of driving forces such as concentration gradients, temperature, and physical contact by sitting or bouncing on the seat. This might cause the release of sorbed organic compounds into the indoor environment.

References

1. Akinrinade, O. E.; Stubbings, W.; Abou-Elwafa Abdallah, M. A.; Ayejuyo, O.; Alani, R.; Harrad, S. Status of Brominated Flame Retardants, Polychlorinated Biphenyls, and Polycyclic Aromatic Hydrocarbons in Air and Indoor Dust in AFRICA: A Review. *Emerg. Contam.* **2020**, *6*, 405–420, DOI: 10.1016/j.emcon.2020.11.005.
2. Civan, M. Y. Y.; Kara, U. M. Risk Assessment of PBDEs and PAHs in House Dust in Kocaeli, Turkey: Levels and Sources. *Environ. Sci. Pollut. Res. Int.* **2016**, *23* (23), 23369–23384, DOI: 10.1007/s11356-016-7512-5.
3. Weschler, C. J.; Nazaroff, W. W. Semivolatile Organic Compounds in Indoor Environments. *Atmos. Environ.* **2008**, *42* (40), 9018–9040, DOI: 10.1016/j.atmosenv.2008.09.052.
4. Balci, E.; Genisoglu, M.; Sofuoglu, S. C.; Sofuoglu, A. Indoor Air Partitioning of Synthetic Musk Compounds: Gas, Particulate Matter, House Dust, and Window Film. *Sci. Total Environ.* **2020**, *729*, 138798, DOI: 10.1016/j.scitotenv.2020.138798.
5. Genisoglu, M.; Sofuoglu, A.; Kurt-Karakus, P. B.; Birgul, A.; Sofuoglu, S. C. Brominated Flame Retardants in a Computer Technical Service: Indoor Air Gas Phase, Submicron (PM1) and Coarse (PM10) Particles, Associated Inhalation Exposure, and Settled Dust. *Chemosphere* **2019**, *231*, 216–224, DOI: 10.1016/j.chemosphere.2019.05.077.
6. Mandin, C.; Mercier, F.; Ramalho, O.; Lucas, J.; Gilles, E.; Blanchard, O.; Bonvallot, N.; Glorennec, P.; Le Bot, B. Semi-Volatile Organic Compounds in the Particulate Phase in Dwellings: A Nationwide Survey in France. *Atmos. Environ.* **2016**, *136*, 82–94, DOI: 10.1016/j.atmosenv.2016.04.016.
7. Usepa, I. *Integrated Risk Information System*. https://www.epa.gov/iris (accessed Aug 11, 2021), 2021.
8. Ali, N.; Dirtu, A. C.; Van den Eede, N.; Goosey, E.; Harrad, S.; Neels, H.; 't Mannetje, A.; Coakley, J.; Douwes, J.; Covaci, A. Occurrence of Alternative Flame Retardants in Indoor Dust from New Zealand: Indoor Sources and Human Exposure Assessment. *Chemosphere* **2012**, *88* (11), 1276–1282, DOI: 10.1016/j.chemosphere.2012.03.100.
9. Darnerud, P. O. Toxic Effects of Brominated Flame Retardants in Man and in Wildlife. *Environ. Int.* **2003**, *29* (6), 841–853, DOI: 10.1016/S0160-4120(03)00107-7.
10. Drage, D. S.; Newton, S.; de Wit, C. A.; Harrad, S. Concentrations of Legacy and Emerging Flame Retardants in Air and Soil on a Transect in the UK West Midlands. *Chemosphere* **2016**, *148*, 195–203, DOI: 10.1016/j.chemosphere.2016.01.034.
11. Morris, A. D.; Muir, D. C. G.; Solomon, K. R.; Teixeira, C. F.; Duric, M. D.; Wang, X. Bioaccumulation of Polybrominated Diphenyl Ethers and Alternative Halogenated Flame Retardants in a Vegetation-Caribou-Wolf Food Chain of the Canadian Arctic. *Environ. Sci. Technol.* **2018**, *52* (5), 3136–3145, DOI: 10.1021/acs.est.7b04890.
12. Gottschall, N.; Topp, E.; Edwards, M.; Payne, M.; Kleywegt, S.; Lapen, D. R. Brominated Flame Retardants and Perfluoroalkyl Acids in Groundwater, Tile Drainage, Soil, and Crop Grain Following a High Application of Municipal Biosolids to a Field. *Sci. Total Environ.* **2017**, *574*, 1345–1359, DOI: 10.1016/j.scitotenv.2016.08.044.
13. Tao, F.; Abdallah, M. A.; Harrad, S. Emerging and Legacy Flame Retardants in UK Indoor Air and Dust: Evidence for Replacement of PBDEs by Emerging Flame Retardants? *Environ. Sci. Technol.* **2016**, *50* (23), 13052–13061, DOI: 10.1021/acs.est.6b02816.

14. Vorkamp, K.; Bossi, R.; Rigét, F. F.; Skov, H.; Sonne, C.; Dietz, R. Novel Brominated Flame Retardants and Dechlorane plus in Greenland Air and Biota. *Environ. Pollut.* **2015**, *196*, 284–291, DOI: 10.1016/j.envpol.2014.10.007.
15. Ali, N.; Harrad, S.; Goosey, E.; Neels, H.; Covaci, A. 'Novel' Brominated Flame Retardants in Belgian and U.K. Indoor Dust: Implications for Human Exposure. *Chemosphere* **2011**, *83* (10), 1360–1365, DOI: 10.1016/j.chemosphere.2011.02.078.
16. Eljarrat, E.; Barceló, D. *The Handbook of Environmental Chemistry: Brominated Flame Retardants*; Springer, 2011, pp 1–287.
17. USEPA. *Technical Fact Sheet – Polybrominated Diphenyl Ethers (PBDEs) and Polybrominated Biphenyls (PBBs)*. https://www.epa.gov/fedfac/technical-fact-sheet-polybrominated-diphenyl-ethers-pbdes-and-polybrominated-biphenyls-pbbs (accessed Aug 15, 2021); Vols. 1–5, 2012.
18. Kemmlein, S.; Hahn, O.; Jann, O. Emissions of Organophosphate and Brominated Flame Retardants from Selected Consumer Products and Building Materials. *Atmos. Environ.* **2003**, *37* (39–40), 5485–5493, DOI: 10.1016/j.atmosenv.2003.09.025.
19. Marklund, A.; Andersson, B.; Haglund, P. Screening of Organophosphorus Compounds and Their Distribution in Various Indoor Environments. *Chemosphere* **2003**, *53* (9), 1137–1146, DOI: 10.1016/S0045-6535(03)00666-0.
20. Gullett, B. K.; Wyrzykowska, B.; Grandesso, E.; Touati, A.; Tabor, D. G.; Ochoa, G. S. PCDD/F, PBDD/F, and PBDE Emissions from Open Burning of a Residential Waste Dump. *Environ. Sci. Technol.* **2010**, *44* (1), 394–399, DOI: 10.1021/es902676w.
21. Cetin, B.; Odabasi, M. Polybrominated Diphenyl Ethers (PBDEs) in Indoor and Outdoor Window Organic Films in Izmir, Turkey. *J. Hazard. Mater.* **2011**, *185* (2–3), 784–791, DOI: 10.1016/j.jhazmat.2010.09.089.
22. Kurt-Karakus, P. B.; Alegria, H.; Jantunen, L.; Birgul, A.; Topcu, A.; Jones, K. C.; Turgut, C. Polybrominated Diphenyl Ethers (PBDEs) and Alternative Flame Retardants (NFRs) in Indoor and Outdoor Air and Indoor Dust from Istanbul-Turkey: Levels and an Assessment of Human Exposure. *Atmos. Pollut. Res.* **2017**, *8* (5), 801–815, DOI: 10.1016/j.apr.2017.01.010.
23. Meeker, J. D.; Johnson, P. I.; Camann, D.; Hauser, R. Polybrominated Diphenyl Ether (PBDE) Concentrations in House Dust Are Related to Hormone Levels in Men. *Sci. Total Environ.* **2009**, *407* (10), 3425–3429, DOI: 10.1016/j.scitotenv.2009.01.030.
24. SC. Listing of Hexabromodiphenyl Ether and Heptabromodiphenyl Ether. In *Stockh. Convention Decision S.C.-4/14 Fourth Meet. conference Parties to Stock. Conv.* May 4–8, Geneva, Switzerland, 2009.
25. SC. Listing of Tetrabromodiphenyl Ether and Pentabromodiphenyl Ether. In *Stockh. Convention Decision S.C.-4/18 Fourth Meet. conference Parties to Stock. Conv.* May 4–8, Geneva, Switzerland, 2009.
26. SC. Listing of Decabromodiphenyl Ether. In *Conv. 14 April Stockh. Convention Decision S.C.-8/10 Eighth Meet. conference Parties to Stock*; May; Geneva, Switzerland, 2017.
27. Toms, L. M. L.; Hearn, L.; Kennedy, K.; Harden, F.; Bartkow, M.; Temme, C.; Mueller, J. F. Concentrations of Polybrominated Diphenyl Ethers (PBDEs) in Matched Samples of Human Milk, Dust and Indoor Air. *Environ. Int.* **2009**, *35* (6), 864–869, DOI: 10.1016/j.envint.2009.03.001.

28. Bramwell, L.; Harrad, S.; Abou-Elwafa Abdallah, M. A.; Rauert, C.; Rose, M.; Fernandes, A.; Pless-Mulloli, T. Predictors of Human PBDE Body Burdens for a UK Cohort. *Chemosphere* **2017**, *189*, 186–197, DOI: 10.1016/j.chemosphere.2017.08.062.

29. Li, Y.; Chen, L.; Ngoc, D. M.; Duan, Y. P.; Lu, Z. B.; Wen, Z. H.; Meng, X. Z. Polybrominated Diphenyl Ethers (PBDEs) in PM2.5, PM10, TSP and Gas Phase in Office Environment in Shanghai, China: Occurrence and Human Exposure. *PLOS ONE* **2015**, *10* (3), e0119144, DOI: 10.1371/journal.pone.0119144.

30. Besis, A.; Samara, C. Polybrominated Diphenyl Ethers (PBDEs) in the Indoor and Outdoor Environments - A Review on Occurrence and Human Exposure. *Environ. Pollut.* **2012**, *169*, 217–229, DOI: 10.1016/j.envpol.2012.04.009.

31. Blum, A.; Behl, M.; Birnbaum, L. S.; Diamond, M. L.; Phillips, A.; Singla, V.; Sipes, N. S.; Stapleton, H. M.; Venier, M. Organophosphate Ester Flame Retardants: Are They a Regrettable Substitution for Polybrominated Diphenyl Ethers? *Environ. Sci. Technol. Lett.* **2019**, *6* (11), 638–649, DOI: 10.1021/acs.estlett.9b00582.

32. Pantelaki, I.; Voutsa, D. Organophosphate Flame Retardants (OPFRs): A Review on Analytical Methods and Occurrence in Wastewater and Aquatic Environment. *Sci. Total Environ.* **2019**, *649*, 247–263, DOI: 10.1016/j.scitotenv.2018.08.286.

33. Yu, X.; Yin, H.; Peng, H.; Lu, G.; Liu, Z.; Dang, Z. OPFRs and BFRs Induced A549 Cell Apoptosis by Caspase-Dependent Mitochondrial Pathway. *Chemosphere* **2019**, *221*, 693–702, DOI: 10.1016/j.chemosphere.2019.01.074.

34. Vallack, H. W.; Bakker, D. J.; Brandt, I.; Broström-Lunden, E.; Brouwer, A.; Bull, K. R.; Gough, C.; Guardans, R.; Holoubek, I.; Jansson, B.; Koch, R.; Kuylenstierna, J.; Lecloux, A.; Mackay, D.; McCutcheon, P.; Mocarelli, P.; Taalman, R. D. F. Controlling Persistent Organic Pollutants–What Next? *Environ. Toxicol. Pharmacol.* **1998**, *6* (3), 143–175, DOI: 10.1016/s1382-6689(98)00036-2.

35. Xu, C.; Niu, L.; Zou, D.; Zhu, S.; Liu, W. Congener-Specific Composition of Polychlorinated Biphenyls (PCBs) in Soil-Air Partitioning and the Associated Health Risks. *Sci. Total Environ.* **2019**, *684*, 486–495, DOI: 10.1016/j.scitotenv.2019.05.334.

36. Gioia, R.; Nizzetto, L.; Lohmann, R.; Dachs, J.; Temme, C.; Jones, K. C. Polychlorinated Biphenyls (PCBs) in Air and Seawater of the Atlantic Ocean: Sources, Trends and Processes. *Environ. Sci. Technol.* **2008**, *42* (5), 1416–1422, DOI: 10.1021/es071432d.

37. Cetin, B.; Yurdakul, S.; Gungormus, E.; Ozturk, F.; Sofuoglu, S. C. Source Apportionment and Carcinogenic Risk Assessment of Passive Air Sampler-Derived PAHs and PCBs in a Heavily Industrialized Region. *Sci. Total Environ.* **2018**, *633*, 30–41, DOI: 10.1016/j.scitotenv.2018.03.145.

38. Frederiksen, M.; Meyer, H. W.; Ebbehøj, N. E.; Gunnarsen, L. Polychlorinated Biphenyls (PCBs) in Indoor Air Originating from Sealants in Contaminated and Uncontaminated Apartments Within the Same Housing Estate. *Chemosphere* **2012**, *89* (4), 473–479, DOI: 10.1016/j.chemosphere.2012.05.103.

39. Johansson, N.; Biologkonsult, M.; Hanberg, A.; Institutet, K.; Defence, S.; Tysklind, M. PCB in building sealant is influencing PCB levels in blood of residents. *Organohalo. Compd* **2003**, *63*, 381–384.

40. Liu, X.; Zhao, D.; Peng, L.; Bai, H.; Zhang, D.; Mu, L. Gas–Particle Partition and Spatial Characteristics of Polycyclic Aromatic Hydrocarbons in Ambient Air of a Prototype Coking Plant. *Atmos. Environ.* **2019**, *204*, 32–42, DOI: 10.1016/j.atmosenv.2019.02.012.
41. Moreau-Guigon, E.; Alliot, F.; Gaspéri, J.; Blanchard, M.; Teil, M.-J.; Mandin, C.; Chevreuil, M. Seasonal Fate and Gas/Particle Partitioning of Semi-Volatile Organic Compounds in Indoor and Outdoor Air. *Atmos. Environ.* **2016**, *147*, 423–433, DOI: 10.1016/j.atmosenv.2016.10.006.
42. Hammel, S. C.; Hoffman, K.; Lorenzo, A. M.; Chen, A.; Phillips, A. L.; Butt, C. M.; Sosa, J. A.; Webster, T. F.; Stapleton, H. M. Associations Between Flame Retardant Applications in Furniture Foam, House Dust Levels, and Residents' Serum Levels. *Environ. Int.* **2017**, *107*, 181–189, DOI: 10.1016/j.envint.2017.07.015.
43. Zhang, X.; Diamond, M. L.; Ibarra, C.; Harrad, S. Multimedia Modeling of Polybrominated Diphenyl Ether Emissions and Fate Indoors. *Environ. Sci. Technol.* **2009**, *43* (8), 2845–2850, DOI: 10.1021/es802172a.
44. Okeme, J. O.; Webster, E. M.; Parnis, J. M.; Diamond, M. L. Approaches for Estimating PUF-Air Partitions Coefficient for Semi-Volatile Organic Compounds: A Critical Comparison. *Chemosphere* **2017**, *168*, 199–204, DOI: 10.1016/j.chemosphere.2016.10.001.
45. Shoeib, M.; Harner, T. Characterization and Comparison of Three Passive Air Samplers for Persistent Organic Pollutants. *Environ. Sci. Technol.* **2002**, *36* (19), 4142–4151, DOI: 10.1021/es020635t.
46. Zhao, D.; Little, J. C.; Cox, S. S. Characterizing Polyurethane Foam as a Sink for or Source of Volatile Organic Compounds in Indoor Air. *J. Environ. Eng.* **2004**, *130* (9), 983–989, DOI: 10.1061/(ASCE)0733-9372(2004)130:9(983).
47. Bennett, D. H.; Furtaw, E. J. Fugacity-Based Indoor Residential Pesticide Fate Model. *Environ. Sci. Technol.* **2004**, *38* (7), 2142–2152, DOI: 10.1021/es034287m.
48. Kamprad, I.; Goss, K. U. Systematic Investigation of the Sorption Properties of Polyurethane Foams for Organic Vapors. *Anal. Chem.* **2007**, *79* (11), 4222–4227, DOI: 10.1021/ac070265x.
49. La Guardia, M. J.; Hale, R. C. Halogenated Flame-Retardant Concentrations in Settled Dust, Respirable and Inhalable Particulates and Polyurethane Foam at Gymnastic Training Facilities and Residences. *Environ. Int.* **2015**, *79*, 106–114, DOI: 10.1016/j.envint.2015.02.014.
50. Carignan, C. C.; Heiger-Bernays, W.; McClean, M. D.; Roberts, S. C.; Stapleton, H. M.; Sjödin, A.; Webster, T. F. Flame Retardant Exposure among Collegiate United States Gymnasts. *Environ. Sci. Technol.* **2013**, *47* (23), 13848–13856, DOI: 10.1021/es4037868.
51. Ceballos, D. M.; Broadwater, K.; Page, E.; Croteau, G.; La Guardia, M. J. Occupational Exposure to Polybrominated Diphenyl Ethers (PBDEs) and Other Flame Retardant Foam Additives at Gymnastics Studios: Before, During and After the Replacement of Pit Foam with PBDE-Free Foams. *Environ. Int.* **2018**, *116*, 1–9, DOI: 10.1016/j.envint.2018.03.035.
52. Gaylor, M. O.; Harvey, E.; Hale, R. C. House Crickets Can Accumulate Polybrominated Diphenyl Ethers (PBDEs) Directly from Polyurethane Foam Common in Consumer Products. *Chemosphere* **2012**, *86* (5), 500–505, DOI: 10.1016/j.chemosphere.2011.10.014.

Chapter 7

Halogen-Based Flame Retardants in Polyurethanes

Nycolle G. S. Silva,[1] Noelle C. Zanini,[2] Alana G. de Souza,[2] Rennan F. S. Barbosa,[2] Derval S. Rosa,[1] and Daniella R. Mulinari*,[1]

[1]Department of Mechanic and Energy, State University of Rio de Janeiro (UERJ), Resende, CEP 27537-000, Brazil
[2]Center for Engineering, Modeling, and Applied Social Sciences (CECS), Federal University of ABC (UFABC), Santo André, CEP 09210-580, Brazil
*Email: dmulinari@hotmail.com

Polyurethanes are polymeric materials widely used in all consumer goods and industrial markets. However, their flammability is a serious problem that, associated with the various tragedies that occur annually involving fires, has highlighted the urgent need to reduce the material's flammability. Flame retardants (FRs) are added to the polymers to improve their fire performance and can be divided into three main types: halogen-based FRs (HFRs), halogen-free FRs, and nano FRs. The HFR is the most common class and includes a broad range of structures, such as brominated FRs (BFRs), chlorinated FRs (CFRs), fluorine, and iodine, where the primary element inhibits fire propagation by substituting •OH and •H free radicals. BFRs and CFRs have broad industrial applications because of their low cost and high effectiveness. More than 75 different BFR compounds are commercially available, and the tetrabromobisphenol A and hexabromocyclododecanes are the most used materials. After thermal decomposition, these materials generate gaseous bromine forms (HBr, Br, and Br_2) and char. The main CFRs are straight-chain paraffin, cycloaliphatic hydrocarbons, aromatic hydrocarbons, and organophosphate compounds, generating HCl and a cross-linked chlorinated residue. In this chapter, the HFR mechanisms in polyurethane foams are reviewed in detail, focusing on the main brominated and chlorinated compounds, and their benefits, harms, suppliers, future trends, and new technologies developed.

© 2021 American Chemical Society

Introduction

The use of polyurethanes (PUs) is common in products of modern society, from mattresses, car seats, and strollers, to furniture, building insulation, refrigeration, and energy storage, among others (1, 2). This application flexibility is reflected in PU materials' high annual production, exceeding 18 million tons (3). This class of materials has important properties, such as low density, superior mechanical and physical properties, abrasion resistance, high hardness, low water absorption, and low thermal conductivity (4). However, PU flammability is a serious problem because this material rapidly decomposes, releasing a large amount of heat when exposed to fire. Besides the heat, large amounts of toxic smoke and molten dripping are generated during combustion, representing a considerable threat to people's lives (5). Because of this, improvements in flame retardancy and smoke suppression for PU products are highly requested and have attracted worldwide attention.

Flame-retardant (FR) additives are chemicals added to the polymeric material that protect ignition sources, improve their fire performance, and are easily used. These additives are used to decrease the burning rate and consequently minimize the fire danger because smoke inhalation is responsible for 80% of deaths in this type of accident. Flammable gases mix with air and generate ignition, which causes combustion (endothermic pyrolysis) of polymers (i.e., the polymer thermally decomposes through chain or bond scission, generating volatile fragments that diffuse in the air to form a combustible gaseous mixture) (6). This gaseous mixture initiates a fire under a determined temperature, and the flame propagation leads to the heat release.

Besides the intrinsic thermal resistance attributed to polymeric linkages, the thermal stability is directly assigned to the materials' fire behavior. Fire behavior or flammability can be determined by some fire tests that simulate a realistic fire scenario and rate the risks or hazards involved for each material composition. The fire stages are ignitability, flammability, flame spread, heat release, and fire penetration. The ignition stage is the onset of flaming combustion, given by an ignition source (e.g., flame, cigarette, glow wire) in a small length scale (cm) with high ventilation and ambient temperatures in the range of 300–400 °C, and the endset of a flame classifies the flammability under ambient conditions. The fire development is assigned with the flame spread and heat release (fire load) phases, where the fire growth or flame spread occurs under the following conditions: heat flux (20–60 kW m^{-2}), larger length scales (dm to m), ambient temperatures higher than the ignition temperature (400–600 °C), and continuous ventilation. A fully developed fire classifies the last phase. It is associated with the heat and fire penetration by the following conditions: heat flux (>50 kW m^{-2}), larger length scales (more than m), ambient temperatures higher than autoignition temperatures (above 600 °C), and low ventilation (7).

The FR acts on the combustion process by scavenging and terminating active radicals in the gas and condensed phases and forming a carbonaceous layer through cyclization, dehydration, and cross-linking reactions between additives and the PU matrix (8). Figure 1 illustrates the polymeric thermal decomposition without and with the use of FRs.

Historically, the first chemical additives used for FR purposes, small halogen molecules, were tested in 1970 (1). In 1964, Boyer and Vajda (9) reported on the chemical structure of halogen–phosphorous compounds, their fire resistance, and their addition to a preformed polymer. After that, mainly between 1968 and 1972, many works described the use of several additives for PU flame retardancy, involving different chemical groups, such as bromine, chlorine, phosphorus, tetrabromophthalic anhydride, and others (10–13). The organic additives are mainly composed of halogens that significantly contribute to PU's FR properties and are attractive because of the low contents needed to obtain a positive antiflammability (14).

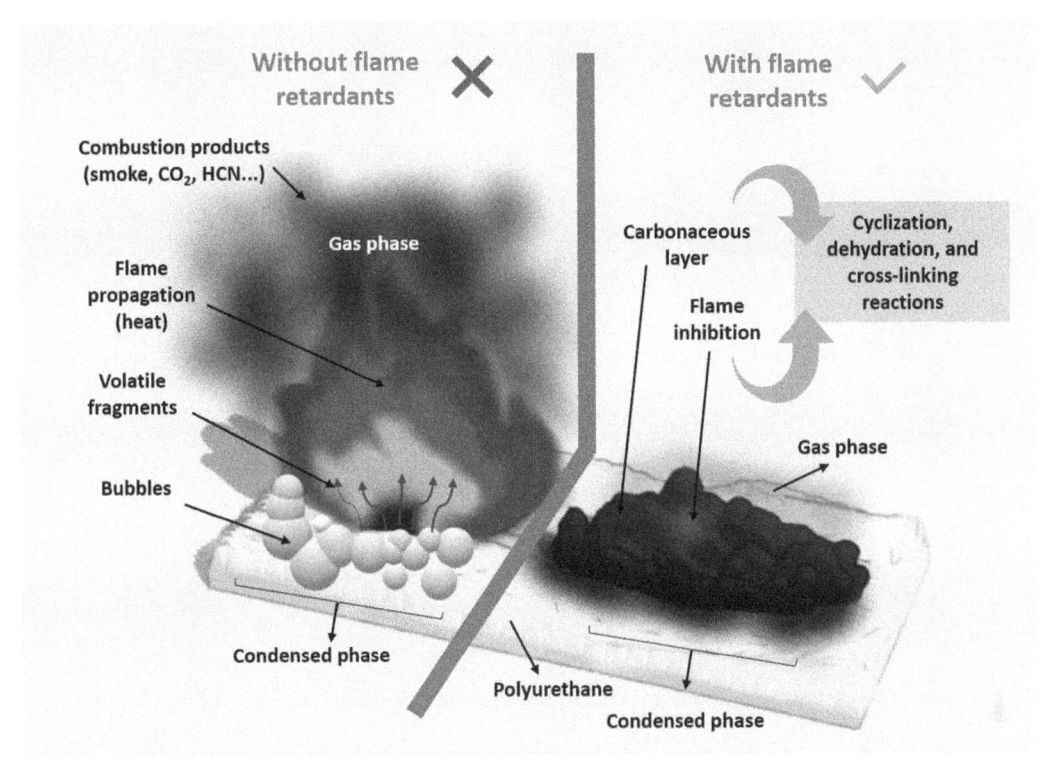

Figure 1. Schematic representation for a typical PU combustion process without and with FR additives.

Halogen-based FRs (HFRs) are widely used in commercial products to improve their fire resistance and meet security regulations. This class includes a broad range of structures, and their effectiveness depends on the main elements, which could be brominated FRs (BFRs), chlorinated FRs (CFRs), fluorine, and iodine, as shown in Figure 2. Active •OH and •H free radicals are released during polymeric combustion, leading to decomposition and burning. The HFR reacts with these radicals to produce others with lower reactivity, inhibiting fire propagation. The easiest way to create this chemical is using chlorinated or brominated alcohols, which can be in aliphatic or aromatic form (*13, 15*). Although the literature suggests that halogenate compounds generate toxic compounds after burning, BFRs are among the widely used compounds of this class because of their high effectiveness, followed by chlorine- and fluorine-based additives (*16*). The global FR market size was evaluated at $7 billion in 2019, with an annual expected growth rate of 3.6% from 2020 to 2027. According to the global FR market size report, the brominated compounds dominate, responsible for approximately 38% of the market (*17*).

Because of environmental and toxicity concerns associated with additives containing bromine and chlorine, they have received particular attention. However, not all bromine or chlorine derivatives have high toxicity, and some natural compounds contain these elements and have biological activities. Because of this, there is a search for new developments focused on developing bromine and chlorine derivatives with high antiflame efficiency and low migration to the environment (*18*). The literature has highlighted FRs based on renewable and sustainable materials, in which the bromine and chlorine groups are synthetically added. In addition to halogenated compounds, two other classes also stand out as FR additives: halogen-free FRs and nano additives (*8*).

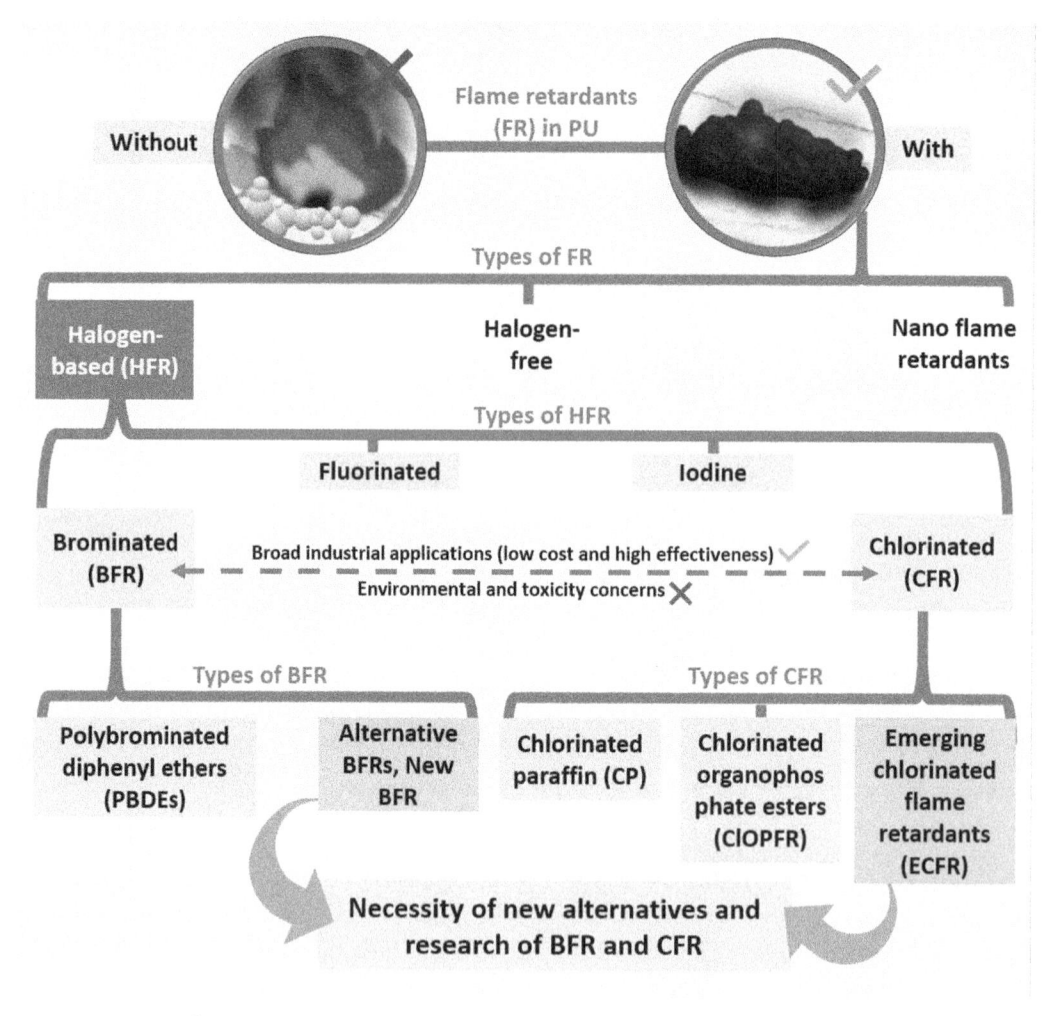

Figure 2. Schematic representation of the main FR classes, highlighting HFRs and their types.

Among the halogen-free FRs (Figure 2), the literature reports the use of organophosphate esters, alumina trihydrate, triphenyl phosphate, and other compounds containing multiple elements, such as phosphorus–silicon, phosphorus–nitrogen, and phosphorus–nitrogen–silicon (*19*). Phosphorus and nitrogen are interesting compounds because of their low toxicity, low smoke, and high FR efficiency. These compounds also form a residual carbonated char that protects the polymeric substrate from the flame (*20*). Lin et al. (*19*) proposed using a phosphorus–nitrogen halogen-free FRs in rigid PU (PUR) foam and reported an effective improvement in PU flame retardancy with 10 wt % of additive. The enhanced thermal property was associated with a new char layer that retarded the heat permeation and volatile degradation products during combustion. The P and N elements acted in the barrier and quenching effects. Cui et al. (*21*) reported using a triazine-based FR combined with ammonium polyphosphate to enhance waterborne PU's thermal resistance. The authors verified the formation of a compact and graphitized char layer that improved fire resistance and smoke suppression. However, halogen-free FRs' big challenge is that they must be used in very high quantities to achieve good flame retardancy, resulting in high costs and industrial unviability (*22*).

The three main types of nano FRs consist of layered nanostructures, such as nanoclays or graphene, nanofibrous materials, carbon nanotubes, and oxide or hydroxide nanometals (23). The layered and fibrillar nanomaterials delay the thermal decomposition by their migration to the polymer surface to create a barrier that inhibits heat, combustible gases, and oxygen transfer during combustion. Yu et al. (24) prepared clay/graphene oxide/montmorillonite networks to increase PU flame resistance. They verified excellent structural stability in a flame because of the compact and interconnected char formed after combustion. Nanometals and their oxides act as FRs by reducing the PU's fire hazard. Despite the promising results presented in the literature, nanoparticles have some disadvantages and limitations, such as involving complex and high costs (25). Even with numerous available FR options, HFRs remain an excellent alternative for application in PU materials.

PU Thermal and Fire Resistance

Polymers' flammability has long been discussed because of the importance of protecting human health and the environment against polymer combustion's hazardous effects (26). Studying the mechanistic details involved in these reactions is necessary to determine and understand PU flame retardation using additives. Depending on the PU composition (e.g., isocyanate or polyol contents), the degradation occurs at different temperatures because of the specific bond energy for each polymer's linkage. Polymer combustion induces highly active •OH and •H free radicals, leading to decomposition and burning (27).

The combustion reaction for PU is highlighted with fast-spreading flames, high thermal emission, toxic smoke production, and complex thermal decomposition with several partial decomposition reactions (28, 29). Thermal decomposition of PU generally occurs in two or three steps, where the first thermal event (~200–350 °C) starts with PU hard segment degradation, including isocyanates, alcohols, amines, olefins, and CO_2, and the second thermal event (~350–600 °C) is related to the soft segments (polyol) and small fragments such as CO_2, amine, and water, which were produced in the first step (28, 30). These decomposition stages involve the linkages of aliphatic and aromatic allophanate, aliphatic and aromatic biuret, and aliphatic, aromatic, and disubstituted urea, which degrade at 85–105 °C, 100–120 °C, 100–110 °C, 115–125 °C, 140–180 °C, 160–200 °C, and 230–250 °C, respectively (7, 28, 31, 32).

According to Singh and Jain (29), PU foam (PUF) combustion occurs only in the presence of a sufficient amount of oxygen, and it produces combustible gases, noncombustible gases, entrained solid particles, and carbonaceous char. The char yield is related to the smoke production once the char layer formation maintains the combustion products in the condensed phase and fewer fragments are released into the gas phase (28). An enhancement in the thermal stability and a smoke release reduction could also be assigned to foams with some groups such as aromatic backbone, carbodiimide, and isocyanurate, which degrade at 160–200 °C, 250–280 °C, and 270–300 °C, respectively (28, 33, 34).

Higher thermal performance and flammability reduction can be achieved either by inserting modifiers or changing the formulation recipe, mainly when the polymerization reaction produces isocyanurate rings (PIRs) (35). The isocyanate cyclotrimerization produces foams based on PIRs, a stable network structure with enhanced mechanical and thermal properties (28, 33–35). According to Schiavoni et al. (36), PIR materials are characterized by a higher fire resistance than PURs. Among foam plastics, PIR materials are the ones with the best fire resistance. Compared to PUFs, PIRs are characterized by a lower thermal conductivity, between 0.018 and 0.027 W m^{-1} K^{-1}, and by similar density values (15–45 kg m^{-3}) and specific heat (about 1.4 kJ kg^{-1} K^{-1}).

PIR foams modified by urethane linkages were first commercialized in 1968 in England by Haggis of I. C. I, in 1969 in the United States by the Upjohn Co., and in Tokyo, Japan, in 1996 by Nisshhinbo Inc. using the Ashida patent. These materials have been extensively applied as heat-insulating materials in civil engineering, pipes and tubing, and the refrigeration industry because of their low thermal conductivity (λ) and superior service temperature (Ts > 149 °C) over PU (Ts ~ 93 °C) (35, 37–39). One of the main advantages of these fire protection foams is the extensive use of aromatic isocyanates rather than aliphatic polyols (38). The main general reactions of PUR, PIR, urea, and carbodiimide are represented in Figure 3.

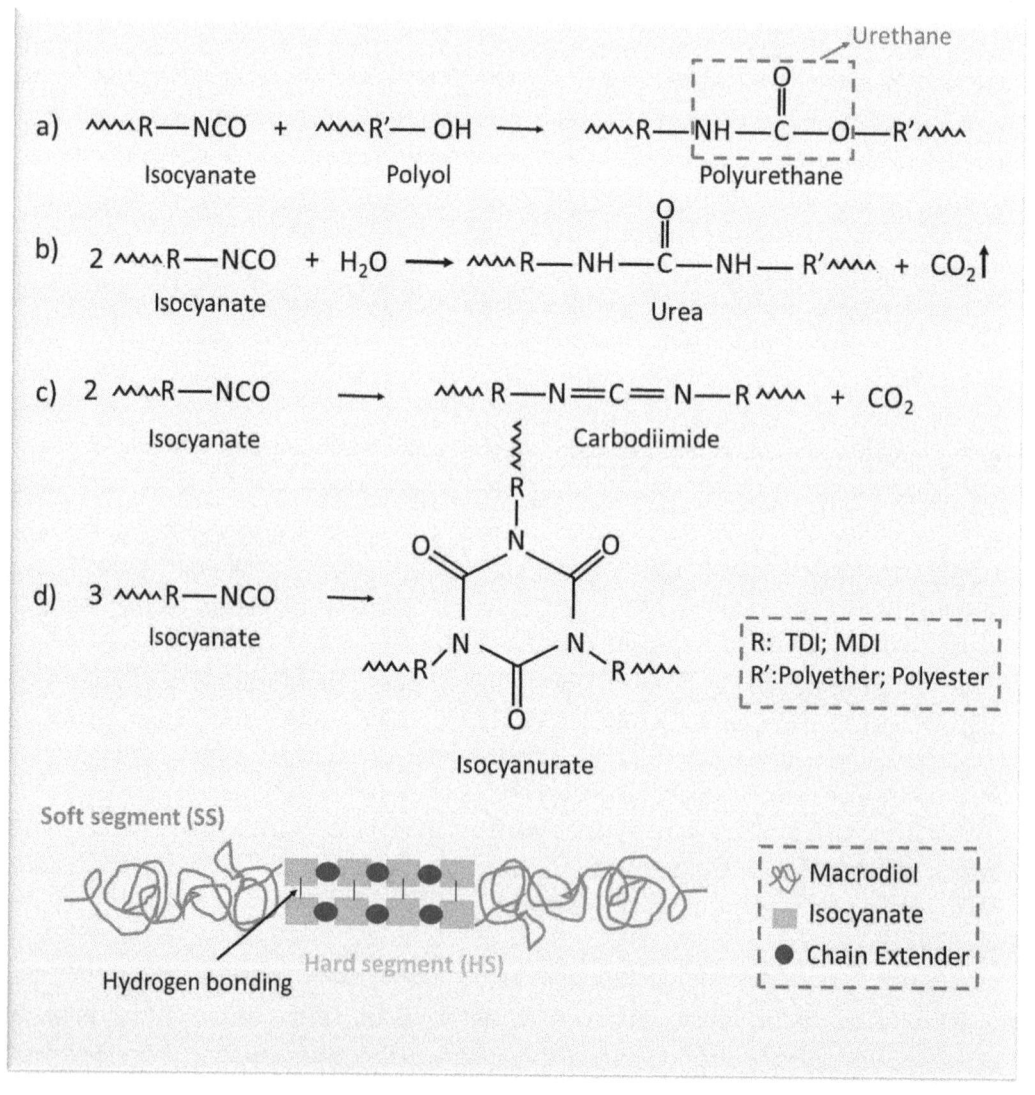

Figure 3. Main reactions of PIR–PUR foams to flame retardance: (a) Reaction between the polyol and isocyanate groups (PUR); (b) Reaction between isocyanate and water (urea); (c) Condensation of two isocyanate groups to form carbodiimide linkage; (d) Cyclotrimerization of isocyanate (PIR).

Even with higher fire retardance, PIR foams are limited because of the price and technological difficulties because of higher volume ratios of isocyanate (OH/NH: 1/1.8–1/6) and higher temperature conditions needed in the polymerization (*31*). Further, unmodified PIR foams are highly brittle. Because of this, some modifications by thermally stable linkages such as amide, imide, and carbodiimide compared to urethane-modified polyisocyanurate foams could improve the materials' flame retardancy and foams' cross-linked density, contributing to mechanical and thermal properties (*34, 37*).

By comparing PIRs and PURs for their fire resistance, some tests and classifications are commonly adopted. For that, there are limiting oxygen index and cone calorimeter apparatuses, which expose the sample under specific fire and heat conditions and measure several parameters such as ignition time (s), mass loss (g s^{-1}), heat release rate (kW m^{-2}), combustion effective heat (MJ kg^{-1}), and carbon monoxide and carbon dioxide production (kg kg^{-1}) (*34, 39*). The technical standards that classify the fire behavior vary according to each country. For the European classification system, building materials are divided into seven classes based on the best fire resistance: A1 and A2 for inorganic materials (noncombustible materials); B, C, D, and E for combustible materials; and F for materials with no determined performance. PU is classified from B to E under fire conditions, depending on its formulation and the fire retardancy additives present (*36*). PIR foams are classified as B, and PUR foams are classified as E (i.e., PIR foams are better thermal insulators because of their stable structure and chemical structure) (*36*).

Although PIR foams present superior fire retardancy to PUR/flexible PUFs, they are susceptible to fire even in the presence of phosphorus or HFRs (*38*). PIR's rigid and brittle character also prevents and limits its use for furniture, automotives, acoustics, and other applications compared to PURs and flexible PUFs (*39*). Consequently, the study of mechanisms involved in HFR and foam reactions is necessary to prevent fire disasters and maintain the foam's widespread usage.

HFRs

Depending on the polymer's nature, the combustion process can be extinguished by physical and/or chemical mechanisms in the solid, liquid, or gaseous combustion phase (*40*). Physical action mechanisms of FR additives can be classified as (1) cooling, which, as an endothermic process, cools the material to a temperature that influences the combustion interruption (such as aluminum hydroxide); (2) the formation of a protective layer that prevents material contact with the gas phase, forming a physical barrier, and; (3) dilution, where inert substances are added (fillers or additives), reducing oxidants in the gaseous phases to limit the ignition level of the gas mixture. Alternatively, the chemical action mechanisms interfere in the combustion process in the solid and gaseous phases. In the gas phase, the exothermic combustion process's mechanism is interrupted, cooling the system. Chemical-acting compounds inhibit the formation of free radicals, generating less energetic and less flammable by-products that delay combustion but can release toxic gases. In the solid phase, there are some possible scenarios: (1) polymer collapse acceleration because of the retardant interrupting the flame influence; (2) formation of the carbonized layer on the polymeric surface, because of cyclization and cross-linking (such as brominated and chlorinated compounds), and (3) inert gas production that dilutes the air supply in the region where the flame occurs (*41, 42*).

When it comes to PUs, in addition to changes in foam formulations (as mentioned in the previous topic), another way to decrease their flammability is by including FRs in their structure. HFRs block the gas phase free radical chain propagation during combustion (*26*). HFR decomposition releases hydrogen halides, which react with •OH and •H radicals from plastics and decrease the energy propagation during the fire ignition and combustion (27, 42). Because HFR activity occurs in the gas phase, it shows efficiency in a solid or liquid phase, where the heavy halogenated fragments (especially brominated fragments) assist in the char layer formation inhibiting the oxygen during combustion (26, 31). The halogenated compounds include iodine, bromine, chlorine, and fluorine. Nevertheless, bromine and chlorine are the most useful in plastics' flame retardancy. While the iodine compounds are the most effective, they present a weak bond attached to carbon (i.e., low thermal stability that prevents FR applications). Furthermore, although fluorine compounds demonstrated excellent thermal stability and fire resistance combined with carbon (fluorocarbons), the C–F strong bond is highly stable, limiting the flame retardancy. Also, fluorine in hydrogen fluoride or radical is highly reactive in the condensed phase, not preventing the start of combustion. The sequence of flame retardance for halogenated compounds is as follows: I > Br > Cl > F (29, 42).

Considering the commercialized HFR, there are numerous halogen additives such as aliphatic, alicyclic, and aromatic materials. Aliphatic ones are the most effective, and aromatic ones are the least effective, indicating that more available halogen is related to flame retardancy efficiency (27, 42). Among BFRs and CFRs, bromine is more effective because it is less attached to carbon than chlorine, facilitating the HFR's action (29). Moreover, chlorine and bromine can be used alone or associated with synergists, such as metal oxides, metal salts, phosphorus-containing agents, and high-charring agents (e.g., antimony trioxide) (27, 29).

Because HFRs are the most common in global production, it is essential to understand the mechanistic details involved in PUs with these compounds' addition. The FR mechanism follows a similar schematic for all brominated compounds, even with different molecular structure performances, as presented in the following sections.

BFRs

BFRs are halogenated derivatives of organic compounds with high fire retardancy and wide industrial application because of their low cost and high effectiveness. BFRs are preferential because of their many different chemicals with different properties (about 75 options), universal applicability, comprehensive knowledge about this compound class, and easy recyclability (43). BFRs also have a low impact on polymers' mechanical properties and a significant potential in ignition delaying. Over 200,000 tons are produced annually worldwide (18), and this abundant use is attributed to BFRs' high efficiency for trapping free radicals at low decomposition temperatures (44). BFRs are divided into two classes: (1) polybrominated diphenyl ethers (PBDEs), such as tetrabromobisphenol A (TBBA) and hexabromocyclododecanes (HBCDs), and (2) alternative FRs, such as 1,2-bis(pentabromodiphenyl) ethane, hexachlorocyclopentadienyldibromo-cyclooctane, tetrabromo bisphenol A-bis(2,3-dibromopropylether), and others.

This last category reflects efforts to produce new variants of retardants with reduced toxicity. However, studies dealing with the toxicity of novel BFRs (NBFRs) are not recurrent. Because of this, NBFR toxicity data is limited, demonstrating the need for a different approach to those new substances in scientific works because they can be as harmful as their previous options of PBDEs. Other examples of NBFRs include decabromodiphenyl ethane (DBDPE), 1,2-bis(2,4,6-tribromophenoxy)ethane (BTBPE), 2-ethylhexyl-2,3,4,5-tetrabromobenzoate (TBB), bis(2-ethylhexyl) tetrabromophtalate (TBPH), hexabromobenzene (HBB), 2,3,4,5,6-pentabromoethyl benzene (PBEB), 2,3,4,5,6-pentabromotoluene (PBT), and 1,2-dibromo-4-(1,2-dibromoethyl) cyclohexane (TBECH), and others (45), as illustrated in Figure 4.

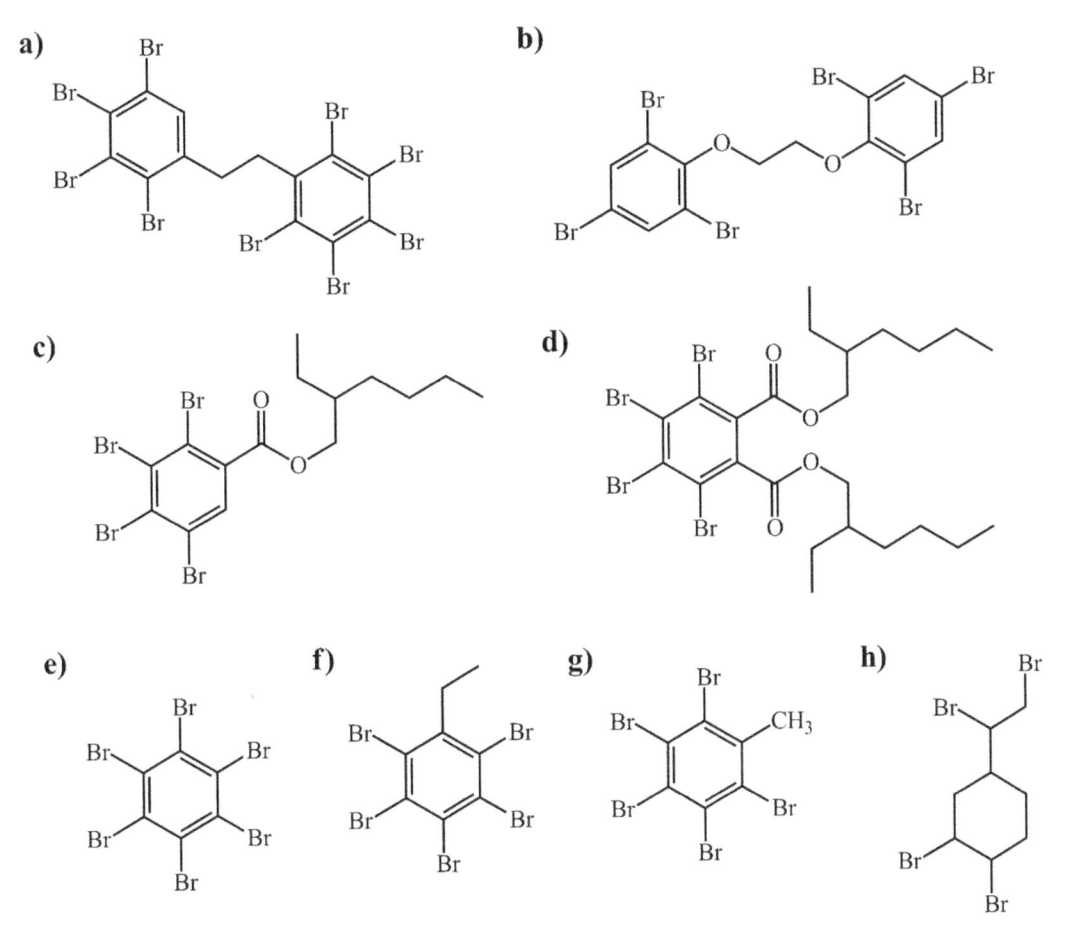

Figure 4. Main BFRs and their chemical structures: (a) DBDPE; (b) BTBPE; (c) TBB; (d) TBPH; (e) HBB; (f) PBEB; (g) PBT; (h) TBECH.

The brominated compounds can be incorporated into the polymers in two ways. The first is in the form of an additive, during processing, in which the compound is not strongly bound to the matrix, having an easier migration to the environment. The other approach is reactive and occurs by forming new covalent bonds between the polymer and the BFR through a new reaction (46). The thermal decomposition of polymers containing BFRs depends on the incorporation method and occurs by the coupling of structurally related precursors and carbon matrix burning (47, 48).

Polymeric combustion occurs in four steps: preheating, decomposition and volatiles evolution, ignition, and propagation, as discussed before. The halogens are highly efficient in capturing free radicals during combustion, eliminating flame propagation. When using BFRs, their heating generates HBr, which inhibits fire propagation by substitution of •OH and •H free radicals for •Br radicals, which are less reactive (47). The generic mechanism of BFR is presented in Equations 1 and 2, where •Br replaces the active chain carriers, and the slow energy rate promotes flame extinguishment.

$$•H + HBr \rightarrow H_2 + •Br \tag{1}$$

$$•OH + HBr \rightarrow H_2O + •Br \tag{2}$$

The weak aliphatic and aromatic C–Br bonds facilitate the Br atoms' release in the early stages of combustion, and adding oxygen to bromine hydrocarbons also releases Br atoms through inversion substitution. So, Equations 1 and 2 can be better described using the gaseous bromine forms (HBr, Br, and Br$_2$) to consume the •H and •OH radicals, according to Equations 3 to 10.

$$•Br + •H \rightarrow HBr \tag{3}$$

$$•Br + •OH \rightarrow HOBr \tag{4}$$

$$HBr + •H \rightarrow H_2 + •Br \tag{5}$$

$$HBr + •OH \rightarrow H_2O + •Br \tag{6}$$

$$Br_2 + •H \rightarrow HBr + •Br \tag{7}$$

$$Br_2 + •OH \rightarrow •Br + HOBr \tag{8}$$

$$•Br + HO_2 \rightarrow HBr + O_2 \tag{9}$$

$$•Br + •Br \rightarrow Br_2 \tag{10}$$

According to Dixon-Lewis et al. (*49*), the presented reactions occur at different rates, and the Br$_2$ and HBr act differently as inhibitors. While Equation 5 rapidly ends the HBr, in Equation 7, there is H atoms' consumption in the reaction zone because of the competition with the propagation reaction (H + O$_2$ → O + OH). Considering the activation energy, the values were 0, 203900, 682, −1290, −7553, 0, 1960, and −7118 J mol^{-1} for Equations 3, 4, 5, 6, 7, 8, 9, and 10, respectively (*50*). These reactions and their intrinsic characteristics, such as activation energy and reaction velocity, in addition to the Br atom's reaction with the formyl radical (HCO) (Equation 11), significantly decrease the material's burning velocity (*50, 51*).

$$•Br + HCO \rightarrow HBr + CO \tag{11}$$

Currently, the most used BFRs are TBBA and HBCDs. TBBA is one of the most common BFRs used in PUs. Its thermal decomposition is complex, occurring predominantly in the condensed phase and involving more than 900 physical processes and chemical reactions. Luda et al. (*52*) investigated TBBA decomposition using thermogravimetric analysis and reported two weight loss events: (1) 200–290 °C (radical polymerization and cross-linking reactions), and (2) 290–500 °C. The authors investigated structural changes in the samples before and after the thermal degradation using IR spectra. The authors reported aromatic wagging vibration displacement and stretched phenol–OH bond disappearance during the first degradation event, indicating fissions in the aromatic ring. Figure 5 shows the different reactions to TBBA during thermal degradation, such as C–Br scission hydrogen abstraction from HBr and phenols with minor bromine content.

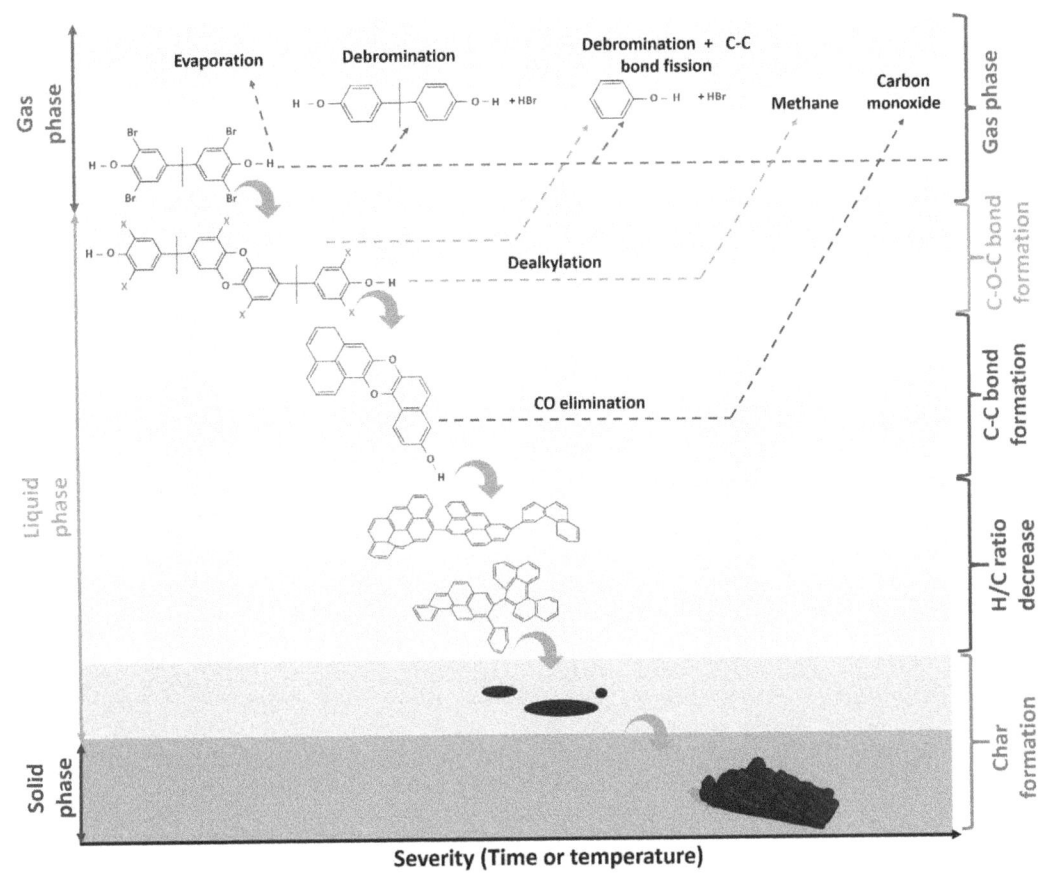

Figure 5. Decomposition steps involved in the TBBA pyrolysis process.

TBBA decomposition products include brominated phenols, HBr, brominated derivatives of bisphenol A, brominated benzenes, and char. The nonbrominated compounds include alkanes and alkylates benzenes (53). Several reaction categories control the TBBA thermal degradation mechanism:

1. Br atoms' bimolecular displacement with phenolic' H atoms, or intramolecular H migration phenolic' H (Figure 6a and 6b) induces HBr elimination and formation of bisphenol A derivatives with a low bromination index via C–Br scission. It is considered an initiation reaction, and HBr loss occurs through bimolecular condensation reactions. The HBr elimination produces char.
2. Isopropylidene bond scissions: Above 300 °C the isopropylidene linkages suffer fissions (Figure 6c and 6d); this reaction can be confirmed by detecting 2,6-dibromophenol and indicates the scissions of aromatic $(H_3C)_2C$ bonds (54).
3. Reactions lead to soot evolution resulting from the C–C and C–O bond scissions.

It is expected that TBBA degradations co-occur with vaporization through five global reactions, as illustrated in Figure 7. Marongiu et al. (55) proposed nine reactions for TBBA degradation: R1 (initiation reactions and bromine radical formation), R2 (heavy component initiation reactions), R3 (bimolecular initiation reactions), R4 (bromine radical substitutive additions), R5 (radical intramolecular additions), R6 (CO elimination), R7 (radical decomposition), R8 (molecular dehydrobromination), and R9 (termination reactions) (54, 55). After that, Font et al. (56) proposed a simplified TBBA degradation model that considers initiation and primary recombination of global reactions after the growth of the first heavy molecules.

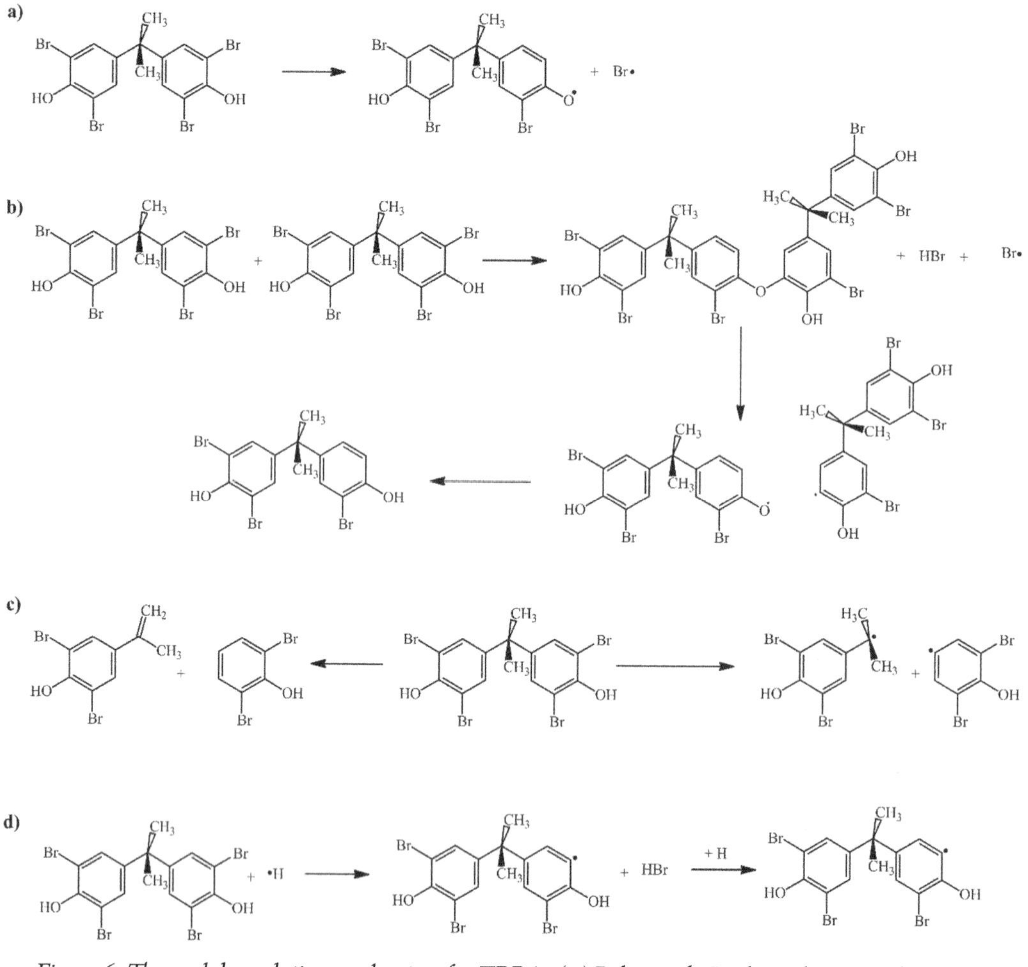

Figure 6. Thermal degradation mechanism for TBBA: (a) Release of •Br through intramolecular H migration; (b) Release of HBr via bimolecular displacement of Br atoms; (c) Scission of isopropylidene bonds; (d) Formation of lower brominated isomers of bisphenol A under a pyrolytic environment.

A common approach to improving TBBA efficiency in polymeric flame retardancy is introducing metal oxides as halogen fixers or absorbers (57). Al-Harahsheh et al. (58) reported that oxide metals, such as Fe_3O_2 and ZnO, capture the HBr. After 400 °C, the emitted quantity is significantly lower because of the easy HBr dissociative adsorption over metal–oxygen bonds. Other studies indicate that calcium hydroxide shows excellent potential to suppress the metal bromides' evaporation and increase the ash residue (59). The HBr adsorption into metal oxides results in volatile metallic halides and molecular bromine that allow the regeneration of Br_2 followed by water elimination (60).

HBCDs are the second most used class of BFRs, and there are 16 stereoisomers (61), all of them with similar degradation profiles, starting at ~230 °C. The main diastereomers are α-, β-, and γ-HBCD. The degradation mechanism of HBCDs is governed by an autocatalytic process derived from radical species. In this process, bromine radicals are released, and their hydrogen abstraction from the substrate leads to new radicals that can decompose by β-scission (51, 62). Once a carbon–bromine bond is broken, several other halogen radicals can be rapidly formed, such as brominated biphenyls, brominated congeners of cyclododecadiene, brominated cyclododecatriene, and brominated methyl-naphthalene. The decomposition mechanism involves unimolecular arrangements, dehybromination reactions, Diels–Alder reactions, C–C bond breakage, and radical-derived reactions. As pointed out by Barontini et al. (53), intramolecular reactions prevail over intermolecular. Because of this, the loss of three atoms of bromine is commonly observed during HBCD degradation. Figure 8 shows the HBCD thermal decomposition mechanism based on the global reaction steps (63–65).

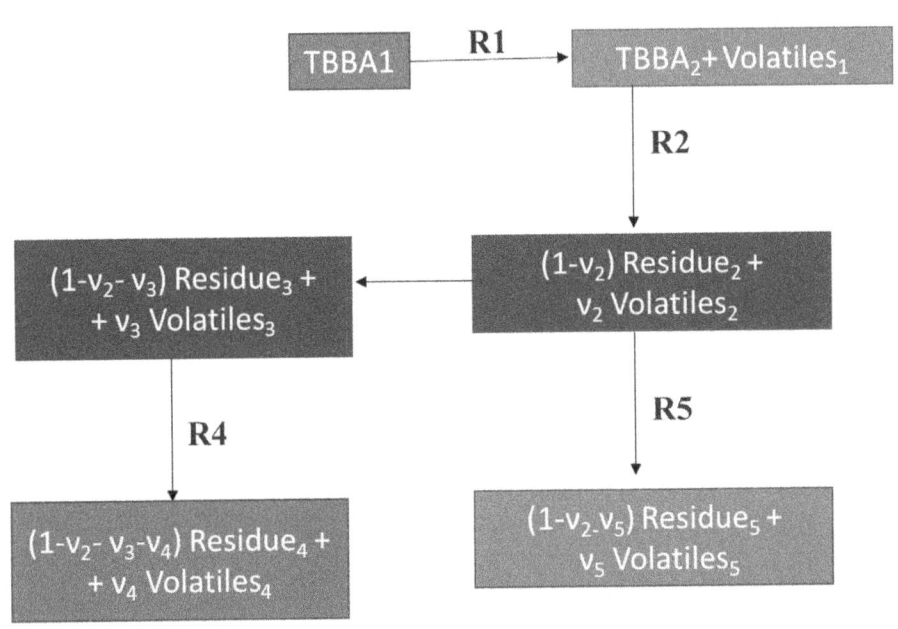

Figure 7. Illustration of the Font et al. (56) model correlation showing how TBBA weight loss degrades and vaporizes through five global reactions.

Pyrolysis studies indicate that HBCD decomposition occurs in one stage, and the introduction of oxygen does not influence the decomposition behavior or the products generated. However, further investigation is needed about the polymeric matrix's effect because C_xH_y radical species may potentialize HBCD degradation before its melting (51). Additionally, iron can promote higher degradation by forming unstable allylic bromides that decompose via radical mechanisms. One alternative to stabilize and retard HBr formation is to include acid scavenging compounds, like calcium octanoate or magnesium oxide, to prevent catalytically active ferric bromide formation. Nevertheless, they do not suppress the radical decompositions and do not affect the rate once the system has turned acidic (62).

In the other BFR compounds, the thermal decomposition is similar and varies according to each additive's chemical structure. The bromine content is generally in the range of 20–200 mg kg^{-1} (51), and its concentration can be achieved at low wt %, which makes this FR class so enjoyable. The Br species are decomposed in gaseous HBr and smaller bromine species at lower temperatures. Vehlow et al. (66) reported that the Br occurs mainly in the gaseous form, mainly at higher contents.

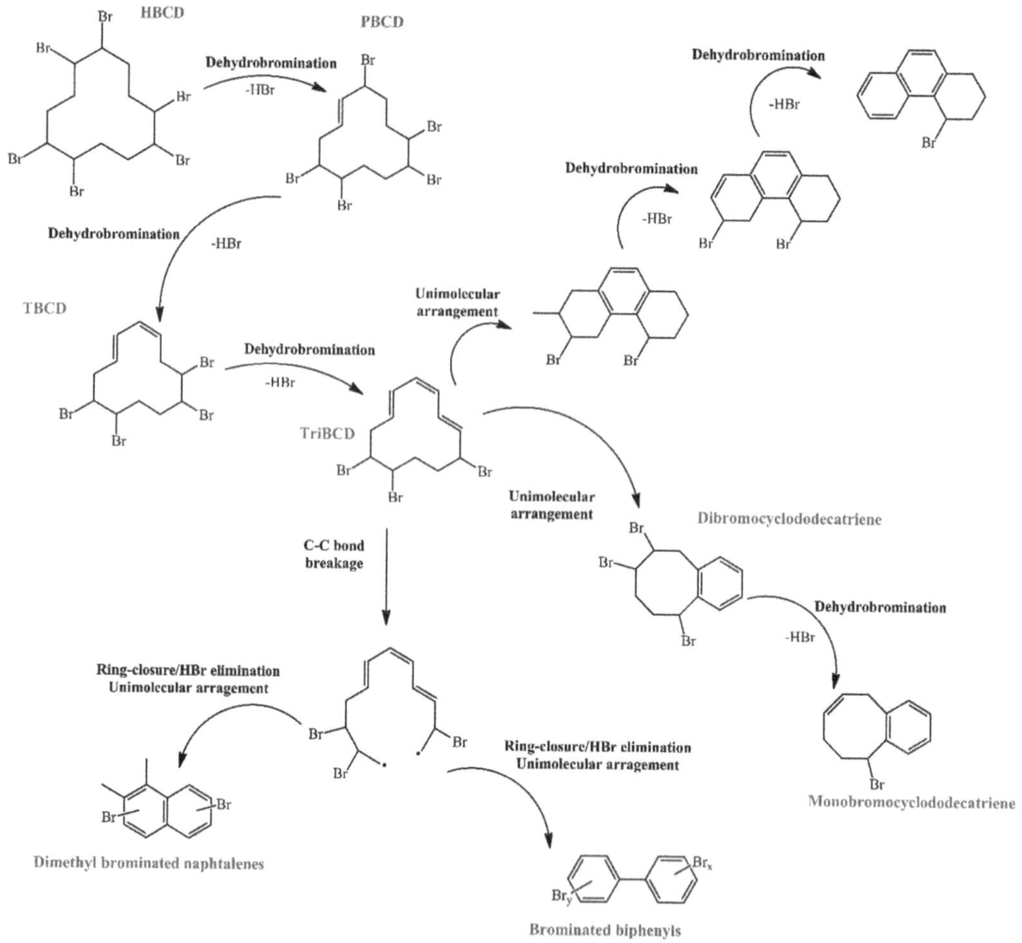

Figure 8. Proposed mechanism for HBCD decomposition.

New BFRs have been used to decrease the toxicity of these compounds, and such BFRs are BTBPE, α-TBECH and β-TBECH, DBDPE or 1,2-bis(pentabromodiphenyl)ethane (DBDPE), and hexachlorocyclopentadienyl–dibromocyclooctane (*48, 67*). However, there are constant efforts to develop new BFRs with low environmental impacts and higher efficiency to guarantee fire protection and human safety.

CFRs

Many CFR compounds have been developed over the years, presenting a wide structure variation from straight-chain paraffin, cycloaliphatic hydrocarbons, aromatic hydrocarbons, and organophosphate compounds. The FRs can present an additive or reactive character, as discussed in the BFR section. While the additive FRs do not present chemical bonds with the products, which allows their migration over time, the reactive FRs present chemical bonds, hindering their release (*68*). This class is used less often for PU flame retardancy than BFRs because of their stronger bonding to carbon, limiting them to interfere favorably in the combustion process. Their main disadvantage is their requirement of a high quantity of PUs, between 18 and 20 wt %, which can negatively affect the foam's properties. However, these compounds are stable until 260 °C, allowing for different applications (*29*).

There is little information about the current annual production of CFR in PUFs with challenges regarding the HFRs' use. According to the biomonitoring program from California's state government, tris(1,3-dichloro-2-propyl) phosphate (TDCPP), an additive FR widely used in PUF, had an annual production in the United States of up to 22.6 million kg between the years 1994 and 2002. However, in 2006, the United States Consumer Product Safety Commission considered TDCPP a probable human carcinogen based on sufficient evidence in animals. Also, Dechlorane Plus (DP) production, a CFR used as an additive, reached 4.5 million kg between 1986 and 2002 in the United States alone. In 2007, traces of DP were found in the air, sediment, fish, eggs, and trees near Niagara Falls (United States) factories (*69*).

In the CFR compounds, flame extinguishment occurs by releasing the HCl generated because of the chlorine radicals' reaction with the reactive radicals •OH and •H (*70*), similar to what was reported in BFRs. CFRs are of three chemical types: aliphatic (mainly chlorinated paraffin (CP)), cycloaliphatic (represented by hexachlorocyclopentadiene Diels–Alder products, such as chlorendic anhydride, with high FR degree, UV stability, and corrosion resistance), and aromatic (chlorinated naphthalenes, tetrachlorophthalic anhydride (low FR degree because of its low chlorine content), and tetrachlorobisphenol A) (*71, 72*), as illustrated in Figure 9.

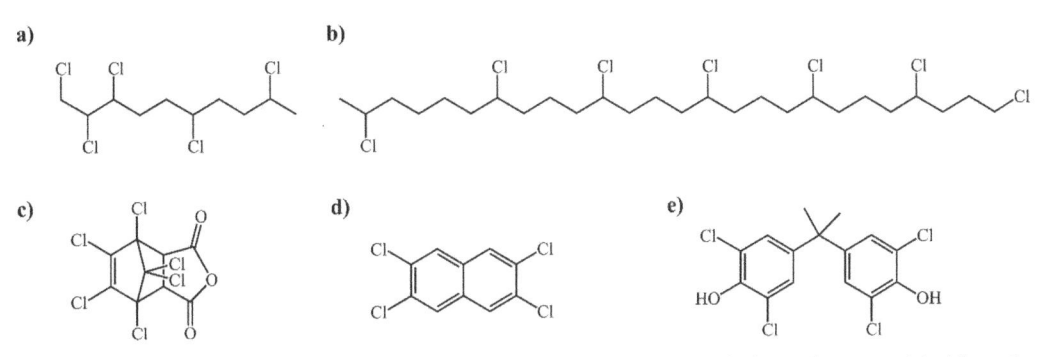

Figure 9. Main CFRs and their chemical structures: (a) short-chain CP; (b) long-chain CP; (c) chlorendic anhydride; (d) chlorinated naphthalenes; (e) tetrachlorobisphenol A.

Within CFRs, there are CP, emerging CFRs (ECFRs), and chlorinated organophosphate esters (ClOPFRs). CP is a complex mixture of synthetic compounds obtained from straight-chained paraffin that is chlorinated and is considered the most common aliphatic CFR. These products present a variety of chlorine content and carbon chain length. Usually, the chlorine content varies between 40 and 70 wt %, while the carbon chain length may be short (C_{10}–C_{13}), medium (C_{14}–C_{17}), or long (C_{18}–C_{30}). CPs are insoluble in water, alcohols, glycerol, and glycols and are soluble in chlorinated solvents, aromatic hydrocarbons, ketones, ester, ethers, and mineral oils. The chain variation impacts their physical state: usually, they are viscous, colorless, or yellowish dense oils, but those with long carbon chain length and high chlorine content are solid materials, acting as polymers' additives. CP is nonflammable and does not quickly evaporate. In the liquid state, CPs are used as plasticizers, FRs in paints, and poly(vinyl chloride) (PVC) formulation, commonly combined with antimony oxide, aiming for a synergic effect (*72, 73*). Camino and Costa (*70*) investigated CP thermal degradation as an FR additive for polymers. The authors reported that HCl was the main volatile degradation product, and the char residue was 35%. These results confirm that CP is an excellent material to act as an FR.

The mechanism observed for CP degradation is related to a dehydrochlorination process in which CP is likely to release HCl and produce double bonds until a charred residue is obtained (*70*).

A similar mechanism is known for PVC, in which the presence of allyl chlorine atoms (i.e., chlorine bonded to an adjacent carbon atom that presents a double bond) are considered weak points that may promote a chain reaction, known as a "zipper reaction" (74). In this process, the weak point is unstable, and radicals formed during thermal degradation may react with chlorine atoms to release HCl. The release of chlorine is responsible for a new double bond in the polymeric chain, conjugated with the previous one. Thus, a new weak point may be formed, and the reaction continues with the release of new HCl molecules. The proposed mechanism is presented in Figure 10.

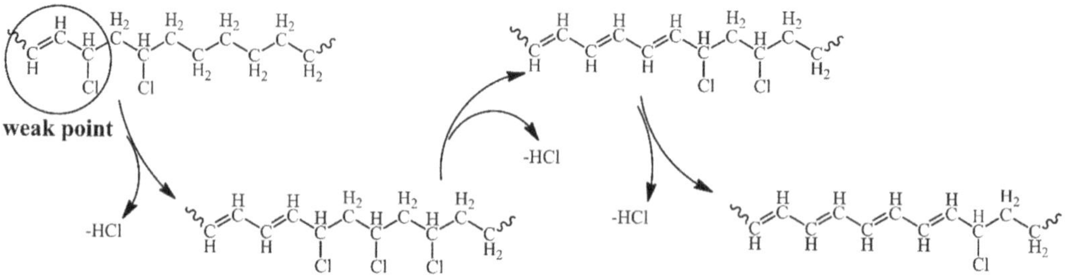

Figure 10. Illustrative process for dehydrochlorination of CP.

Although CP is expected to present only saturated carbon bonds, a first partial release of HCl or the presence of β-chlorine unsaturated structure originated from polymerization may be responsible for the CP chain's weak points (74). Additionally, an intermolecular addition between double bonds could lead to a cross-linked chlorinated residue (70).

The consumption of CPs reached 11.2 million tons in 2015, with an annual production estimated over 1 million tons, for which China is the largest producer and consumer (75). However, the concerns raised about environmental persistence and possible toxic effects of CP by-products led to some of these compounds, especially the short-chain CPs, being included as persistent organic pollutants (POPs) under the Stockholm Convention (76). So, a decrease in its application is expected for the near future. Xin et al. (75) investigated the thermal degradation of highly CP (CP70). They observed the release of large quantities of short and medium–medium chain CP and unsaturated analogs and toxic chlorinated aromatic hydrocarbons, as illustrated in Figure 11. These toxicity observations increase the search for new alternatives to CFRs.

The main ECFR is DP and its structural analogs, such as Dechloranes 602, 603, and 604, and Chlordene Plus. This compound has a high melting temperature (350 °C), its chemical name is dodecachlorododecahydrodimetranodibenzocyclooctene, and its molecular formula is $C_{18}H_{12}Cl_{12}$. DP has a molecular mass of 653.7, is insoluble in water, and is chemically stable. DP manufacturers include Occidental Chemical Corporation (Oxychem, Buffalo, New York, United States) and Anpon Electrochemical Co., Ltd. (Huaian City, Jiangsu, China), with an annual production of ~5000 t (77). Originally, DP's use was in nylons, being later applied in polyolefins and other polymers that require formulation with low smoke content, such as polypropylene, synthetic elastomers, and PUs. The DP action mode involves flame inhibition and ash formation (resulting in less smoke), while brominated additives act mainly to inhibit the flame (78). DP is generally used with antimony oxide because of its synergic effect that enhances the condensed phase because this mixture can form $SbCl_3$ on exposure to heat in a fire environment. Numerous mechanisms were proposed over the years, as presented in Equations 12–14 (79).

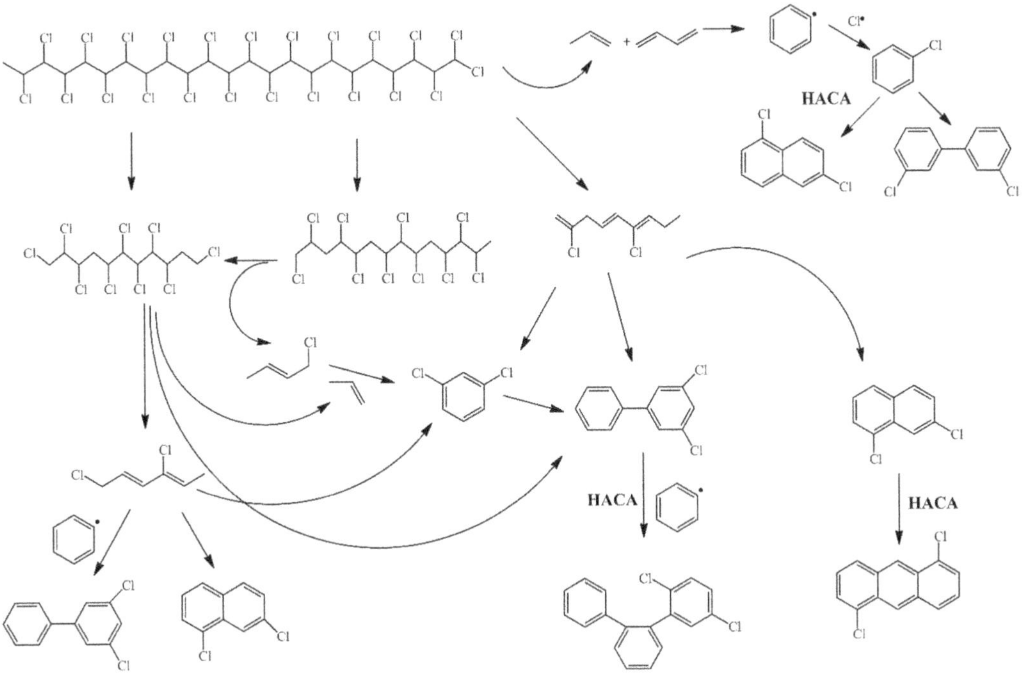

*Where HACA means hydrogen abstraction/acetylene addition

Figure 11. Thermal degradation of long-chain highly CP (CP70), indicating possible main by-products.

$$Sb_2O_{2(s)} + 6HCl_{(g)} \rightarrow 2SbCl_{3(g)} + H_2O \qquad (12)$$

$$Sb_2O_{3(s)} + 2HCl_{(g)} \rightarrow 2SbOCl_{(s)} + H_2O \qquad (13)$$

$$2SbOCl_{(s)} + 2HCl_{(g)} \rightarrow SbCl_{3(s)} + H_2O \qquad (14)$$

Pitts et al. (*80*) proposed the SbOCl thermal disproportionation mechanism, as presented in Equations 15–17.

$$270\text{–}275\ °C: 2SbOCl_{(s)} \rightarrow Sb_4O_5Cl_{2(s)} + SbCl_{3(g)} \qquad (15)$$

$$405\text{–}475\ °C: 11Sb_4O_5Cl_{2(s)} \rightarrow 5Sb_8O_{11}Cl_{2(s)} + 4SbCl_{3(g)} \qquad (16)$$

$$475\text{–}570\ °C: 3Sb_8O_{11}Cl_{2(s)} \rightarrow 11Sb_2O_3(s) + 2SbCl_{3(g)} \qquad (17)$$

The antimony trioxide also acts as a catalyst, improving flame retardancy properties. When HCl is released, this gas reacts with antimony oxides to initiate $SbCl_{3(g)}$ formation, which improves the ashes that extinguish the fire and minimize the toxic HCl released to the environment. However, from the 2000s onward, this compound started to be found in natural environments, with signs of high contamination and toxicity capacity, and is currently considered a highly chlorinated emerging FR from its being detected in several internal matrices and the human body (*81, 82*). For critical human and environmental safety reasons, its use is rarely reported in the literature, especially for PUFs.

The ClOPFRs are a large group of chemicals used as additive FRs with applications in PUFs, PVC, and textiles that emerged in the 1990s, with annual production in the United States over 38,000 tons in 2012 (*83*). ClOPFRs have a central phosphate molecular group, and when used as an FR additive, there is no chemical bond involved in the product (*84*). In a fire event, ClOPFR results in a char layer that suppresses the fire and leads to its extinguishment and decomposition into gases, such as HCl. The main compounds in these class are tris(chloroisopropyl) phosphate (TCPP), TDCPP, tris(2-chloroethyl) phosphate (TCEP), and tetrakis (2-chloroethyl) dichloroisopentyldiphosphate. Among these, TCPP and TDCPP are the most expressive ClOPFRs applied, especially in PUFs. Their chemical structure is presented in Figure 12.

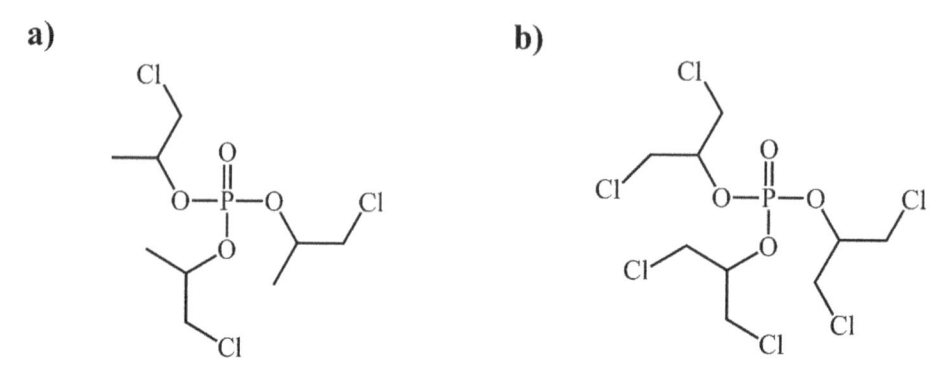

Figure 12. Chemical structure of major ClOPFRs used as PU FRs: (a) TCPP; (b) TDCPP.

Jianzong et al. (*85*) investigated the TCPP FR mechanism on epoxy and reported that a TCPP addition improved the polymer thermal stability and affected the thermo-oxidative process. Samples containing the FR were more carbonized and did not show spectral bands of the epoxy resin. Also, the chlorine atoms were lost during the heating, and the phosphorous atoms were still present in the polymer, probably in the form of phosphoric acid or pyrophosphoric acid, which induces char formation and confirms TCPP efficiency. FR activity is promoted by the phosphorous in the solid phase and the chlorine in the gas phase. As a decomposition product, the emitted gases are carbon monoxide, carbon dioxide, phosphorous compounds, and hydrochloric acid (*86*). The proposed thermal degradation is presented in Figure 13 (*87*).

Matuschek (*87*) evaluated the thermal degradation of different ClOPFRs using mass spectrometry and evaluated the degradation pathway of TDCPP, where the main products are organic phosphoresters and some chlorinated hydrocarbons presented in Figure 14. Additionally, as Levchik and Weil (*88*) observed, the retardation on which TDCPP acts is related to the barrier properties of a phosphorous-containing carbonaceous layer formed beneath the flame, producing a physical obstruction for flame propagation.

The main advantage of these compounds is that the FR action mechanism is responsible for suppressing decomposition gases. However, one concern about these compounds is their water solubility, which has raised concerns after observations of their presence in drinking water (*89*). Thus, some researches have been made to propose its degradation using TiO_2 and H_2O_2/UV systems to ensure its complete degradation and guarantee human safety (*90, 91*).

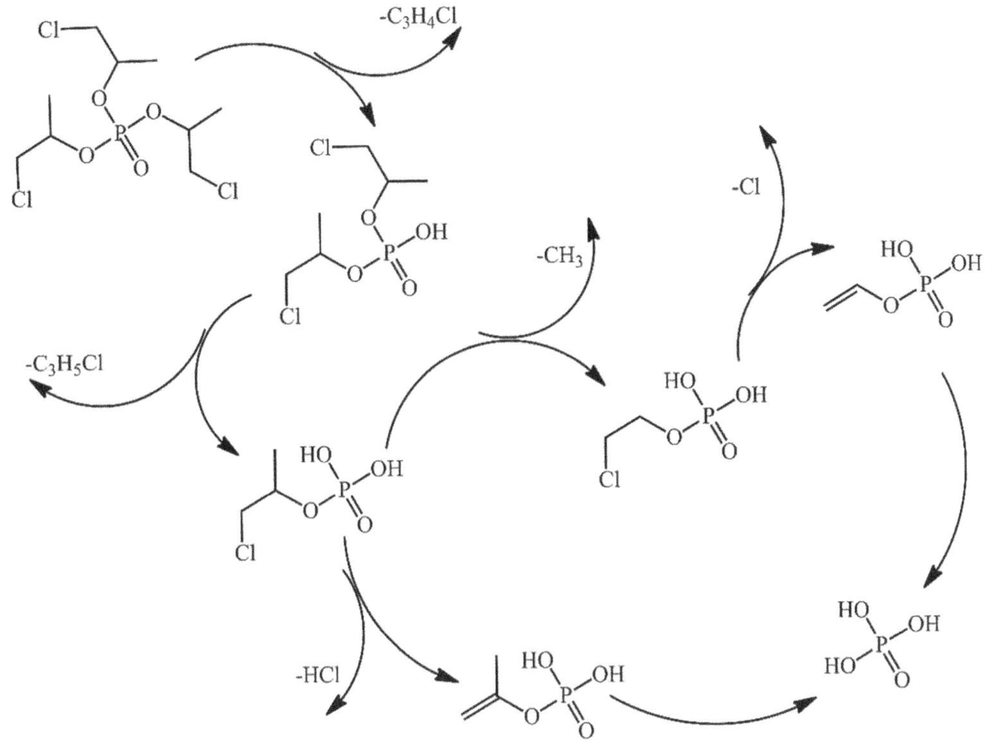

Figure 13. TCPP thermal degradation pathway.

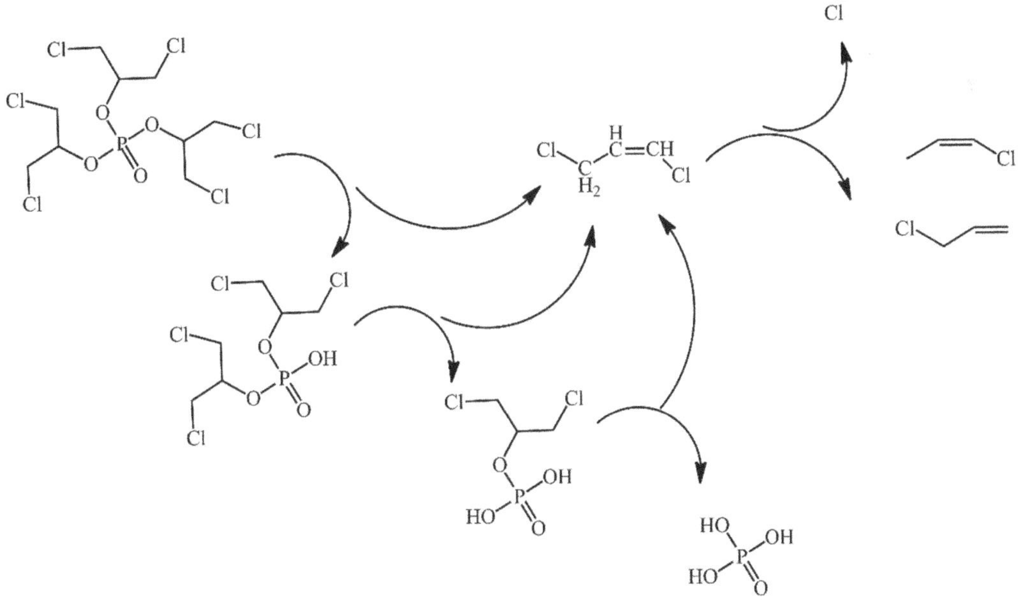

Figure 14. TDCPP thermal degradation pathway.

Main Halogen Compounds and Suppliers

The halogenated compounds market for FR applications in PUs can be classified based on BFRs and CFRs. The diversity and variety of those FRs are vast, having a range of different products in companies spread worldwide. End-use industries (transportation, construction, and electronics) are the ones that most stimulate the FR market, with compound annual growth rate (CAGR) of 3.6% by 2027. Countries such as Japan, Canada, and the European Union have restrictions on (or have even banned) the use of halogenated compounds for ignition delay, promoting challenges in the development of new halogenated materials (brominated and chlorinated) aimed at reducing toxicity or even stimulating the production of nonhalogenated compounds like aluminum hydroxide-based compounds (92).

According to the Global Market Trend report for brominated compounds as FRs, there will be a CAGR of 4% until the year 2023. The most stimulating fact for such retardants' production is the application in electronic and electrical equipment. One factor that hinders the market's growth is the high demand for environmentally friendly FRs and the brominated compounds' toxicity (93). Despite brominated compounds carrying the stigma of possible toxicity as FRs, they are an essential product segment in the United States. Demands and regulations of use can differ from country to country in Europe, Africa, and the Middle East (92). Moreover, some companies continue to manufacture brominated compounds that were once used as FRs for other purposes. For example, the compound BDE-47 (PBDE class) can be used as a test to study genes related to thyroid hormones' homeostasis (94). Big companies such as Akzo Nobel (the Netherlands), Albemarle (United States), ICL (Israel), Jiangsu Yoke Technology (China), LANXESS (Germany), and Velsicol Chemical (United States) are the key players in the BFR market (93). Tables 1 and 2 present other suppliers and compounds available in the market by the year 2021.

Table 1. PBDE Compounds and Their Varieties, Names, Abbreviations, Commercial Names, Chemical Abstract Service (CAS) Numbers, Suppliers, and Respective Nationalities

Compound	Abbreviation	Commercial name	CAS number	Supplier	Country
2,2′,4,4′-Tetrabromodiphenyl ether	BDE 47	BDE 47	5436-43-1	Merck	Germany
Polybrominated diphenyl ethers	PBDE	Chemical FR powder 8010	–	Shanghai King Chemicals Co., Ltd.	China

In addition to the mentioned BFRs for PU, chlorine is also widely used. A report on the chlorinated PU market forecasted a CAGR of 8% by 2024. It was also mentioned that one end use of those chlorinated foams was indeed the ignition retardants, along with impact modification, wire and cable jacketing, tubing, adhesives, among others. Companies such as Showa Denko KK (Japan), the DOW Chemical Company (United States), Hangzhou Keli Chemical Co. Ltd. (China), Novista Group Co. Ltd. (China), and Sundow Polymers Co. (China) are examples of the key players in this market (95). Table 3 shows other chlorinated suppliers.

Table 2. NBFR Compounds and Their Varieties, Names, Abbreviations, Commercial Names, CAS Numbers, Suppliers, and Respective Nationalities

Compound	Abbreviation	Commercial name	CAS number	Supplier	Country
1,1'-(ethane-1,2-diyl) bis(pentabromobenzene)	DBDPE	DPDPE	84852-53-9	TDS Hunan Chemical BV	The Netherlands
Mixture of brominated and nonhalogen compounds: Benzoic acid, 2,3,4,5-tetrabromo-, 2-ethylhexyl ester; Di(2-ethylhexyl) tetrabromophthalate; isopropylated triphenyl phosphate; triphenyl phosphate	TBB, TBPH, ITPP, TPP	Firemaster® 550	Mixture of: 183658-27-7; 26040-51-7; 68937-41-7; and 115-86-6, respectively	Parchem - Fine & Specialty Chemicals	United States
A reactive diol that contains both halogen and phosphorous	—	SAYTEX RB-7001	—	Albemarle Corporation	United States
Mixture: ester of tetrabromophthalic anhydride with diethylene glycol and propylene glycol	—	SAYTEX RB-79	77098-07-8	Albemarle Corporation	United States
A blend of SAYTEX RB-79 FR, polyol, and a liquid phosphate ester	—	SAYTEX RB-7980	—	Albemarle Corporation	United States
Aromatic bromine-containing diol with high bromine content	—	SAYTEX RB-9170	77098-07-8	Albemarle Corporation	United States

Table 3. Chlorinated Compounds for PU Flame Retardancy

Compound	Abbreviation	Commercial name	CAS number	Supplier	Country
Tris(2-chloro-1-methylethyl) phosphate	TCPP	TCPP	13674-84-5; 6145-73-9	Arpadis Benelux NV	Belgium
Tris(2-chloro-1-methylethyl) phosphate)	TCPP	AMIFLAME RFP 1201	13674-84-5; 6145-73-9	AMINO Química	Brazil
Tris(2-dichloro isopropylphosphate)	TDCP	AMIFLAME AMF 1205	—	AMINO Química	Brazil
Tris(2-chloroethyl) phosphate	TCEP	TCEP	115-96-8	WEGO	United States
Tris(1,3-dichloro-2-propyl) phosphate	TDCPP	PESTANAL®	13674-87-8	Merck	Germany

Future Trends and Possible New Technologies

Several factors propagate the trends that dictate FRs' production and technological and scientific development today. One of them is social pressure because one of the ideal living standards is achieved through a fire-safe environment, with materials that do not stimulate eventual flame spread with catastrophic consequences. However, the most decisive factor in trends can be the legislation in each territory worldwide (96), affecting the market and the production, supply, and demand of retardants. Generally, legislation is stimulated by movements from health or environmental organizations, research demonstrating the advantages or disadvantages of their use and the possible consequences for organisms and nature, and cases where the FR power of such materials has been put to the test in accidents.

Because of the adverse effects of toxicity, persistence, and bioaccumulation, conventional HFRs are gradually being replaced by more environmentally friendly and health-safe alternatives. Global conventions, such as the Stockholm Convention (2009), carried out by the United Nations Environment Program, delimited POPs, including some brominated compounds commonly used as FRs in PUs such as PBDEs, BCD, and hexabromobiphenyl. As a relevant event in the international scenario, the convention encouraged brominated compounds' restriction (or even prohibition) in countries participating in the multilateral treaty. Such determinations affect the global economy, where countries that adhere to such measures commit themselves by taking administrative and legal actions against production through adaptation periods (97, 98).

In the European Union, the primary regulation for POPs is the 2019/2021 Regulation, stemming from the Stockholm Convention and the Multilateral Environmental Agreement of the Long-Range Transboundary Air Pollution of POPs. But also taken into account is the 2011/65/ European Union Directive on the "restriction of the use of certain hazardous substances in electrical and electronic equipment" or "RoHS". These provisions delineate the POPs that need prohibition or restriction, HFRs included, especially in their brominated form (98). As for South and North America, the situation regarding BFRs is more complex. Brazil, for example, continues to rely on exemptions for the use of PBDEs, having a timeframe to restrict their use starting in 2030, as it considers points defined by the Stockholm Convention. Still, there is no specific Brazilian legislation regulating the use or commercialization of PBDE or HBCD. Alternatively, the United States, according to Sharkey et al. (98), has not adopted the Stockholm provisions but has internal restrictions that limit the use of BFRs. In the Asia–Pacific region, countries like Japan are looking at alternatives to produce new FRs for insulator materials (98), of which PUs can be an example.

Although finding ideal substitutes for banned HFRs has been a challenge for countries, researchers, institutions, and the business market, collaborative countries encourage others to commit to a sustainability agenda, guarantee a good international reputation, and share goals for trading technology and resources (98). Once the Stockholm Convention (2001) prohibited the circulation of 23 organo-halogen FRs (all BFRs), they fomented challenges to find FRs with higher flame retardancy, lower cost, higher thermal stability, lower melt-dripping, and slower deterioration in mechanical properties (99). As a result of the restriction, ban, or phase-out of the so-called legacy emerging HFRs, it is the responsibility of the scientific community to evaluate and purpose the development of new and safer alternatives, which is the reason for increased interest in alternative studies with a new generation of HFRs (100).

The international concern for sustainability to reduce the ecological footprint is to produce new FR types, especially novel HFRs and halogen-free FRs (45, 99, 101). Because HFRs demonstrated excellent fire performance, a new generation of environmentally friendly BFRs and CFRs with additional benefits has been slowly incorporated into the market. This new generation includes FRs with appropriate softening temperatures such as brominated indan, tris(tribromophenyl)cyanurate, and tris(tribromoneopentyl)phosphate, which provide processing aid effects and better flow properties. In addition to that, developments have also occurred with CFRs offering cost-efficient systems with minimal environmental impact (without diphenyl oxide) and better fire performance (27). The representative emerging HFR compounds are commonly reported in scientific literatures, and they are listed according to their respective groups as follows: (1) NBFRs: DBDPE, BTBPE, TBB, HBB, PBEB, bis(2-ethylhexyl)-3,4,5,6-tetrabromophthalate, PBT, and 2,3-diphenylpropyl-2,4,5-tribromophenylether; (2) ECFRs: DP and its structural analogs, such as Dechlorane 602 (Dec 602, $C_{14}H_4Cl_{12}O$), Dechlorane 603 (Dec 603, $C_{17}H_8Cl_{12}$), Dechlorane 604 (Dec 604, $C_{13}H_4Br_4Cl_6$), and Dechlordene Plus (DP, $C_{18}H_{12}Cl_{12}$); and (3) ClOPFRs: TCPP, TDCPP, TCEP, and tetrakis(2-chloroethyl)dichloroisopentyldiphosphate (*100, 102, 103*).

Figure 15 shows the relation between the publication number and publication year of HFRs, BFRs, CFRs, and halogen-free FRs. The data used in discussing HFR research trends were based on the online database "Web of Science" in March 2021.

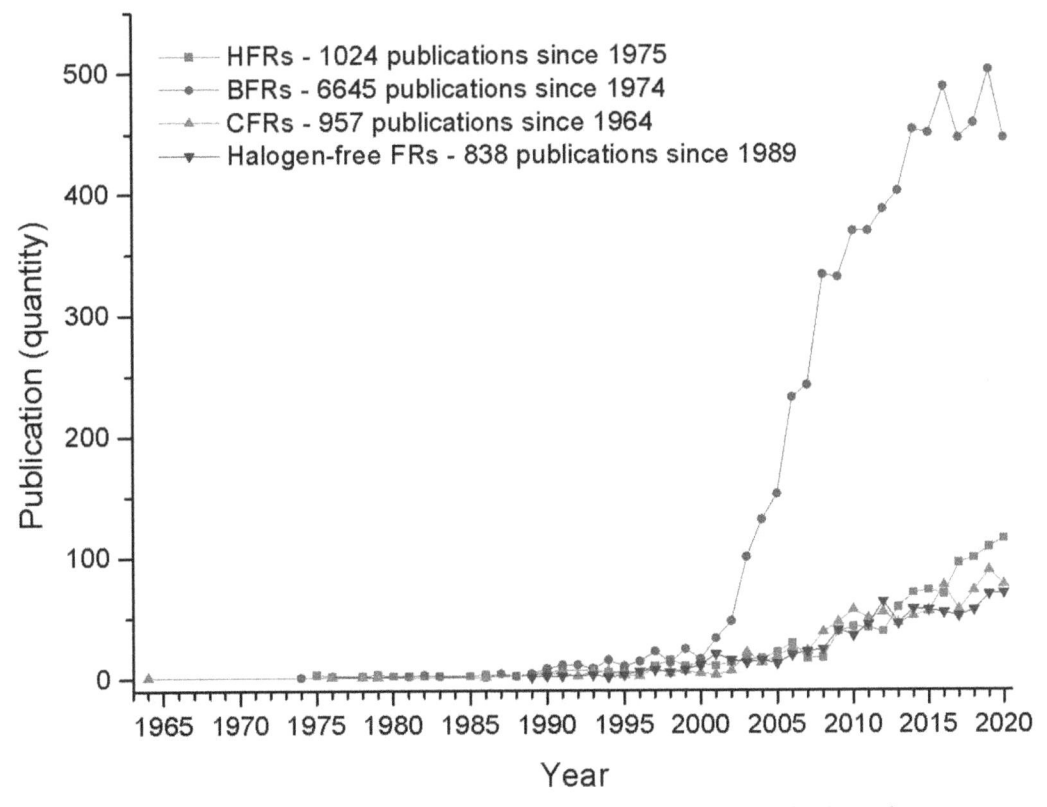

Figure 15. Summary of publications per year on HFRs, BFRs, CFRs, and halogen-free FRs.

The mention of "chlorinated flame retardants" (CFRs) in the database presented older reports than the keyword "brominated flame retardants" (BFRs) (1963 and 1974, respectively); while CFRs accumulated 957 articles as of March 2021, BFRs had 6645. From the 2000s on, growth in the number of publications occurred, mainly for BFRs, reaching 502 publications in 2019 and 446 publications in 2020. This progression regarding BFRs demonstrates their popularity and relevance because of their widespread usage, and the number of scientific reports detailing their environmental occurrence, analysis, and toxicity increased over the years. The same did not occur with CFRs, which reached 77 and 89 publications in 2019 and 2020, respectively, showing that chlorine's more pronounced toxicity influenced such retardants' production. The term "halogenated flame retardants" (HFRs) has been used in publications since 1973 and contains a total of 1024 publications in 47 years (up to March 2021). However, the term "halogen-free flame retardants" was first used in 1989, and in 31 years, it has accumulated 838 publications (up to March 2021), evidencing that it is on the rise.

As mentioned previously, besides the development of NHFRs with reduced toxicity, halogen-free FRs can be encouraged over conventional HFRs for application in flexible PU and PUR foams (22, 104–106). There are various classifications of halogen-free FRs (99), such as phosphorous-based FRs (107), nitrogen-based FRs (108), mineral-based FRs (109), intumescent FRs (21), nanomaterials (8), nanocomposites (110), silicon FRs (111), and biomacromolecular FRs (101).

Despite the demand to replace HFRs, this process takes a long time to find all-new eco-friendly alternative modifications. The FRs' scientific reports are moving toward creating new solutions. Nevertheless, a small number of periodic table atoms have been applied as FRs (99), indicating that there are many possibilities yet to explore. Other possible new technologies could also include catalysts (metal compounds that convert the flammable polymer to graphite under fire conditions) (112), low-melting glasses and ceramics (inorganic materials that are noncombustible) (113), and new vapor phase combustion inhibitors (oxygen scavengers or even free radical inhibitors) (114).

Conclusion

The concerns related to PU flammability are recurrent because they release high energy during burning, besides generating toxic smoke. During the burning of these materials, the following steps are observed: ignitability, flammability, flame spread, heat release, and fire penetration, in which highly reactive radical species of •OH and •H are released and aid in flame propagation. Intrinsically, PUs can be developed as PUR, and the PIRs are responsible for increasing thermal resistance. However, they are still not sufficient to prevent material flammability, and the development and use of FRs have been stimulated to reduce PU flammability. Among them, HFRs have gained much prominence because of their high efficiency. BFRs and CFRs are highlighted because they can release hydrogen halides that react with •OH and •H radicals and form a layer that reduces the spread of fire. FRs can be of additive or reactive type, impacting possible migration effects from the polymer matrix. Among BFRs, TBBA and HBCDs are the most applied materials. They present a complex thermal decomposition that may include HBr elimination, isopropylidene, and C–C and C–O bond scissions with unimolecular arrangements that lead to gaseous bromine forms and char. For the CFRs, the main compounds include straight-chain paraffin, cycloaliphatic hydrocarbons, aromatic hydrocarbons, and organophosphate compounds, for which the thermal degradation releases HCl and chlorinated residues. However, the HCl generated and observation of water solubility for some CFRs has raised some concerns about these additives. Some countries present restrictions for halogenated compounds, but they are still widely applied worldwide, with over 200,000 tons produced annually, especially for BFRs, which present higher efficiency and are continuously investigated. Nevertheless, it has stimulated the development of novel HFRs or halogen-free FRs that present minimal environmental impacts and better fire performance.

References

1. Morgan, A. B. Revisiting Flexible Polyurethane Foam Flammability in Furniture and Bedding in the United States. *Fire Mater.* **2021**, *45*, 68–80.
2. Jamsaz, A.; Goharshadi, E. K. An Environmentally Friendly Superhydrophobic Modified Polyurethane Sponge by Seashell for the Efficient Oil/Water Separation. *Process Saf. Environ. Prot.* **2020**, *139*, 297–304.
3. Rao, W. H.; Xu, H. X.; Xu, Y. J.; Qi, M.; Liao, W.; Xu, S.; Wang, Y. Z. Persistently Flame-Retardant Flexible Polyurethane Foams by a Novel Phosphorus-Containing Polyol. *Chem. Eng. J.* **2018**, *343*, 198–206.
4. Visakh, P. M.; Semkin, A. O.; Rezaev, I. A.; Fateev, A. V. Review on Soft Polyurethane Flame Retardant. *Constr. Build. Mater.* **2019**, *227*, 116673–116679.
5. Liu, L.; Xu, Y.; Li, S.; Xu, M.; He, Y.; Shi, Z.; Li, B. A Novel Strategy for Simultaneously Improving the Fire Safety, Water Resistance and Compatibility of Thermoplastic Polyurethane Composites through the Construction of Biomimetic Hydrophobic Structure of Intumescent Flame Retardant Synergistic System. *Compos. Part B Eng.* **2019**, *176*, 107218–107231.
6. Mouritz, A. P.; Gibson, A. G. *Fire Properties of Polymer Composite Materials*, Vol. 143; Springer Netherlands: Dordrecht, 2006.
7. Schartel, B.; Hull, T. R. Development of Fire-Retarded Materials—Interpretation of Cone Calorimeter Data. *Fire Mater.* **2007**, *31*, 327–354.
8. He, W.; Song, P.; Yu, B.; Fang, Z.; Wang, H. Flame Retardant Polymeric Nanocomposites through the Combination of Nanomaterials and Conventional Flame Retardants. *Prog. Mater. Sci.* **2020**, *114*, 100687–100735.
9. Boyer, N. E.; Vajda, A. E. Fireproofing of Polymers with Derivatives of Phosphines and with Halogen–Phosphorus Compounds. *Polym. Eng. Sci.* **1964**, *4*, 45–55.
10. Hilado, C. J.; Burgess, P. E.; Proops, W. R. Bromine, Chlorine, and Phosphorus Compounds as Flame Retardants in Rigid Urethane Foam. *J. Cell. Plast.* **1968**, *4*, 67–78.
11. Pape, P. G.; Sanger, J. E.; Nametz, R. C. Tetrabromophthalic Anhydride in Flame-Retardant Urethane Foams. *J. Cell. Plast.* **1968**, *4*, 438–442.
12. Hilado, C. J. Flame Retardant Urethane Foams. *J. Cell. Plast.* **1968**, *4*, 339–344.
13. Wang, S. X.; Zhao, H. B.; Rao, W. H.; Huang, S. C.; Wang, T.; Liao, W.; Wang, Y. Z. Inherently Flame-Retardant Rigid Polyurethane Foams with Excellent Thermal Insulation and Mechanical Properties. *Polymer (Guildf).* **2018**, *153*, 616–625.
14. Zhang, G.; Lin, X.; Zhang, Q.; Jiang, K.; Chen, W.; Han, D. Anti-Flammability, Mechanical and Thermal Properties of Bio-Based Rigid Polyurethane Foams with the Addition of Flame Retardants. *RSC Adv.* **2020**, *10*, 32156–32161.
15. Papa, A. J.; Proops, W. R. Influence of Structural Effects of Halogen and Phosphorus Polyol Mixtures on Flame Retardancy of Flexible Polyurethane Foams. *J. Appl. Polym. Sci.* **1972**, *16*, 2361–2373.
16. Bhoyate, S.; Ionescu, M.; Radojcic, D.; Kahol, P. K.; Chen, J.; Mishra, S. R.; Gupta, R. K. Highly Flame-Retardant Bio-Based Polyurethanes Using Novel Reactive Polyols. *J. Appl. Polym. Sci.* **2018**, *135*, 46027–46038.
17. *Flame Retardant Market Size, Share & Trends Analysis Report By Product*, 2020. https://www.reportlinker.com/p05930639/Flame-Retardant-Market-Size-Share-Trends-Analysis-Report-

By-Product-By-Application-By-End-Use-By-Region-And-Segment-Forecasts.html?utm_source=GNW (accessed July 2021).

18. Shishlov, O.; Dozhdikov, S.; Glukhikh, V.; Eltsov, O.; Kraus, E.; Orf, L.; Heilig, M.; Stoyanov, O. Synthesis of Bromo-Cardanol Novolac Resins and Evaluation of Their Effectiveness as Flame Retardants and Adhesives for Particleboard. *J. Appl. Polym. Sci.* **2017**, *134*, 45322–45331.

19. Lin, Z.; Zhao, Q.; Fan, R.; Yuan, X.; Tian, F. Flame Retardancy and Thermal Properties of Rigid Polyurethane Foam Conjugated with a Phosphorus–Nitrogen Halogen-Free Intumescent Flame Retardant. *J. Fire Sci.* **2020**, *38*, 235–252.

20. Zhang, Z.; Li, D.; Xu, M.; Li, B. Synthesis of a Novel Phosphorus and Nitrogen-Containing Flame Retardant and Its Application in Rigid Polyurethane Foam with Expandable Graphite. *Polym. Degrad. Stab.* **2020**, *173*, 109077–109088.

21. Cui, M.; Li, J.; Qin, D.; Sun, J.; Chen, Y.; Xiang, J.; Yan, J.; Fan, H. Intumescent Flame Retardant Behavior of Triazine Group and Ammonium Polyphosphate in Waterborne Polyurethane. *Polym. Degrad. Stab.* **2021**, *183*, 109439–109450.

22. Akdogan, E.; Erdem, M.; Ureyen, M. E.; Kaya, M. Rigid Polyurethane Foams with Halogen-Free Flame Retardants: Thermal Insulation, Mechanical, and Flame Retardant Properties. *J. Appl. Polym. Sci.* **2020**, *137*, 47611–47625.

23. Luo, Y.; Miao, Z.; Sun, T.; Zou, H.; Liang, M.; Zhou, S.; Chen, Y. Preparation and Mechanism Study of Intrinsic Hard Segment Flame-Retardant Polyurethane Foam. *J. Appl. Polym. Sci.* **2021**, *138*, 49920–49936.

24. Yu, Z. R.; Mao, M.; Li, S. N.; Xia, Q. Q.; Cao, C. F.; Zhao, L.; Zhang, G. D.; Zheng, Z. J.; Gao, J. F.; Tang, L. C. Facile and Green Synthesis of Mechanically Flexible and Flame-Retardant Clay/Graphene Oxide Nanoribbon Interconnected Networks for Fire Safety and Prevention. *Chem. Eng. J.* **2021**, *405*, 126620–126631.

25. Wan, L.; Deng, C.; Chen, H.; Zhao, Z.; Huang, S.; Wei, W. Flame-Retarded Thermoplastic Polyurethane Elastomer: From Organic Materials to Nanocomposites and New Prospects. *Chem. Eng. J.* **2021**, *417*, 129314–129328.

26. Papa, A. J. Reactive Flame Retardants for Polyurethane Foams. *Ind. Eng. Chem. Prod. Res. Dev.* **1970**, *9*, 478–496.

27. Georlette, P. Applications of Halogen Flame Retardants. In *Fire Retardant Materials*; Horrocks, R.; Price, D., Eds.; Woodhead Publishing Limited: England; CRC Press: Boca Raton, FL, 2001; pp 264–292.

28. Liu, X.; Hao, J.; Gaan, S. Recent Studies on the Decomposition and Strategies of Smoke and Toxicity Suppression for Polyurethane Based Materials. *RSC Adv.* **2016**, *6*, 74742–74756.

29. Singh, H.; Jain, A. K. Ignition, Combustion, Toxicity, and Fire Retardancy of Polyurethane Foams: A Comprehensive Review. *J. Appl. Polym. Sci.* **2009**, *111*, 1115–1143.

30. Silva, N. G. S.; Cortat, L. I. C. O.; Orlando, D.; Mulinari, D. R. Evaluation of Rubber Powder Waste as Reinforcement of the Polyurethane Derived from Castor Oil. *Waste Manag.* **2020**, *116*, 131–139.

31. Kirpluks, M.; Cabulis, U.; Avots, A. Flammability of Bio-Based Rigid Polyurethane Foam as Sustainable Thermal Insulation Material. In *Insulation Materials in Context of Sustainability*; Almusaed, A.; Almssad, A., Eds.; Intech: Croatia, 2016; pp 87–111.

32. Stirna, U.; Cabulis, U.; Beverte, I. Water-Blown Polyisocyanurate Foams from Vegetable Oil Polyols. *J. Cell. Plast.* **2008**, *44*, 139–160.
33. Kurańska, M.; Cabulis, U.; Auguścik, M.; Prociak, A.; Ryszkowska, J.; Kirpluks, M. Bio-Based Polyurethane–Polyisocyanurate Composites with an Intumescent Flame Retardant. *Polym. Degrad. Stab.* **2016**, *127*, 11–19.
34. Modesti, M.; Lorenzetti, A. Improvement on Fire Behaviour of Water Blown PIR–PUR Foams: Use of an Halogen-Free Flame Retardant. *Eur. Polym. J.* **2003**, *39*, 263–268.
35. Hejna, A.; Kosmela, P.; Klein, M.; Gosz, K.; Formela, K.; Haponiuk, J.; Piszczyk, Ł. Rheological Properties, Oxidative and Thermal Stability, and Potential Application of Biopolyols Prepared via Two-Step Process from Crude Glycerol. *Polym. Degrad. Stab.* **2018**, *152*, 29–42.
36. Schiavoni, S.; D'Alessandro, F.; Bianchi, F.; Asdrubali, F. Insulation Materials for the Building Sector: A Review and Comparative Analysis. *Renew. Sustain. Energy Rev.* **2016**, *62*, 988–1011.
37. Ashida, K. *Polyurethane and Related Foams: Chemistry and Technology*; CRC Press: Boca Raton, FL, 2007.
38. Song, D.; Gupta, R. K. The Use of Thermosets in the Building and Construction Industry. In *Thermosets: Structure, Properties and Applications*; Guo, Q., Ed.; Woodhead Publishing Limited: England, 2012; pp 165–188.
39. Günther, M.; Lorenzetti, A.; Schartel, B. Fire Phenomena of Rigid Polyurethane Foams. *Polymers (Basel)* **2018**, *10*, 1166–1188.
40. Bourbigot, S.; Duquesne, S. Fire Retardant Polymers: Recent Developments and Opportunities. *J. Mater. Chem.* **2007**, *17*, 2283–2300.
41. Pestana, C. R.; Borges, K. B.; Da Fonseca, P.; De Oliveira, D. P. Risco Ambiental Da Aplicação de Éteres de Difenilas Polibromadas Como Retardantes de Chama. *Rev. Bras. Toxicol.* **2008**, *21*, 41–48.
42. Green, J. Mechanisms for Flame Retardancy and Smoke Suppression—A Review. *J. Fire Sci.* **1996**, *14*, 426–442.
43. Nnorom, I. C.; Osibanjo, O. Sound Management of Brominated Flame Retarded (BFR) Plastics from Electronic Wastes: State of the Art and Options in Nigeria. *Resour. Conserv. Recycl.* **2008**, *52*, 1362–1372.
44. Vojta, Š.; Bečanová, J.; Melymuk, L.; Komprdová, K.; Kohoutek, J.; Kukučka, P.; Klánová, J. Screening for Halogenated Flame Retardants in European Consumer Products, Building Materials and Wastes. *Chemosphere* **2017**, *168*, 457–466.
45. Xiong, P.; Yan, X.; Zhu, Q.; Qu, G.; Shi, J.; Liao, C.; Jiang, G. A Review of Environmental Occurrence, Fate, and Toxicity of Novel Brominated Flame Retardants. *Environ. Sci. Technol.* **2019**, *53*, 13551–13569.
46. Stubbings, W. A.; Harrad, S. Extent and Mechanisms of Brominated Flame Retardant Emissions from Waste Soft Furnishings and Fabrics: A Critical Review. *Environ. Int.* **2014**, *71*, 164–175.
47. Zhang, M.; Buekens, A.; Li, X. Brominated Flame Retardants and the Formation of Dioxins and Furans in Fires and Combustion. *J. Hazard. Mater.* **2016**, *304*, 26–39.

48. Kodavanti, P. R. S.; Stoker, T. E.; Fenton, S. E. Brominated Flame Retardants. In *Reproductive and Developmental Toxicology*, 2nd ed.; Gupta, R. C., Ed.; Academic Press: Cambridge, MA, 2017; pp 681–710.
49. Dixon-Lewis, G.; Marshall, P.; Ruscic, B.; Burcat, A.; Goos, E.; Cuoci, A.; Frassoldati, A.; Faravelli, T.; Glarborg, P. Inhibition of Hydrogen Oxidation by HBr and Br 2. Combust. *Flame* **2012**, *159*, 528–540.
50. Babushok, V.; Tsang, W. Inhibitor Rankings for Alkane Combustion. *Combust. Flame* **2000**, *123*, 488–506.
51. Altarawneh, M.; Saeed, A.; Al-Harahsheh, M.; Dlugogorski, B. Z. Thermal Decomposition of Brominated Flame Retardants (BFRs): Products and Mechanisms. *Prog. Energy Combust. Sci.* **2019**, *70*, 212–259.
52. Luda, M. P.; Balabanovich, A. I.; Hornung, A.; Camino, G. Thermal Degradation of a Brominated Bisphenol a Derivative. *Polym. Adv. Technol.* **2003**, *14*, 741–748.
53. Barontini, F.; Marsanich, K.; Petarca, L.; Cozzani, V. The Thermal Degradation Process of Tetrabromobisphenol A. *Ind. Eng. Chem. Res.* **2004**, *43*, 1952–1961.
54. Altarawneh, M.; Dlugogorski, B. Z. Mechanism of Thermal Decomposition of Tetrabromobisphenol A (TBBA). *J. Phys. Chem. A* **2014**, *118*, 9338–9346.
55. Marongiu, A.; Bozzano, G.; Dente, M.; Ranzi, E.; Faravelli, T. Detailed Kinetic Modeling of Pyrolysis of Tetrabromobisphenol A. *J. Anal. Appl. Pyrolysis* **2007**, *80*, 325–345.
56. Font, R.; Moltó, J.; Ortuño, N. Kinetics of Tetrabromobisphenol A Pyrolysis. Comparison between Correlation and Mechanistic Models. *J. Anal. Appl. Pyrolysis* **2012**, *94*, 53–62.
57. Terakado, O.; Ohhashi, R.; Hirasawa, M. Bromine Fixation by Metal Oxide in Pyrolysis of Printed Circuit Board Containing Brominated Flame Retardant. *J. Anal. Appl. Pyrolysis* **2013**, *103*, 216–221.
58. Al-Harahsheh, M.; Altarawneh, M.; Aljarrah, M.; Rummanah, F.; Abdel-Latif, K. Bromine Fixing Ability of Electric Arc Furnace Dust during Thermal Degradation of Tetrabromobisphenol: Experimental and Thermodynamic Analysis Study. *J. Anal. Appl. Pyrolysis* **2018**, *134*, 503–509.
59. Kumagai, S.; Grause, G.; Kameda, T.; Yoshioka, T. Thermal Decomposition of Tetrabromobisphenol-A Containing Printed Circuit Boards in the Presence of Calcium Hydroxide. *J. Mater. Cycles Waste Manag.* **2017**, *19*, 282–293.
60. Grabda, M.; Oleszek-Kudlak, S.; Shibata, E.; Nakamura, T. Vaporization of Zinc during Thermal Treatment of ZnO with Tetrabromobisphenol A (TBBPA). *J. Hazard. Mater.* **2011**, *187*, 473–479.
61. Becher, G. The Stereochemistry of 1,2,5,6,9,10-Hexabromocyclododecane and Its Graphic Representation. *Chemosphere* **2005**, *58*, 989–991.
62. Larsen, E. R.; Ecker, E. L. Thermal Stability of Fire Retardants: V. Decomposition of Haloalkyl Phosphates under Polyurethane Processing Conditions. *J. Fire Sci.* **1988**, *6*, 363–379.
63. Peled, M.; Dec, R. S.; Sondack, D.; Compounds, B.; Sheva, B. Thermal Rearrangement of Hexabromo-Cyclododecane (HBCD). *Ind. Chem. Libr.* **1995**, *7*, 92–99.
64. Larsen, E. R.; Ecker, E. L. Thermal Stability of Fire Retardants: II. Pentabromochlorocyclohexane. *J. Fire Sci.* **1987**, *5*, 215–227.

65. Barontini, F.; Cozzani, V.; Petarca, L. Thermal Stability and Decomposition Products of Hexabromocyclododecane. *Ind. Eng. Chem. Res.* **2001**, *40*, 3270–3280.
66. Vehlow, J.; Bergfeldt, B.; Jay, K.; Seifert, H.; Wanke, T.; Mark, F. E. Thermal Treatment of Electrical and Electronic Waste Plastics. *Waste Manag. Res.* **2000**, *18*, 131–140.
67. Covaci, A.; Harrad, S.; Abdallah, M. A. E.; Ali, N.; Law, R. J.; Herzke, D.; de Wit, C. A. Novel Brominated Flame Retardants: A Review of Their Analysis, Environmental Fate and Behaviour. *Environ. Int.* **2011**, *37*, 532–556.
68. Weil, E. D.; Levchik, S. V. *Flame Retardants for Plastics and Textiles: Practical Applications*; Hanser Publications: Ohio, 2015.
69. *California Environmental Contaminant Biomonitoring. Brominated and Chlorinated Organic Chemical Compounds Used as Flame Retardants*, 2008. https://biomonitoring.ca.gov/downloads/brominated-and-chlorinated-organic-chemical-compounds-used-flame-retardants-0 (accessed June 2021).
70. Camino, G.; Costa, L. Thermal Degradation of a Highly Chlorinated Paraffin Used as a Fire Retardant Additive for Polymers. *Polym. Degrad. Stab.* **1980**, *2*, 23–33.
71. Liepins, R.; Pearcet, E. M. Chemistry and Toxicity of Flame Retardants for Plastics. *Environ. Health Perspect.* **1976**, *17*, 55–63.
72. Mack, A. G. Flame Retardants. In *Kirk-Othmer Encyclopedia of Chemical Technology*; Kirk, R. E.; Othmer, D. F., Eds.; John Wiley & Sons: New Jersey, 2004; pp 454–483.
73. Fiedler, H. Short-Chain Chlorinated Paraffins: Production, Use and International Regulations. In *The Handbook of Environmental Chemistry*; de Boer, J., Ed.; Springer Berlin Heidelberg: Berlin, 2010; pp 1–40.
74. Baum, B.; Wartman, L. H. Structure and Mechanism of Dehydrochlorination of Polyvinyl Chloride. *J. Polym. Sci.* **1958**, *28*, 537–546.
75. Xin, S.; Gao, W.; Wang, Y.; Jiang, G. Identification of the Released and Transformed Products during the Thermal Decomposition of a Highly Chlorinated Paraffin. *Environ. Sci. Technol.* **2018**, *52*, 10153–10162.
76. Yuan, S.; Wang, M.; Lv, B.; Wang, J. Transformation Pathways of Chlorinated Paraffins Relevant for Remediation: A Mini-Review. *Environ. Sci. Pollut. Res.* **2021**, *28*, 9020–9028.
77. Xian, Q.; Siddique, S.; Li, T.; Feng, Y. lai; Takser, L.; Zhu, J. Sources and Environmental Behavior of Dechlorane Plus—A Review. *Environ. Int.* **2011**, *37*, 1273–1284.
78. Weil, E. D.; Levchik, S. V. Flame Retardants in Commercial Use or Development for Polyolefins. *J. Fire Sci.* **2008**, *26*, 5–43.
79. Costa, L.; Goberti, P.; Paganetto, G.; Camino, G.; Sgarzi, P. Thermal Behaviour of Chlorine-Antimony Fire-Retardant Systems. *Polym. Degrad. Stab.* **1990**, *30*, 13–28.
80. Kuryla, W. C.; Papa, A. J. *Flame Retardancy of Polymeric Materials*; M. Dekker: New York, 1973.
81. Sun, J.; Xu, Y.; Zhou, H.; Zhang, A.; Qi, H. Levels, Occurrence and Human Exposure to Novel Brominated Flame Retardants (NBFRs) and Dechlorane Plus (DP) in Dust from Different Indoor Environments in Hangzhou, China. *Sci. Total Environ.* **2018**, *631–632*, 1212–1220.
82. Hansen, K. M.; Fauser, P.; Vorkamp, K.; Christensen, J. H. Global Emissions of Dechlorane Plus. *Sci. Total Environ.* **2020**, *742*, 140677–140685.
83. Schreder, E. D.; Uding, N.; La Guardia, M. J. Inhalation a Significant Exposure Route for Chlorinated Organophosphate Flame Retardants. *Chemosphere* **2016**, *150*, 499–504.

84. Greaves, A. K.; Letcher, R. J. A Review of Organophosphate Esters in the Environment from Biological Effects to Distribution and Fate. *Bull. Environ. Contam. Toxicol.* **2017**, *98*, 2–7.
85. Jianzong, L.; Shiyuan, C.; Xiaoming, X. Mechanism of Flame-Retardant Action of Tris(2,3-dichloropropyl) Phosphate on Epoxy Resin. *J. Appl. Polym. Sci.* **1990**, *40*, 417–426.
86. van der Veen, I.; de Boer, J. Phosphorus Flame Retardants: Properties, Production, Environmental Occurrence, Toxicity and Analysis. *Chemosphere* **2012**, *88*, 1119–1153.
87. Matuschek, G. Thermal Degradation of Different Fire Retardant Polyurethane Foams. *Thermochim. Acta* **1995**, *263*, 59–71.
88. Levchik, S. V.; Weil, E. D. Thermal Decomposition, Combustion and Fire-Retardancy of Polyurethanes—A Review of the Recent Literature. *Polym. Int.* **2004**, *53*, 1585–1610.
89. Yu, X.; Yin, H.; Peng, H.; Lu, G.; Dang, Z. Oxidation Degradation of Tris-(2-Chloroisopropyl) Phosphate by Ultraviolet Driven Sulfate Radical: Mechanisms and Toxicology Assessment of Degradation Intermediates Using Flow Cytometry Analyses. *Sci. Total Environ.* **2019**, *687*, 732–740.
90. Chen, Y.; Ye, J.; Chen, Y.; Hu, H.; Zhang, H.; Ou, H. Degradation Kinetics, Mechanism and Toxicology of Tris(2-Chloroethyl) Phosphate with 185 nm Vacuum Ultraviolet. *Chem. Eng. J.* **2019**, *356*, 98–106.
91. Son, Y.; Lee, Y. M.; Zoh, K. D. Kinetics and Degradation Mechanism of Tris(1-Chloro-2-Propyl) Phosphate in the UV/H2O2 Reaction. *Chemosphere* **2020**, *260*, 127461–127468.
92. Research and Markets. *Flame Retardant Market Size, Share & Trends Analysis Report By Product (Halogenated, Non-Halogenated), By Application, By End Use, By Region, and Segment Forecasts, 2020–2027*; Research and Markets, 2020.
93. Research and Markets. *Global Brominated Flame Retardants Market 2019–2023*; Research and Markets, 2018.
94. Kang, H. M.; Lee, Y. H.; Kim, B. M.; Kim, I. C.; Jeong, C. B.; Lee, J. S. Adverse Effects of BDE-47 on in vivo Developmental Parameters, Thyroid Hormones, and Expression of Hypothalamus-Pituitary-Thyroid (HPT) Axis Genes in Larvae of the Self-Fertilizing Fish Kryptolebias Marmoratus. *Chemosphere* **2017**, *176*, 39–46.
95. Research and Markets. *Chlorinated Polyethylene Market Report: Trends, Forecast and Competitive Analysis*; Research and Markets, 2021.
96. Green, J.; Israel, S. C.; Krasny, J. F.; Lawson, D. F.; Tewarson, A. *Flame-Retardant Polymeric Materials*; Plenum Press: New York, 1975.
97. Pieroni, M. C.; Leonel, J.; Fillmann, G. Retardantes de Chama Bromados: Uma Revisão. *Quim. Nova* **2016**, *40*, 317–326.
98. Sharkey, M.; Harrad, S.; Abou-Elwafa Abdallah, M.; Drage, D. S.; Berresheim, H. Phasing-Out of Legacy Brominated Flame Retardants: The UNEP Stockholm Convention and Other Legislative Action Worldwide. *Environ. Int.* **2020**, *144*, 106041–106053.
99. Choudhury, A. K. R. Advances in Halogen-Free Flame Retardants. *Latest Trends Text. Fash. Des.* **2018**, *1*, 70–74.
100. Ekpe, O. D.; Choo, G.; Barceló, D.; Oh, J. E. Introduction of Emerging Halogenated Flame Retardants in the Environment. In *Comprehensive Analytical Chemistry*; Barceló, D.; Milacic, R.; Scancar, S.; Goenaga-Infante, H.; Vidmar, J., Eds.; Elsevier B.V.: Amsterdam, 2020; pp 1–39.
101. Costes, L.; Laoutid, F.; Brohez, S.; Dubois, P. Bio-Based Flame Retardants: When Nature Meets Fire Protection. *Mater. Sci. Eng. R Reports* **2017**, *117*, 1–25.

102. Shen, L.; Reiner, E. J.; MacPherson, K. A.; Kolic, T. M.; Helm, P. A.; Richman, L. A.; Marvin, C. H.; Burniston, D. A.; Hill, B.; Brindle, I. D.; McCrindle, R.; Chittim, B. G. Dechloranes 602, 603, 604, Dechlorane Plus, and Chlordene Plus, a Newly Detected Analogue, in Tributary Sediments of the Laurentian Great Lakes. *Environ. Sci. Technol.* **2011**, *45*, 693–699.

103. Von Eyken, A.; Pijuan, L.; Martí, R.; Blanco, M. J.; Díaz-Ferrero, J. Determination of Dechlorane Plus and Related Compounds (Dechlorane 602, 603 and 604) in Fish and Vegetable Oils. *Chemosphere* **2016**, *144*, 1256–1263.

104. Modesti, M.; Lorenzetti, A. Halogen-Free Flame Retardants for Polymeric Foams. *Polym. Degrad. Stab.* **2002**, *78*, 167–173.

105. Zhao, Q.; Chen, C.; Fan, R.; Yuan, Y.; Xing, Y.; Ma, X. Halogen-Free Flame-Retardant Rigid Polyurethane Foam with a Nitrogen-Phosphorus Flame Retardant. *J. Fire Sci.* **2017**, *35*, 99–117.

106. Wang, C.-Q.; Lv, H.-N.; Sun, J.; Cai, Z.-S. Flame Retardant and Thermal Decomposition Properties of Flexible Polyurethane Foams Filled With Several Halogen-Free Flame Retardants. *Polym. Eng. Sci.* **2014**, *54*, 2497–2705.

107. Arora, S.; Mestry, S.; Naik, D.; Mhaske, S. T. O-Phenylenediamine-Derived Phosphorus-Based Cyclic Flame Retardant for Epoxy and Polyurethane Systems. *Polym. Bull.* **2020**, *77*, 3185–3205.

108. Wang, S.; Du, X.; Jiang, Y.; Xu, J.; Zhou, M.; Wang, H.; Cheng, X.; Du, Z. Synergetic Enhancement of Mechanical and Fire-Resistance Performance of Waterborne Polyurethane by Introducing Two Kinds of Phosphorus–Nitrogen Flame Retardant. *J. Colloid Interface Sci.* **2019**, *537*, 197–205.

109. Guler, T.; Tayfun, U.; Bayramli, E.; Dogan, M. Effect of Expandable Graphite on Flame Retardant, Thermal and Mechanical Properties of Thermoplastic Polyurethane Composites Filled with Huntite&hydromagnesite Mineral. *Thermochim. Acta* **2017**, *647*, 70–80.

110. Ababsa, H. S.; Safidine, Z.; Mekki, A.; Grohens, Y.; Ouadah, A.; Chabane, H. Fire Behavior of Flame-Retardant Polyurethane Semi-Rigid Foam in Presence of Nickel (II) Oxide and Graphene Nanoplatelets Additives. *J. Polym. Res.* **2021**, *28*, 87–101.

111. Chen, C. H.; Chiang, C. L. Preparation and Characteristics of an Environmentally Friendly Hyperbranched Flame-Retardant Polyurethane Hybrid Containing Nitrogen, Phosphorus, and Silicon. *Polymers (Basel)* **2019**, *11*, 720–735.

112. Wu, Y.; Zhou, X.; Xing, Z.; Ma, J. Metal Compounds as Catalysts in the Intumescent Flame Retardant System for Polyethylene Terephthalate Fabrics. *Text. Res. J.* **2019**, *89*, 2983–2997.

113. Wu, G. M.; Schartel, B.; Yu, D.; Kleemeier, M.; Hartwig, A. Synergistic Fire Retardancy in Layered-Silicate Nanocomposite Combined with Low-Melting Phenylsiloxane Glass. *J. Fire Sci.* **2012**, *30*, 69–87.

114. Hobbs, C. E. Recent Advances in Bio-Based Flame Retardant Additives for Synthetic Polymeric Materials. *Polymers (Basel)* **2019**, *11*, 224–255.

Chapter 8

Mechanistic Study of Boron-Based Compounds as Effective Flame-Retardants in Polyurethanes

Saptaparni Chanda[1] and Dilpreet S. Bajwa[*,1]

[1]Department of Mechanical and Industrial Engineering Montana State University, Bozeman, Montana 59717, United States
*Email: dilpreet.bajwa@montana.edu

Polyurethane (PU) is one of the most flammable polymeric materials. Therefore, there is an immense need for efficient flame-retardant (FR) materials to prevent fire hazard from polymers. The low toxicity, high thermal stability, and ecofriendly nature of boron-based compounds place them among the most promising FRs in the present research scenario. The boron-based compounds, either as additive or reactive FRs, offer an effective FR behavior for PU. They display high limiting oxygen index values, reduced smoke release rates, and effective smoke suppression, thus improving the FR properties of PU. In this chapter, the working mechanisms of boron-based FRs are reviewed in detail.

Introduction

The growth of fossil-fuel derived polymeric materials is ubiquitous in every sector of modern life: from packaged goods and consumer electronics to transportation, construction, aerospace, and industrial machinery. This progress also brings additional hazards: hydrocarbon-based polymeric materials are inherently highly flammable and release toxic gases during combustion (*1*). The loss of property and lives has become a continuous reminder of the need to find a suitable solution to limit the flammability of such materials (*2*). Polyurethanes (PUs) are one of the most versatile polymers, with industrial applications in the building construction, furniture upholstery, footwear, textile, adhesives, sealants, and aviation industries. PUs can be obtained from the reaction between polyols and diisocyanates. Rigid, elastic, or semirigid PU foams (PUFs) can be obtained by tuning the mechanical properties of the polymer. PUFs have found a variety of applications as insulating materials. However, their large surface area, light weight, and high content of carbon and hydrogen make PUFs more flammable. PUFs can withstand heat up to 90 °C, and their combustion produces highly toxic gases accompanied by dripping (*2–4*).

Several factors govern the flammability of PUFs, including the structure of the polyol and the isocyanurate index, density of the foam, types of blowing agents, and degree of cross-linking of polymer. Aromatic polyols that contain PUFs are significantly less flammable than their aliphatic

© 2021 American Chemical Society

counterparts. The presence of the isocyanurate ring is also responsible for a reduction in the flammability of PUFs. The limiting oxygen index (LOI) increases from 20.4 to 22.0 vol% with an increase in the foam isocyanurate index from 220 to 2080 (*4, 5*). The increase in the degree of cross-linking of the polymer and the density of the foam has significant influences on the reduction of flammability for PUFs.

The combustion of polymers is an endothermic pyrolysis to flammable and toxic gases, which ignite in contact with air and lead to the exothermic processes of flame propagation and heat release. The modes of action of flame retardancy are mainly in the condensed phase and vapor phase. In the condensed phase, the process of char formation is promoted by dehydration of the polymeric structures. Moreover, the formation of carbonaceous char helps to reduce the release of volatile gases, which act as fuel. In some cases, flame retardancy can also be achieved through intumescence. In these cases, a multicellular residue acts as a barrier to the underlying material by slowing heat transfer. The gas phase mode of action usually acts in parallel with the condensed phase mode. In the gas phase mode, noncombustible gases are released during the decomposition of the material, reducing combustion efficiency. H• and OH• radicals are formed during the combustion of hydrocarbon materials, and the fuel combustion cycle is propagated through the following exothermic reaction: OH• + CO → H• + CO_2. Flame retardants (FRs) generally act chemically or physically in the condensed or vapor phase to interfere with the combustion process during heating, pyrolysis, ignition, or flame spread (*1, 6*).

To reduce the flammability of the polymers, FRs can be added to polymers by physical addition, by mechanical mixing of FRs with the foaming composition, or by chemical reaction. FRs used to reduce flammability of PUFs are expected to be environmentally friendly and to meet certain criteria for effective flame retardancy. The efficacy of an FR depends strongly on the interaction between the FR and the polymer matrix but also on the structure–property relationship between the two during thermal degradation (*1*).

The efficiency of nonreactive FRs depends on their content in PUFs. Nonreactive FRs inhibit the combustion process by physical means. Usually, they act as plasticizers or fillers, depending on their compatibility with the polymer matrix. The incorporation of a large quantity of fillers dilutes the polymer and reduces the release of combustible gases. Additionally, hydrated fillers provide a cooling effect to the substrate. However, the physical addition of FRs can adversely affect the physical properties of PUFs. The most popular and effective nonreactive FRs are bromine, phosphorus, or a combination of both (*7*).

Halogen FRs—such as aliphatic, alicyclic, and aromatic chlorine and bromine compounds—can be used as additives to a polymer matrix or they can be added as structural components during the polymerization process. The active free radicals are replaced with less-reactive halogen radicals, thereby slowing the rate of heat production. Unfortunately, the use of halogen FRs is restricted because of their negative impacts on the environment and on human health (*8*).

Nitrogen-based compounds, such as melamine (MEL) and urea, can also act as effective FRs for PUs. The toxic gases released during the combustion of PUs become diluted by the presence of nitrogen formed during the decomposition of MEL. Moreover, the charred layer formed on the surface of the PU material due to the thermal degradation of MEL acts as a barrier to reduce the flammability of the material (*9*).

Inorganic FRs, such as alkali metals, are another group of effective FRs because of their low toxicity and low smoke evolution. Aluminum hydroxide is one of the most effective FRs in this group. It decomposes endothermically, vaporizes water, and releases noncombustible gases during decomposition, reducing the efficiency of combustion. Aluminum hydroxide is used as an FR in phenolic resins, elastomers, thermoplastic resins, and epoxy resins (10, 11).

Boron-based compounds have gained particular attention as combustion inhibitors because of their low toxicity and ecofriendly nature. Boric acid (H_3BO_3) and boroorganic compounds are emerging as promising FRs (12, 13).

The mechanisms of flame retardancy of different FRs are strongly dependent on the decomposition temperatures of polymers as well as the FRs themselves. Therefore, the FR must be chosen to match the polymer processing parameters. In the case of foams, the FRs must not influence the pore structure, thermal stability, or durability of foams, and they must have good foamability. Furthermore, the fiber and textile FRs must maintain their FR properties after undergoing the processes of spinning, weaving, and washing (1, 4).

Because of the aforementioned reasons, the search for novel FRs inevitably tries to meet all the conditions of effective firefighting and eco-friendliness. Boron-based FRs have emerged as a promising class of FRs that is overall effective, safe, economical, and compatible with several polymer matrices.

Significance of Boron-Based FRs

Boron-based compounds are a rapidly developing group of FRs, and their FR properties have been noted for quite some time. Usually, mixtures of H_3BO_3 and borax have been used as FRs for cellulosic materials. The endothermic decomposition of H_3BO_3 occurs in two stages to release water. In the first stage, H_3BO_3 decomposes at 130–200 °C to form HBO_2; in the second stage, boron trioxide (B_2O_3) is formed at about 265 °C. This mixture dissolves in the inherent water of hydration and finally forms a glassy surface coating. H_3BO_3 can dehydrate oxygen-containing polymers, exhibiting a char-forming catalytic effect. The formation of the glassy barrier inhibits the flow of the combustible volatiles to the surface of the material exposed to fire. The inhibition of the mass transfer of the combustible gases then results in the dilution of fuel and the subsequent termination of the combustion cycle.

Borax is known to reduce flame spread, but it can also promote smoldering or glowing. On the other hand, H_3BO_3 can effectively suppress smoldering but has little influence on flame spread. Therefore, these two compounds are generally used synergistically (6, 14). Zinc borates (ZnB) are another group of well-known boron-based compounds that are used as FRs for polymeric materials. The zinc oxide moiety present in ZnB promotes the formation of the B_2O_3 moiety, which suppresses heat flow and stabilizes char.

There are many beneficial effects of boron-based compounds, including their preservative effectiveness, neutral pH, and less impact on the mechanical properties of the substrate. These advantages, along with their low toxicity and environmentally friendly nature, make them one of the most popular group of FRs for polymeric materials (14, 15).

Boron-Based FRs and Their Mechanisms

Currently, most of the FRs used for polymers are halogen free and are used in synergy with other FRs. Boron-based FRs are usually used in synergy with nitrogen- or phosphorus-based FRs. The whole system acts in the condensation phase, where the boron-based compound produces a glassy barrier layer on the substrate, preventing the flow of the fire and decreasing the heat release rate, flame spread rate, and toxic gas production. The nitrogen-based compound promotes the formation of nontoxic nonflammable gases that dilute the toxic flammable gases. Depending on the mode of action, boron-based compounds can be either used during polymerization to produce oligoetharols or used in producing boron-induced PUFs, which show increased resistance toward thermal degradation. In addition, boron-based nanoparticles have gained considerable attention as a new class of FRs because of their superior thermally stable layered structure, excellent thermal conductivity, and superior chemical inertness. These boron-based nanoparticles can be homogeneously dispersed in the polymer matrix to form a continuous barrier network (3, 16).

Boron-Containing PUs

One of the most effective ways to use boron as an FR is to induce boron into the structure of PUs during the synthesis step. Introduction of thermally stable 1,3,5-triazine rings and boron-reactive retardants can significantly reduce the flammability of PUFs. In addition, MEL, MEL polyphosphate, and tris (2-chloro-1-methylethyl) phosphate are also used as additive FRs. Lubczak et. al produced boron-containing PUFs with carbazole that had increased thermal resistance and reduced flammability. H_3BO_3 was used as the source of boron. In the first step, carbazole (Figure 1a) reacted with a 7-fold molar excess of glycidol (GL) to produce a multifunctional alcohol (Figure 1b) (3).

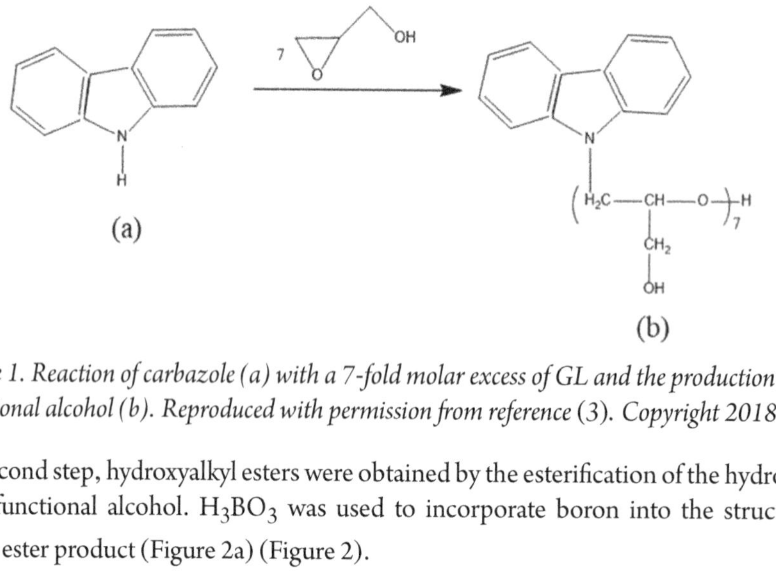

Figure 1. Reaction of carbazole (a) with a 7-fold molar excess of GL and the production of a multifunctional alcohol (b). Reproduced with permission from reference (3). Copyright 2018 Elsevier.

In the second step, hydroxyalkyl esters were obtained by the esterification of the hydroxyl groups of the multifunctional alcohol. H_3BO_3 was used to incorporate boron into the structure of the hydroxyalkyl ester product (Figure 2a) (Figure 2).

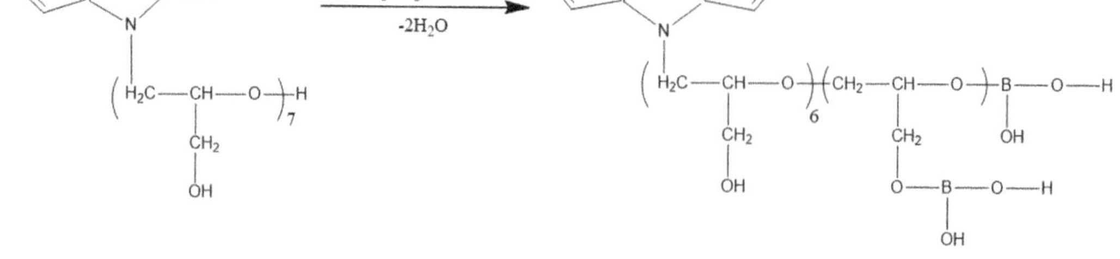

Figure 2. Incorporation of boron into the structure during synthesis using H_3BO_3. Reproduced with permission from reference (3). Copyright 2018 Elsevier.

The oligoetherol was obtained by hydroxyalkylation of product (Figure 2a). The reaction occurred with the alkylene carbonate, ethylene carbonate (EC), in the presence of potassium carbonate at 145–150 °C (Figure 3).

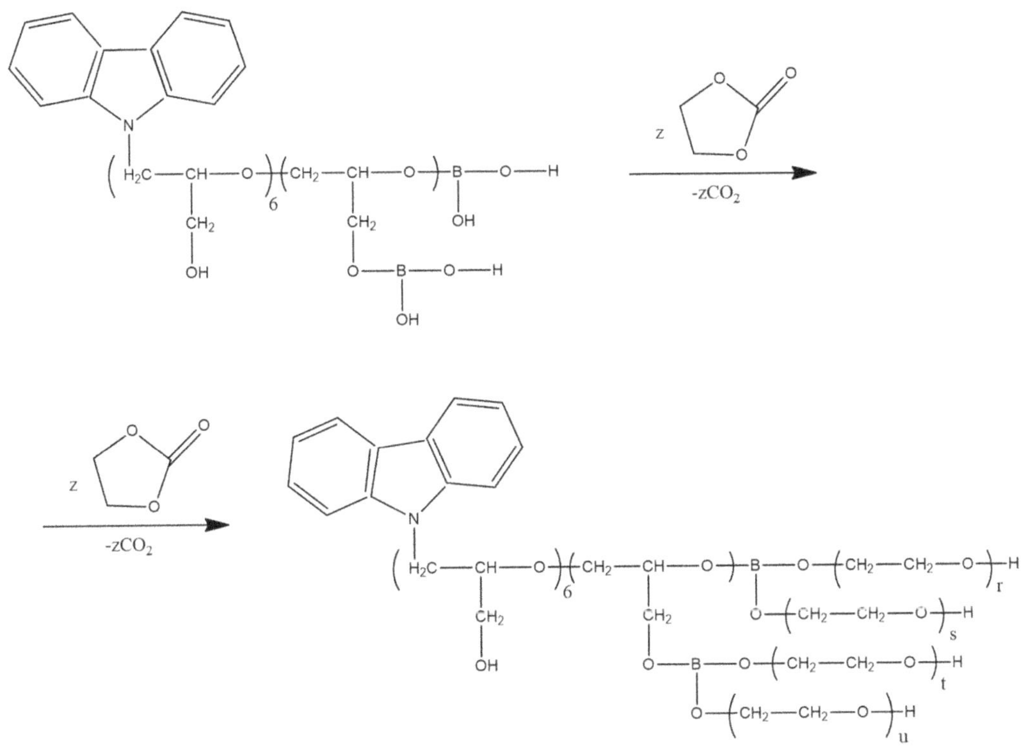

Figure 3. Hydroxyalkylation to produce boron-induced oligoetherol. Reproduced with permission from reference (3). Copyright 2018 Elsevier.

The PUFs with additive FRs and reactive FRs (e.g., H_3BO_3) exhibited increased apparent density. PUFs obtained from oligoetherols with additive FRs showed apparent densities within 45.5–52.9 kg/m³, whereas PUFs that contain H_3BO_3 showed a considerable increase in apparent density, to within 91.5–113.6 kg/m³. PUFs with H_3BO_3 indicated a much higher water uptake, facilitating coordination of the water molecules. The unmodified PUFs were flammable, having an oxygen index (LOI) of 20.2 and a flame spread rate of 5.3–5.4 mm/s in the horizontal flame test. The modified PUFs that contain additive FRs were nonflammable and ceased burning immediately after the removal of the flame source. Their LOI was between 22.8 and 25.3, indicating the self-extinguishing phenomena. The highest LOI (24.7%) was observed in the case with boron-containing oligoetherol and MEL as additive FRs.

Chmiel et. al discussed the production of two types of oligoetherols. One was obtained from metasilicic acid, GL, and EC. The other oligoetherol, with a 1,3,5-triazine ring, was obtained from MEL diborate (MDB) and EC (17). These two types of oligoetherols were mixed and then foamed to obtain PUFs. The thermal resistance of the PUFs was studied by both static and dynamic tests. In the case of the static test at 150 and 175 °C, the mass loss was 13.0–14.7% and 29.4–32.6%, respectively, after a 30-d exposure. The formation of the protective borate glassy layer on the surface of PUFs contributed toward the decrease in mass loss at 150 °C with increasing boron concentration. However, at the higher temperature of annealing, this effect was not observed owing to the thermolysis of the B–O bonds. In the case of the dynamic test, the PUFs started decomposing at temperatures above 200 °C. The largest mass loss was observed within 290–305 °C. In the first step, at 200 °C, the thermolysis of the urethane and urea bond occurred, and at higher temperatures, the allophanate and biuret bond breakage occurred. In the second step of thermal degradation, between 360 and 410 °C, polyol fragment degradation occurred. PUFs obtained from the mixture of oligoetherols are flammable, having OIs within 20.3–20.5%. After thermal exposure at 150 °C, PUFs became self-extinguishing (OI = 24.4–27.6%). They did not flame upon burner ignition after exposure at 175 °C. Annealed PUFs released less fume compared with nonannealed PUFs. The same PUFs showed an LOI of between 46.5 and 48.0% after exposure for 1 month at 175 °C. Thus, the long-term heating exposure increased carbonization and the formation of a more heat-resistant carbon structure that had a higher ignition temperature or higher oxygen availability for flame ignition.

Zarzyka et. al synthesized polyols by the reaction (4) of N,N'-bis(2-hydroxyethyl) oxamide esterified with H_3BO_3 and various excesses of EC. Rigid PU foams were produced on the basis of these new polyols. The boron and nitrogen content of these PUFs were 0.8–1.2 wt% and 7.9–8.5 wt%, respectively. The OI values of the PUFs were in the range of 21.8–23.5 vol%. The OI values increased with the increase in boron and nitrogen content of the foams, indicating their self-extinguishing nature. The foams obtained from polyols that contain the oxamide group showed an OI value of 20.7 vol%. Therefore, the influence of boron on the flammability of the PUFs was noticeable. In general, the higher the boron and nitrogen content, the higher the OI value: indicating the synergistic effect of boron and nitrogen on the flammability of PUFs. Nitrogen compounds decreased the oxygen concentration in the combustion zone. A superficial char layer was created on the surface of the burning material due to the endothermic decomposition of the nitrogen compound. Boron compounds acted in the condensed phase, participating in an endothermic reaction to release water and to create a protective glassy layer. The horizontal flame test results indicated that the foams with the highest nitrogen (8.5 wt%) and boron amounts (1.1 wt%) were characterized by the slowest burning rates. Decreasing amounts of the elements (boron and nitrogen) increased the burning rates. In the case of the vertical test (Butler test), the burning residue in the vertical chimney decreased from 86.3% to 82.7%, decreasing with the decreasing amount of boron and nitrogen. The foams were subjected to annealing for 30 d, at 150 and 175 °C. Borate groups present in the foam structure started to decompose at temperatures above 130 °C. The weight loss of the foams was highly dependent on the boron content and decreased with the increase in boron concentration.

Chmiel et. al synthesized boron-containing oligoetherol (*18*) from MDB with EC (Figure 4), and PUFs were obtained using these oligoetherols. Additional MDB was added at the foaming step as an additive FR. The obtained PUFs were subjected to annealing at 150 and 175 °C.

The PUF synthesized from oligoetherol that contain MDB and EC showed an OI of 21.2. To decrease its flammability, MDB was added as the additive FR. A nonflammable PUF was obtained with the addition of 11.5% MDB. In the horizontal flammability test, this PUF ceased burning immediately after the flame source was removed. The PUF obtained from a higher boron-containing oligoetherol showed a decreased rate of flame spread and a lower mass loss compared with the lower amount of the lower boron-containing one. The OI was 21.2, indicating that it was flammable in atmospheric exploration conditions. The further addition of MDB into the PUF resulted in a higher

OI, 23, which indicated that the PUF was self-extinguishing. When the PUF was annealed at 150 °C, the OI increased to 25.4%, whereas annealing at 175 °C resulted in a significant increase in OI, 55%, which is the value for nonflammable foams. High-temperature annealing was accompanied by carbonization of the foams, indicating a higher oxygen requirement for combustion.

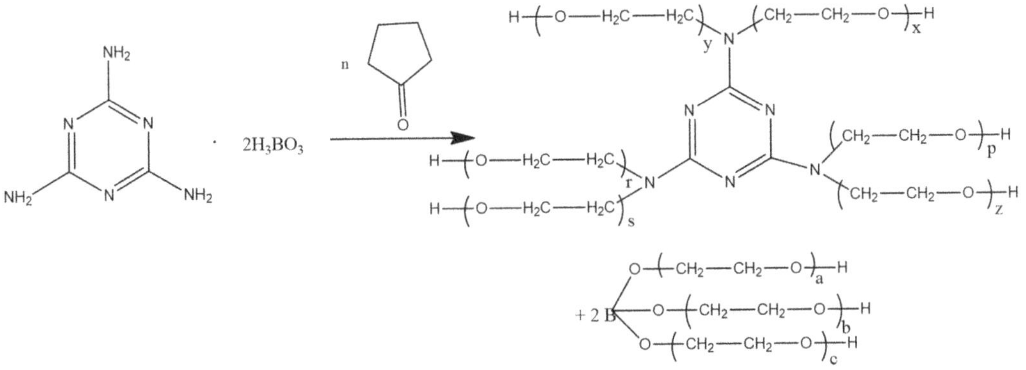

Figure 4. Schematic illustration of the MEL and H_3BO_3 reaction. Reproduced with permission from reference (18). Copyright 2019 Taylor & Francis.

A boron-containing UV-curable oligomer was produced by Patil et. al from linseed oil, phenylboronic acid, and glycidal methacrylate (19). This boron-containing epoxidized linseed oil oligomer (BELO) was added to a conventional PU acrylate (PUA) at varying concentrations of from 10 to 40 wt% for FR-coating application. The thermal stability of the coating was strongly dependent on the number of hard and soft segments present in its structure. The urethane linkage and isocyanate structure were responsible for the hard segments, whereas the long aliphatic chains were responsible for the soft segments. From the thermogravimetric analysis (TGA) curves, it can be inferred that the PUA40 coating had a higher thermal resistance compared with the others. The possible reason behind the increased thermal resistance was the influence of the increased percentage of BELO on the cross-linking density of the cured coating. Additionally, the presence of an aromatic ring in the structure enhanced the thermal resistance of the coating. The LOI value increased from 19 to 25, increasing with the increase in BELO content in the coating. The highest LOI of 25 was obtained for the PUA40 coating (Figure 5). The improvement in flame retardancy was probably due to the presence of boron in the coating. During firing, the heat of combustion was absorbed due to the endothermic dehydration reaction of boron. The released water vapor diluted the oxygen and the other inflammable gases. This flame inhibition mechanism occurred in the vapor phase, whereas the formation of boron oxide acted as a barrier in the condensed phase. The protective layer of boron oxide prohibited the heat transfer and slowed the spreading of flame. The residual char yield of the coatings was also increased with the increase of BELO oligomer content in the coating, indicating the improvement in flame retardancy of the coating (19).

Xia et. al derived diglyceride borate from H_3BO_3 with glycerin by esterification reaction (16). Polyether polyol, 4,4′-diphenylmethane diisocyanate, and diglyceride borate were reacted together to obtain boron-containing PUs (Figure 6).

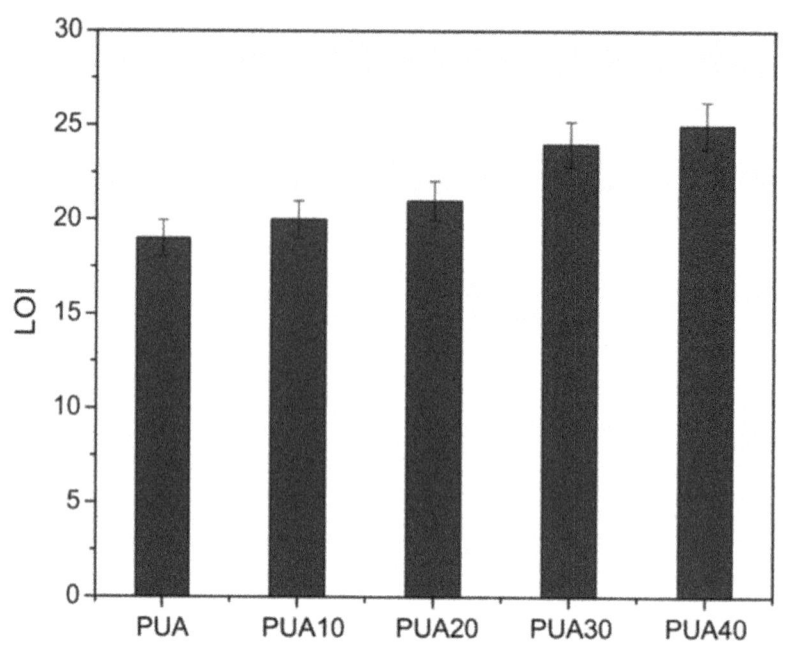

Figure 5. LOI analysis of PUA coatings. Adapted with permission from reference (19). Copyright 2018 Springer.

The TGA test showed that the with the increase in the content of diglyceryl borate, the residual carbon amount also increased. At 600 °C, the residual carbon amount increased from 13.7 to 17.9%, increasing with the increase in the diglyceride borate from 13.6 to 25.7%. The B–O–C char layer covered the PU surface, thereby insulating against heat and oxygen and promoting initial carbon residue formation. It was observed that with the increase in boron content, the LOI value of the PU also increased from 23.8% to 30.0%. During the combustion process, a dense nitrogen-rich carbon residue was formed on the polymer surface, preventing the polymer from exposure to heat and oxygen. The boron atom has an empty orbital, and its electron-deficient structure facilitates its combination with a nitrogen atom, which contains a lone pair of electrons, by a coordination bond. The free radicals, O• and OH•, produced by the cracking or pyrolysis during the combustion process were the cause of the end of the combustion chain reaction. The boron FR promoted the formation of carbon and a glass-like protective layer, preventing the escape of the volatile combustible gases. Moreover, a new compound, B_2O_3, was also formed during the combustion process. The char layer produced during the combustion process effectively prevented the entrance of oxygen, inhibiting further combustion of the PU and improving the FR performance. Additionally, the inhibition of the tight coal char layer facilitated the pyrolysis products to remain in the condensed phase, resulting in a denser, larger coal char residue.

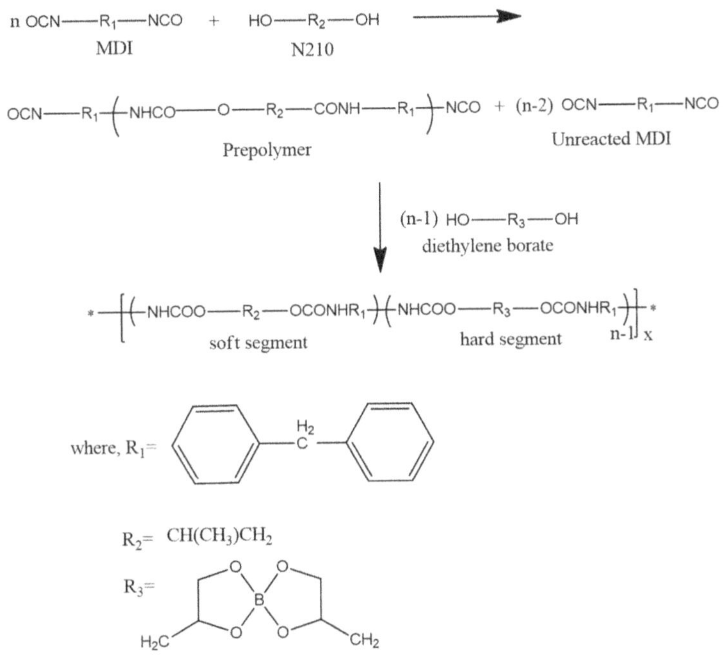

Figure 6. Schematic diagram for the synthesis of boron-containing PU. MDI, methane diisocyanate, 4,4'-diphenylmethane diisocyanate. Reproduced with permission from reference (16). Copyright 2020 Taylor & Francis.

Boron Nitride

Hexagonal boron nitride (h-BN) has gained enormous attention as an FR because it can act as a rigid barrier on the polymer surface, thus improving the fire safety of polymers. However, the homogeneous dispersion of h-BN can be challenging because of the chemically inert surface and strong Van der Waals interaction between the h-BN nanoplates. The exfoliation and functionalization of the bulky h-BN can be a potential solution for these problems. A copper phytate decorated polypyrrole shell enwrapped boron nitride nanosheet (CPBN) (20) was successfully prepared by Wang et. al (Figure 7). The thermal stability of pure thermoplastic PU (TPU) and its composites with h-BN and CPBN were investigated using the TGA test. TPU/h-BN showed a higher mass loss compared with the TPU-CPBN composites. The higher thermal conductivity of h-BN promoted heat transfer from the heat source to the polymer matrix. Additionally, the agglomeration of h-BN in the polymer matrix created an undesirable barrier network that hindered heat diffusion. On the other hand, the continuous CPBN barrier network provided efficient inhibition of heat transfer. In the cone calorimeter test, it was observed that the peak heat release rate (PHRR) was the highest in the case of the TPU/CPBN composites, indicating the highest flame retardancy. TPU/CPBN composites also exhibited the highest char residues. The formation of phosphor carbonaceous char was facilitated by the decorated phytate components. Moreover, the barrier network of CPBN held the decomposed products in the condensed phase to effectively promote char formation.

The char residue that was formed blocked the diffusion of the degraded products, thereby protecting the underlying polymer matrix from burning (Figure 8). In addition, the TPU/CPBN composites exhibited the lowest total smoke production values, suggesting the smoke suppression function of CPBN. The composites with CPBN showed increased fire performance index values and reduced fire growth index values compared with pure TPU. Therefore, it can be inferred that the addition of CPBN to the polymer matrix probably postponed the occurrence of flashover, that is, the rapid and simultaneous ignition of the combustible surface. A discrete and inefficient barrier network was created owing to the poor compatibility and nonhomogeneous dispersion of h-BN in the polymer matrix. In the case of CPBN, a continuous and efficient barrier network was generated owing to the homogenous dispersion. This barrier network hindered the mass transfer of combustible volatiles and toxic effluents. The char residues composed of boron nitride and carbonaceous layers acted as a protective shield for the underlying polymer and an additional barrier for mass diffusion.

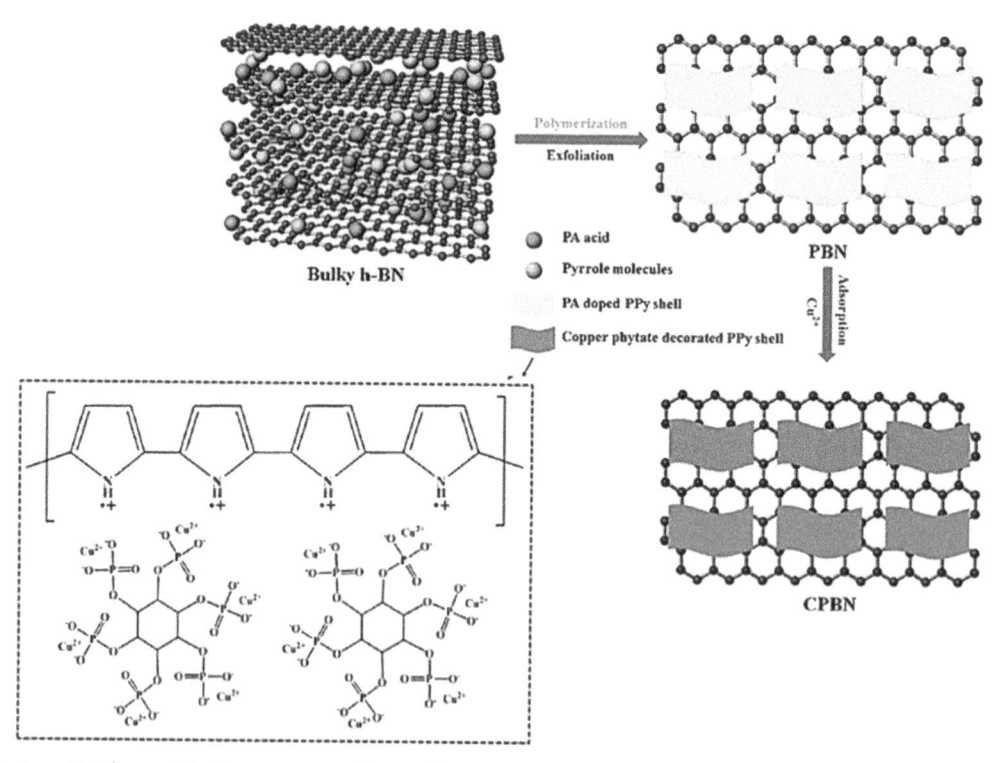

Figure 7. The graphical preparation scheme of CPBN. PA, phytic acid, polypyrrole (PPy). Adapted with permission from reference (20). Copyright 2019 Elsevier.

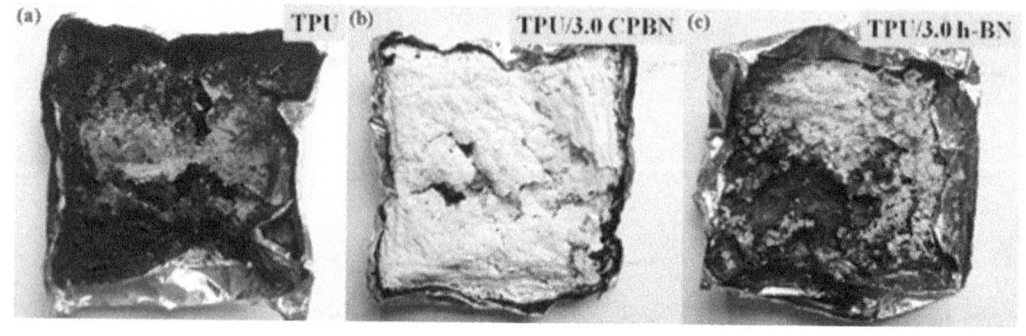

Figure 8. Digital pictures of the char residues: (a) pure TPU, (b) TPU/3.0 wt% CPBN, and (c) TPU/3.0 wt% h-BN. Adapted with permission from reference (20). Copyright 2019 Elsevier.

In h-BN, the boron and nitrogen atoms alternate in the hexagonal lattice, constructing a honeycomb-like layered structure. A partial ionic characteristic exists in h-BN because of the asymmetric boron and nitrogen electron pairs dispersed within the B–N domain. Boron atoms can interact with neighboring nitrogen atoms and form peculiar B–N stacking characteristics. This chemical interaction is commonly known as a "lip–lip" interaction. The functionalization of h-BN is particularly difficult because of the partial ionic characteristic of B–N bonds and the lip–lip interaction between the adjacent layers of h-BN. Cai et. al formed an operable platform by using a silicon dioxide coating—via Lewis acid–base interactions—onto the surface of h-BN nanosheets, introducing phytic acid (PA) into the layered structure (Figure 9) (21).

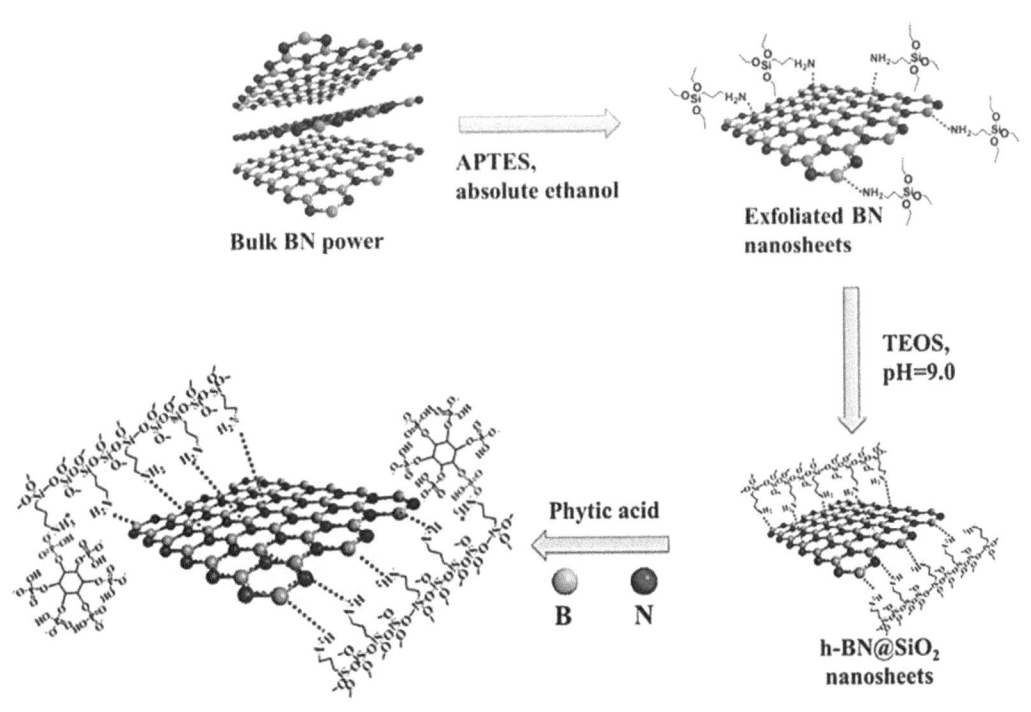

Figure 9. Schematic illustration for the preparation of h-BN nanosheets with incorporation of silicon dioxide and PA. APTES, 3-aminopropyl)triethoxysilane; TEOS, tetraethyl orthosilicate. Adapted with permission from reference (21). Copyright 2019 Elsevier.

The resultant h-BN nanohybrids offered an improved flame retardancy for the TPU. The TPU composites showed an obvious reduction in the values of PHRR (−23.5%) and total heat release (−22.1%). Moreover, the smoke production rate and total smoke release of TPU composites were also reduced by 29.2% and 8.6%, respectively.

The layered structure of h-BN nanosheets acted as the first barrier line to prevent the diffusion of the flammable pyrolysis gases (Figure 10). The pyrolysis gases reacted with pre-graded PA and produced the char residue. The mechanical robustness of the char was also improved owing to the presence of the silicon element. The final char layer that was produced acted as both a heat shield and a barrier, inhibiting the permeation of the flammable products. The high temperature of the flame facilitated the production of electron hole onto the surface of h-BN and assisted in oxidizing the toxic reducing gases, such as CO, CH_4, C_2H_6.

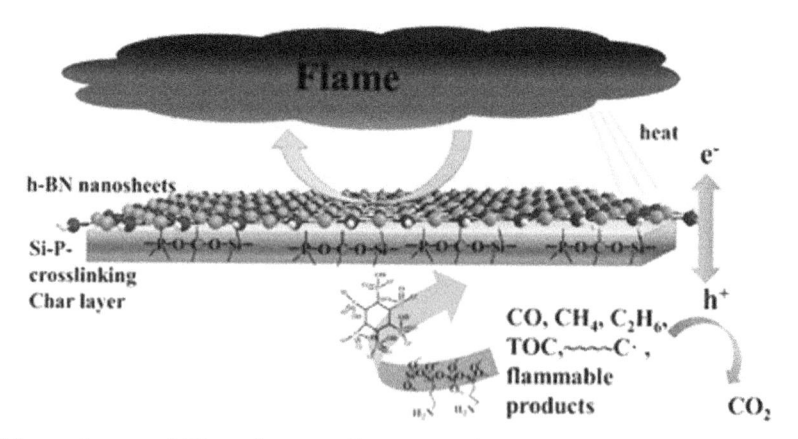

Figure 10. Scheme of proposed FR mechanism of h-BN nanosheets for TPU. Adapted with permission from reference (21). Copyright 2019 Elsevier.

Davesne et. al prepared boron nitride nanoplatelets via an aqueous route and produced hydroxylated h-BN nanosheets. These exfoliated nanosheets were then paired with polyethylenimine (PEI) and deposited as a single bilayer on a PU foam by using layer-by-layer assembly (Figure 11) (2).

Figure 11. (a) Digital image, (b) transmission electron microscopy cross-sectional micrograph of one bilayer of PEI/h-BN–coated PUF. Adapted with permission from reference (2). Copyright 2019 American Chemical Society.

The thermal stability of the produced foam was investigated using TGA under nitrogen and air atmospheres. In both conditions, the nanocoating aided in delaying the oxidation, but the degradation pathway remained unchanged. In this case, the coating might be very thin to accelerate heat transfer and modify the degradation pathway. Under both nitrogen and air, the control foam (i.e., pure PUF) left no residue at 800 °C, whereas a residual mass of 7% was observed in the case of coated foams. In a butane torch test, the control foam completely degraded with melt dripping. The coated samples were protected by the PEI/h-BN inflammable shield and exhibited a degradation gradient that was dependent on the foam thickness. In the cone calorimeter test, the coated samples showed a reduction in PHRR of 54% and total heat release of 20%, along with the reduction in the amount of gas emitted.

Other Boron-Based Compounds

ZnB, an inorganic FR, is widely used either solely or in combination with other FRs for its synergistic effect. Dike et. al studied the effect of ZnB as an FR on the FR and thermal properties of a TPU that contain huntite–hydromagnesite (HH) (22). The addition of 50% ZnB to TPU increases the LOI value of the composite to 24.1%, compared with the LOI (21.2%) of the pure TPU. HH and ZnB each promoted endothermic decomposition and showed a dilution effect in the gas phase by water release and the formation of a protective char layer in the condensed phase. There was no remarkable change observed in the FR properties when HH and ZnB were used together. In the mass calorimeter test, the PHRR, average heat release rate, and total heat evolved were decreased with increases in the ZnB amount in the composite. Therefore, it can be concluded that the addition of ZnB increased the barrier effect by the formation of the glass-like char residue.

Liu et. al investigated the effect of the catalysis of boron phosphate (BP) on the thermal properties and char forming ability of the FR-PU–polyisocyanurate foams (FPUR-PIR) with dimethylmethylphosphonate and tris(2-chloropropyl) phosphate (23). The heat release rate, total smoke released, and carbon dioxide production were decreased with the incorporation of 3 mass% BP in FPUR-PIR. Adding BP suppressed the release of smoke and toxic products, such as cyanate compounds, or cyanic acid in the gas phase. The presence of a large number of Bronsted and Lewis acid sites on the surface of BP accelerated the decomposition of FPUR-PIR and promoted the cross-linking of hydroxyl compounds. BP catalyzed the decomposed fragments such as cyanate to a more thermally stable structure by the formation of C=N groups in the condensed phase, thereby increasing the stability of the char and reducing the smoke and toxic gas production.

Conclusion

Boron-based FRs have emerged as an effective class of FRs. The addition of boron-based compounds into the PU structure promotes the formation of a glass-like protective barrier on the polymer surface, preventing the diffusion of toxic volatiles. The functionalized h-BN nanosheets form a continuous and effective barrier network, which act as a heat shield and hinders the permeation of flammable toxic gases. The other boron-based compounds, such as ZnB or BP, promote the formation of a protective char layer in the condensed phase and suppress the evolution of smoke and toxic gas production in the gas phase. There is an increasing market demand for an efficient FR for PUs given that they are among the most flammable polymeric materials and release toxic smoke upon burning. Therefore, the addition of boron-based FRs improves the fire safety standard. Additionally, boron compounds can act synergistically with nitrogen-based compounds and significantly improve firefighting performance. Boron-based compounds help slow the ignition process, increase the thermal stability of PUs, reduce the PHRR and total heat release, and vanquish smoke evolution and toxic gas release. Therefore, in conclusion, boron-based compounds can play a pivotal role in saving human lives and property and in reducing the risk of injuries.

References

1. Velencoso, M. M.; Battig, A.; Markwart, J. C.; Schartel, B.; Wurm, F. R. Molecular firefighting—how modern phosphorus chemistry can help solve the challenge of flame retardancy. *Angewandte Chemie International Edition* **2018**, *57* (33), 10450–67.
2. Davesne, A. L.; Lazar, S.; Bellayer, S.; Qin, S.; Grunlan, J. C.; Bourbigot, S.; Jimenez, M. Hexagonal boron nitride platelet-based nanocoating for fire protection. *ACS Applied Nano Materials* **2019**, *2* (9), 5450–9.
3. Lubczak, R.; Szczęch, D.; Broda, D.; Szymańska, A.; Wojnarowska-Nowak, R.; Kus-Liśkiewicz, M.; Lubczak, J. Preparation and characterization of boron-containing polyurethane foams with carbazole. *Polymer Testing* **2018**, *70*, 403–12.

4. Zarzyka, I. Foamed polyurethane plastics of reduced flammability. *Journal of Applied Polymer Science* **2018**, *135* (4), 45748.
5. Dick, C.; Dominguez-Rosado, E.; Eling, B.; Liggat, J. J.; Lindsay, C. I.; Martin, S. C.; Mohammed, M. H.; Seeley, G.; Snape, C. E. The flammability of urethane-modified polyisocyanurates and its relationship to thermal degradation chemistry. *Polymer* **2001**, *42* (3), 913–23.
6. Kicko-Walczak, E. Novel halogen-free flame retardants—flame retardation of unsaturated polyester resins with use of boron compounds. *Polimery.* **2008**, *53* (2), 126–32.
7. Bastin, B.; Paleja, R.; Lefebvre, J. Fire behavior of polyurethane foams. *Journal of Cellular Plastics* **2003**, *39* (4), 323–40.
8. Ashford, P. *Proceeding of Polyurethanes World Congress 97*; Technomic Publishing Co. Inc.: Amsterdam, 1997; p 612.
9. Price, D.; Liu, Y.; Milnes, G. J.; Hull, R.; Kandola, B. K.; Horrocks, A. R. An investigation into the mechanism of flame retardancy and smoke suppression by melamine in flexible polyurethane foam. *Fire and Materials* **2002**, *26* (4-5), 201–6.
10. Furukawa, M.; Yokoyama, T. Mechanical properties of organic–inorganic polyurethane elastomers. I. Al(OH)3–polyurethane composites based on PPG. *Journal of Applied Polymer Science* **1994**, *53* (13), 1723–9.
11. Bourbigot, S.; Le Bras, M.; Leeuwendal, R.; Shen, K. K.; Schubert, D. Recent advances in the use of zinc borates in flame retardancy of EVA. *Polymer Degradation and Stability* **1999**, *64* (3), 419–25.
12. Paciorek-Sadowska, J. Modification of PUR-PIR foams by boroorganic compound prepared on the basis of di(hydroxymethyl) urea. *Journal of Porous Materials.* **2012**, *19* (2), 161–71.
13. Green, J. Mechanisms for flame retardancy and smoke suppression—a review. *Journal of Fire Sciences* **1996**, *14* (6), 426–42.
14. LeVan, S. L.; Tran, H. C. The role of boron in flame-retardant treatments. In *First International Conference on Wood Protection with Diffusible Preservatives*; Nashville, Tennessee, November 28–30, 1990; Forest Products Research Society: Madison, WI, 1990; pp 39–41, ill. 1990.
15. Ishii, T.; Kokaku, H.; Nagai, A.; Nishita, T.; Kakimoto, M. Calcium borate flame retardation system for epoxy molding compounds. *Polymer Engineering & Science* **2006**, *46* (6), 799–806.
16. Xia, L.; Liu, J.; Li, Z.; Wang, X.; Wang, P.; Wang, D.; Hu, X. Synthesis and flame retardant properties of new boron-containing polyurethane. *Journal of Macromolecular Science, Part A* **2020**, *57* (8), 560–8.
17. Chmiel, E.; Lubczak, J. Polyurethane foams with 1, 3, 5-triazine ring, boron and silicon. *Journal of Cellular Plastics* **2020**, *56* (2), 187–205.
18. Chmiel, E.; Oliwa, R.; Lubczak, J. Boron-containing non-flammable polyurethane foams. *Polymer-Plastics Technology and Material.* **2019**, *58* (4), 394–404.
19. Patil, D. M.; Phalak, G. A.; Mhakse, S. T. Boron-containing UV-curable oligomer-based linseed oil as flame-retardant coatings: Synthesis and characterization. *Iranian Polymer Journal* **2018**, *27* (10), 795–806.
20. Wang, J.; Zhang, D.; Zhang, Y.; Cai, W.; Yao, C.; Hu, Y.; Hu, W. Construction of multifunctional boron nitride nanosheet towards reducing toxic volatiles (CO and HCN)

generation and fire hazard of thermoplastic polyurethane. *Journal of Hazardous Materials* **2019**, *362*, 482–94.

21. Cai, W.; Wang, B.; Liu, L.; Zhou, X.; Chu, F.; Zhan, J.; Hu, Y.; Kan, Y.; Wang, X. An operable platform towards functionalization of chemically inert boron nitride nanosheets for flame retardancy and toxic gas suppression of thermoplastic polyurethane. *Composites Part B: Engineering.* **2019**, *178*, 107462.

22. Dike, A. S.; Tayfun, U.; Dogan, M. Influence of zinc borate on flame retardant and thermal properties of polyurethane elastomer composites containing huntite-hydromagnesite mineral. *Fire and Materials* **2017**, *41* (7), 890–7.

23. Liu, X.; Wang, J. Y.; Yang, X. M.; Wang, Y. L.; Hao, J. W. Application of TG/FTIR TG/MS and cone calorimetry to understand flame retardancy and catalytic charring mechanism of boron phosphate in flame-retardant PUR–PIR foams. *Journal of Thermal Analysis and Calorimetry* **2017**, *130* (3), 1817–27.

Chapter 9

Two-Dimensional Nanomaterials as Smart Flame Retardants for Polyurethane

Emad S. Goda,*,[1,3] Mahmoud H. Abu Elella,[2] Heba Gamal,[4] Sang Eun Hong,[1] and Kuk Ro Yoon[1]

[1]Organic Nanomaterials Lab, Department of Chemistry, Hannam University, Daejeon 34054, Republic of Korea
[2]Chemistry Department, Faculty of Science, Cairo University, Giza 12613, Egypt
[3]Fire Protection Laboratory, National Institute of Standards, Giza 12211, Egypt
[4]Home Economy Department, Faculty of Specific Education, Alexandria University, Alexandria, 21526, Egypt
*Email: emadzidan630@gmail.com

Polyurethanes (PUs) are multilateral polymeric materials utilized widely for various industrial purposes. However, these polymers have the drawbacks of poor thermal stability, high combustion behavior, and emission of toxic gases and smoke. This is because of their characteristic cellular morphology, porosity, and aliphatic segments restricting further applications. Consequently, the research on improving the flammability of polymers is of great concern. In particular, halogen-based flame retardants (FRs) demonstrate effective behavior against fire growth, but they are restricted because of their apparent toxicity. Product specifications such as mechanical properties, durability, and suitable safeness are greatly needed for applications in the market. There is therefore an urgent need to develop new-generation FRs for PU. The usage of 2D nanomaterial FR fillers has been broadly expanded for polymers—especially PUs—because of their outstanding features such as high thermal stability, lower thermal conductivity, and ability to form a strong carbonaceous layer on the polymer. Also, 2D materials can largely boost the mechanical properties of polymers. Commonly, 2D nanomaterials can retard flames in the condensed phase by creating an insulating barrier for suppressing smoke density and heat transfer. In this chapter, the authors conduct an overview of the current research performed on utilizing 2D material FRs for PU and the related mechanism for the retardation of fire growth.

© 2021 American Chemical Society

Introduction

Polyurethanes (PUs) are considered substantial for industrial applications because of their versatile properties such as resistance against chemicals, good wear resistance, excellent abrasion, and significant mechanical strength. Various raw materials and processability techniques are currently used for synthesizing PU materials that can be then converted into final products with the desired features needed for many applications. In the 1930s, the premier PUs were invented by Otto Bayer, opening the door for interest in the preparation of such materials. The research on PUs has been considered a hot topic, because these materials are used frequently in making products that are utilized in daily life such as paint, furniture, automotive seating, thermal insulation, and medical materials (1). They occupy approximately 5% of the worldwide polymer market, which was estimated at 18 million tons in 2016 (2).

Various sources can be used to synthesize PUs with specified properties, offering large areas for specific fields. PUs materials are classified depending on the desired properties into thermoplastic materials, rigid materials, flexible materials, coatings, waterborne materials, adhesives, binders, sealants, and elastomers. In the chemical structure of PUs, the urethane group is considered the major repeating unit that can be formed by the reaction between isocyanate (NCO), and alcohol (–OH), aside from the presence of other functional groups such as esters, ethers, and ureas accompanied by some aromatic compounds (3, 4) (See Scheme 1).

The properties of PUs can possibly be manipulated by controlling the desired flexibility in the structure when selecting raw components such as polyols, di- or triisocyanates, and chain extenders. For instance, polyester polyols, polyethers, and acrylic polyols are the commonly used polyols in the PU industry. Meanwhile, the diisocyanates utilized in the formulation of PUs are usually methylene diphenyl diisocyanate, hydrogenated methylene diphenyl diisocyanate, toluene diisocyanate, 1,5-naphthalene diisocyanate, isophorone diisocyanate, and xylene diisocyanate. Typically, PU polymer can be found in thermosets and thermoplastics. Additionally, PU foam can be formed chemically through two main reactions, as seen in Figure 1. The first is a gelling reaction known as the isocyanate–polyol reaction. It leads to the formation of a urethane group on the polymer backbone and cross-linking of the polymer by the hydroxyl groups of polyol compound. In this reaction, the polymer can also be cross-linked by the reaction of the isocyanate group with the urethane group, resulting in an allophanate.

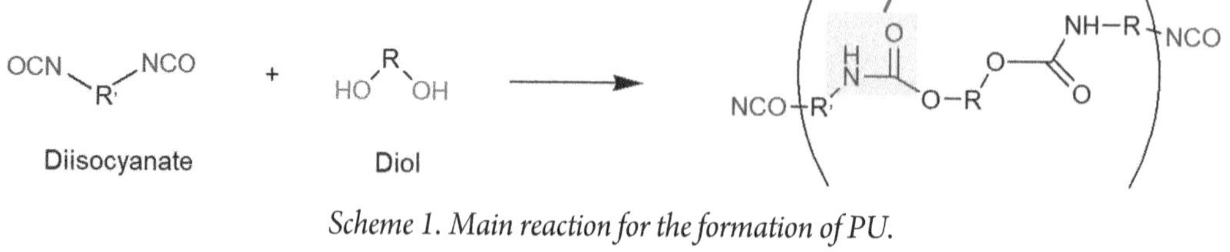

Scheme 1. Main reaction for the formation of PU.

The second reaction of isocyanate–water refers to the blowing reaction by which carbamic acid is formed and decomposed for obtaining carbon dioxide as the gas bubbles, along with the amine. Consequently, the interaction of the obtained amine with another isocyanate group gets the disubstituted urea that helps in extending the chain of the isocyanate molecules. This occurs through the increase of the aromatic groups for producing hard and linear segments. A further step relates to the construction of linkages from an allophanate and biuret, assisting in the covalent cross-linking of the polymer foam. These reactions should be carefully balanced for controlling the stability of the obtained foam and attaining the desired physical properties. Notably, the reaction rates between isocyanate and polyols in the preparation method of PU foam can be controlled using catalysts. Improper evenness between the aforementioned reactions can deteriorate the foam structure by possibly creating inappropriate cells that might be prematurely opened or closed (5–8). Generally, thermoplastic PUs (TPUs) are copolymers of linear chains composed of soft segments and hard segments. The hard segments contain the molecules of diisocyanates and small chains of extenders. They appear mainly with features of rigidity and high polarity, such as diamines and diols. Consequently, the hydrogen bonding that occurs between the urethane/urea groups of the chains is the reason for hard segments being found with a higher interchain interaction. Therefore, they could behave as fillers for reinforcing the soft chains of the PU structure. However, the soft segments are composed of flexible and linear long-chain polyols or diols with a weak polarity. The component of polyols in PUs can be represented by polyfunctional polyethers such as polypropylene glycol, polyethylene glycol, polytetramethylene glycol, and castor oil.

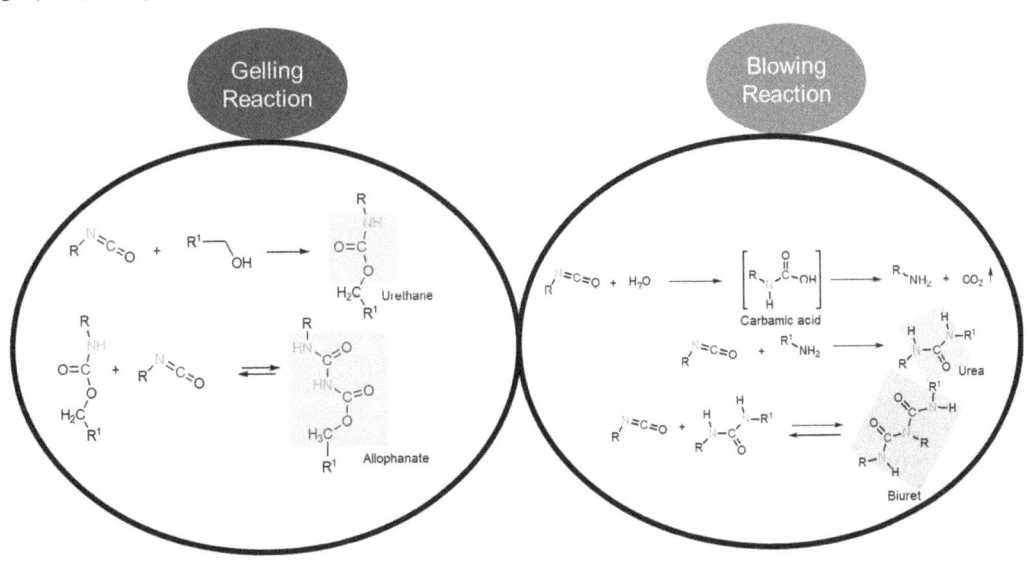

Figure 1. Schematic diagram for gelling and blowing reactions for forming PU foams.

Thermodynamical immiscibility can happen between the hard and soft segments, resulting in the phase separation of TPU chains (3, 9–12). Meanwhile, thermoset PU is usually obtained using different methods, such as polyols or isocyanate compounds individually being used with functionalities of more than 2; a ratio of isocyanate and alcohol being made more than 1; and the addition of cross-linker in the reaction.

As previously described, PUs have been documented as having massive industrial significance in daily use because of their simplicity in terms of implementation. However, PUs foam faces shortcomings, such as poor thermal stability, significant combustion behavior, and emission of toxic gases and smoke because of their characteristic cellular morphology, porosity, and aliphatic segments restricting their further applications (13, 14). Consequently, the research on improving the flammability of polymers is of great concern (15–18). The combustion process was introduced as a complex one, simultaneously combining the diffusion of heat and mass with chemistry decomposition. As shown in Figure 2, it involves the occurrence of four main steps: ignition, pyrolysis, combustion, and feedback. Through application of a sufficient flame source for the polymer surface, decomposition will occur through the pyrolysis step for the evolution of flammable gases that combine with air and start the polymer combustion cycle, as stated in eqs 1 and 2:

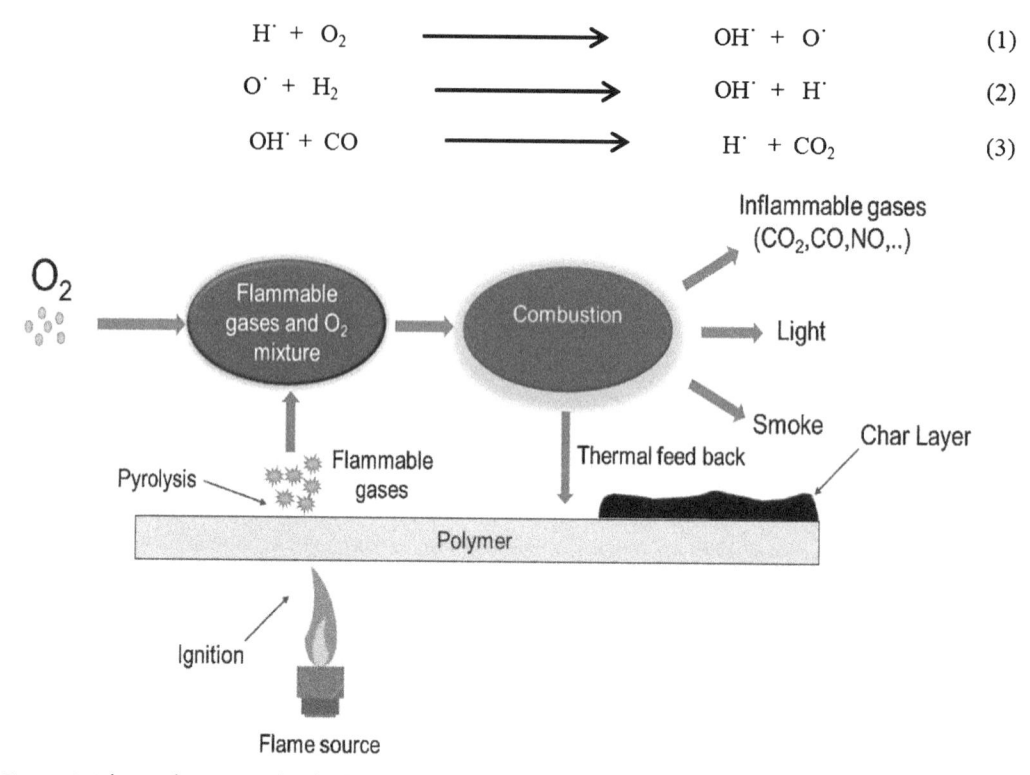

$$H^\cdot + O_2 \longrightarrow OH^\cdot + O^\cdot \tag{1}$$

$$O^\cdot + H_2 \longrightarrow OH^\cdot + H^\cdot \tag{2}$$

$$OH^\cdot + CO \longrightarrow H^\cdot + CO_2 \tag{3}$$

Figure 2. The combustion cycle of polymers. Reproduced with permission from reference (21). Copyright 2018 Elsevier.

The thermal decomposition of polymer using an external fire source creates hydrogen-free radical. This interacts with one molecule of oxygen to produce hydroxyl-free radical as an essential source for the production of exothermic heat required for the sustainability of the polymer combustion, as shown in eq 3.

PU is highly flammable because of its extreme surface area and oxygen permeability, leading to the generation of diverse toxic gases during the fire occurrence. In fire accidents, the main reason for the death is because of these hazardous gases. Generally, the toxic products obtained from the decomposition of PU structure in the air can be classified into two types. One type is made up of asphyxiant gases (such as CO_2 and HCN) that can prohibit the transfer of oxygen to the cells, causing depression of the nervous system and leading to death. The other type is made up of hydrogen halides, which are irritant gases promoting incapacitation that mostly occurs in the upper respiratory tract and eyes (19, 20).

To overcome the combustion dilemma of polymers, especially PUs, an appropriate flame retardant (FR) material must be added into the foam structure to disturb the combustion cycle. The addition of FRs generates a barrier layer that protects the polymer from fire or causes the dilution of the flammable gases created by the decomposition of polymer thermally. Consequently, the role of these materials is to terminate fire growth (21). Halogen-containing FRs have made up the majority on the market because of their efficient flame retardancy behavior. Their decomposed products when exposed to fire are environmentally toxic. A large concern has been replacing these FRs with the nonhalogenated ones in polymers. Nevertheless, the usage of these safe materials—such as metal hydroxides with high loadings for fabricating PU nanocomposites—distorts their physicochemical properties (22). The development of chemically stable, new, and cost-effective materials for inhibiting fire propagation is necessary for obtaining future polymer nanocomposites with high performances to be utilized in more advanced applications. In addition, 2D nanomaterials—such as graphene, layered double hydroxides (LDHs), nanoclays, molybdenum disulfide (MoS_2), and boron nitride (BN)—possess outstanding fire retardancy behavior over other dimensions of nanomaterials when mixed in small loadings with polymers. This is because of the considerable decrease observed in the mass-loss rate and heat release rate (HRR) of the polymers (23–29).

This chapter aims to describe the research done on utilizing various 2D nanomaterials as FRs for PU, aside from their mechanisms of action for retardancy against flame.

FR Actions

When FRs containing polymers are exposed to flame heat, they can degrade thermally with various conceivable mechanisms in the condensed and gas phases, as shown in Figure 3. The retardancy behavior of FRs depends mainly on the chemical structure and the chemical binding of these materials with the polymer.

The condensed phase is composed of three possible scenarios that act through altering the pathways of combustion cycles for forming low-emitted flammable gases and more carbonaceous yields. The scenarios are the construction of an insulator layer on the polymer surface, the cooling effect, and melt dripping. A typical example of FR material working in the condensed phase can proceed through one, two, or three mechanisms in one of or both of the condensed and gas phases together. Regarding the barrier layer, the combustion process can give layers of different forms—such as char (carbonaceous), vitreous, inorganic, and intumescent forms—employed as protective insulators for restricting heat, mass, and oxygen transport between the flame and polymer. For this reason, the entities of flammable gases that feed the flame, rate of polymer decomposition under heat, and oxygen permeation to the polymer are all decreased. Physical and chemical reactions are responsible for the formation of such a layer. The decomposition of some FRs assists in yielding various polyaromatic structures along with intermolecular reactions. This leads to the formation of a carbonaceous layer in what is known as the barrier effect.

The described char layer is very different from the residue obtained from the decomposition of inorganic FRs such as silicones, aluminum hydroxides, and magnesium hydroxides (30, 31). The intumescent char is specially formed after thermal exposure to the systems of intumescent FRs. This forms an expanded carbonaceous char on the polymer surface with a lower thermal conductivity for enclosing heat transfer to the polymer. The composition of these systems at most involves polyols as a carbon source, an acid source for dehydrating the carbon material, and a swelling agent that is under heat and can release a nonflammable gas mixture, creating an expansion of the carbonaceous layer on the material surface.

The cooling action is an endothermic process that happens when FRs such as metal hydroxide are treated by the heat of flames. This results in the absorption of some transmitted energy to the material, causing the decrease in temperature of the condensed phase. The melt-dripping factor is performed by changing the rheology and viscosity of polymer, generating the isolation of polymer molecules from the flame, and therefore the easy dripping of the polymer. As shown in Figure 3, two processes are typically observed for the FRs that act in the gas phase, where inert gases can be released into the region of the flame. This helps to dilute flammable gases, and the free radicals derived from

FRs' decomposition can cause trapping of the energetic radicals (such as hydroxyl) to inhibit fire propagation.

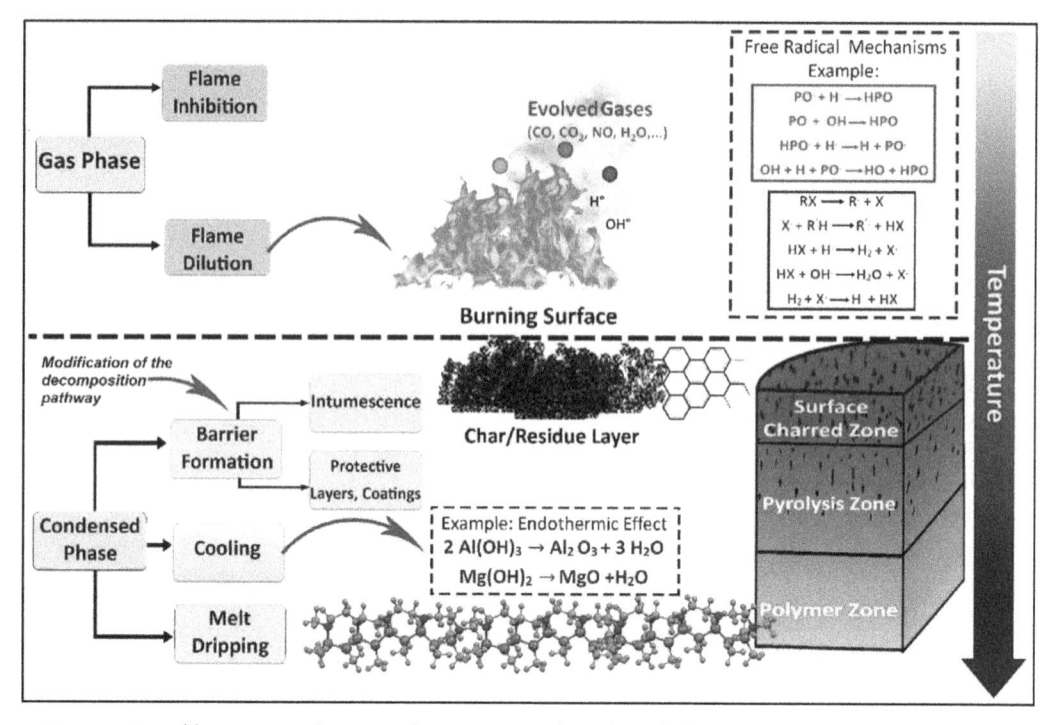

Figure 3. Possible action mechanisms of FRs in gas and condensed phases. Reproduced with permission from reference (31). Copyright 2021 Elsevier.

Advances on 2D FRs as PU Nanocomposites

Generally, nanomaterials with sheet morphology have proven to be efficient FRs because of their high thermal stability, lower thermal conductivity, and ability to form a strong carbonaceous layer on the polymer (32). Various 2D nanomaterials, as shown in Figure 4, are being added to the polymer with a lower weight percent (less than 5 wt %) for fabricating polymer nanocomposites with enhanced mechanical and thermal stability properties. These nanomaterials must be uniformly dispersed in the polymer matrix, as they are likely restacked and agglomerated, which affects the final properties of the polymer. There is also an incompatibility issue between these hydrophilic fillers and the organic polymer, resulting in the bad distribution in the polymer.

There are three common microstructures when the polymer is mixed with 2D materials. First, a phase-separated one occurs through the inability of the polymer to intercalate between the nanosheets, obtaining new conventional composites. Second, intercalated nanocomposite can be yielded when the polymer chains can intercalate through the nanolayers for designing an ordered alternative layered structure from the polymer and 2D layers. The last structure is the exfoliated structure that can be found in a case where there is complete and uniform dispersion of nanosheets in the polymer matrix (33). As a consequence, there are various approaches to performing the beneficial exfoliation, intercalation, and modification with other organic modifiers before mixing with the polymer (21, 32, 34–43).

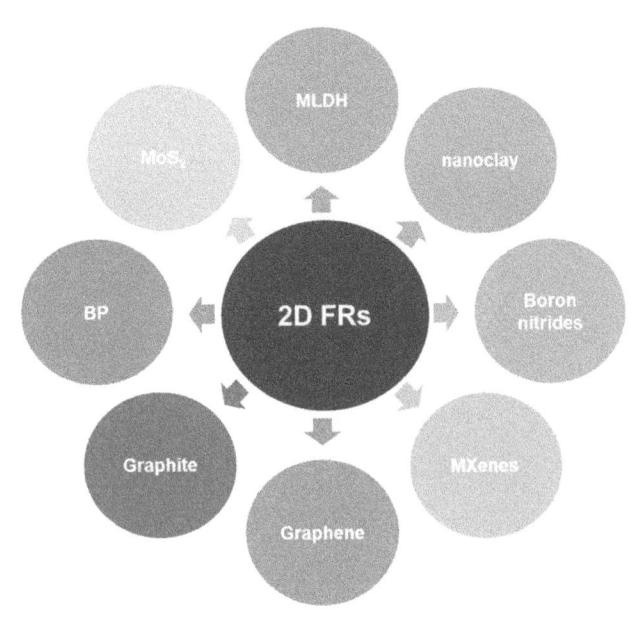

Figure 4. A schematic diagram for the various 2D FRs used for PUs.

Layered Carbon as PU Nanocomposites

Graphene as a PU Nanocomposite

Graphene is an interesting 2D nanomaterial composed of a hexagonal lattice from carbon atoms in the form of thin flat layers. This chemical structure of graphene offers a variety of distinctive features, such as high hydrophobicity, surface area, flexibility, current density, and thermal conductivity. It can be a promising material for energy storage devices, liquid crystals, chemical sensors, and polymer nanocomposites (44–47). Graphene and its derivatives have been broadly used as highly efficient FRs for improving the fire resistance of PUs. It helps in efficaciously decreasing the diffusion of mass and heat during the combustion cycle, as it forms a highly stable, compact, and large adsorptive layer on the polymer surface. As a result of continuous demand for eco-friendly and safe FRs, graphene is expected to be a reliable and promising material specified for such a purpose (16).

However, for the polymer composite, sheets of graphene are able to easily agglomerate by restacking through van der Waals forces, directly affecting the physical properties of the polymer (48). For this dilemma, another derivative of graphene, graphene oxide (GO), has been widely utilized instead of pure graphene in the field of polymer nanocomposites. GO is usually prepared through oxidizing and exfoliating the graphite layers, creating a structure with numerous oxygenated functional groups—such as epoxide, hydroxyl, carboxyl—and is found in edges and basal planes (49, 50). As GO is in an oxidation state for the structure of graphene, it shows numerous extra chemical and physical features over graphene. Unfortunately, there are still certain drawbacks when GO is mixed with PU, such as compatibility and interfacial interactions. This is because of the likely occurring hydrogen bonding between the oxygenated chains of graphene (51). Consequently, it is urgent to find a suitable approach for surface modifying of GO to enhance the compatibility of the polymer and achieve good mechanical properties and flame retardancy (35).

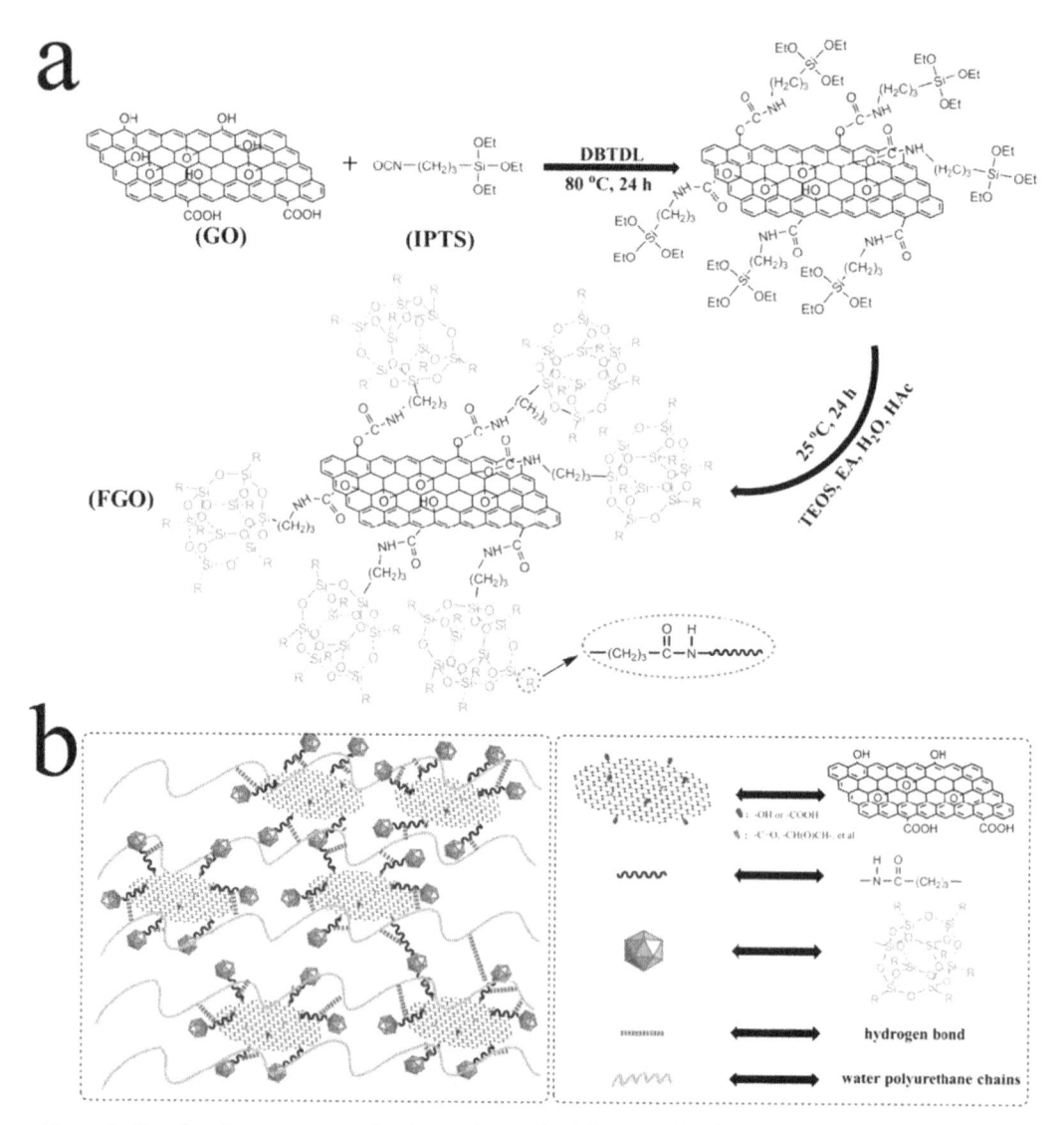

Figure 5. Graphical representation for the synthesis of (a) functionalized GO (FGO) and (b) PU/FGO nanocomposite films. Reproduced with permission from reference (52). Copyright 2019 Elsevier.

Du et al. (52) introduced the functionalization of GO using urethane–silica as an FR for waterborne PUs, as shown in Figure 5. The GO modification involves two steps. First, the isocyanatopropyltriethoxysilane agent was reacted with GO in the presence of dibutyltin dilaurate as a catalyst to create isocyanatopropyltriethoxysilane–functionalized GO (FGO) (see Figure 5a). In the second step, the previous product was treated with a tetraethoxysilane agent to prepare silica nanoparticles on a GO surface using the sol–gel method, as shown in Figure 5b.

The modified graphene was then added to PUs in mass ratios of 1, 2, and 3% using the common solution casting method. The authors used an indirect differential scanning calorimetry tool using the glass-transition temperatures (Tg) for investigating the compatibility of modified GO sheets with PUs matrices. Clearly, a lower Tg was detected for the PUs/untreated GO (−59.8 °C) compared with the pristine PUs (−55.3 °C) as a consequence of the aggregated GO sheets and their lower interfacial bindings with the polymer. Conversely, the addition of modified graphene in PUs caused increase of the Tg value of PUs to −41.1 °C when FGO was used with only 1 wt %. This was an indication for the confinement of the polymer chain near FGO sheets. In addition, the thermal stability for the PU composites was improved by exploiting FGO filler. The essential decomposition peak at 300–400 °C for the pure PU was obviously shifted to a higher temperature when FGO was dispersed in the polymer matrix.

This is because of the tortuous path raised from the modified graphene slowing down the volatile escapes and boosting the strength of the yielded char (53). The limiting oxygen index (LOI) and vertical burning (UL-94) tests were performed to investigate the flammability properties of the composite films. The LOI of PU film increased by 17.8–23.5%, when 3% from FGO was used aside from the composite achieved V-2 class from the UL-94 test.

In another study, Guo et al. (54) fabricated PU foam with excellent hydrophobicity, flame retardancy, and flexibility by the surface decoration with a mixture of GO and inorganic ammonium polyphosphate (APP) through electrostatic interaction, subsequently followed by functionalization with a silane agent. The existence of both APP and silane caused selective distributions on the surface of GO with apparent roughness and less susceptibility to water. Also, a synergism effect was noticed for the ternary coating in the flame resistance of PUs because the phosphorous and silicone elements were uniformly dispersed on GO to obtain a compact layer on the surface of the foam.

A similar dipping method was found by Jamsaz et al. for the fabrication of FR sponges from PUs with superoleophilic and superhydrophobic properties based on reduced GO (rGO) and orthoaminophenol (55). The modified sponge was burned after 60 s, while the virgin PU was completely fired after 6 s, suggesting the enhanced flame retardancy of the composited PUs.

Another interesting study was done for the electrochemical exfoliation and modification of graphene simultaneously with phosphazene rings to obtain sheets with few defects, as the ratio of carbon to oxygen equals 10.4 (56). The anodic graphite slice was exfoliated at 10 V for 6 h under ice with a cathodic copper rod in an electrolyte solution from hex(4-carboxylphenoxy)cyclotri phosphazene to give functionalized graphene nano sheets with the code of FGNS. It was expected that the noncovalent interaction with a cyclophosphazene structure could yield elements such as nitrogen and phosphorous that could assist in ameliorating the flame retardancy of a polymer. For this reason, the incorporation of FGNS in the PU matrix was conducted, and thermogravimetric analysis data in the air for the various composites were added, as shown in in Table 1. A negligible decrease was detected for the temperature at 5 wt % mass loss ($T_{5\%}$) compared with that of pure PU because of the thermal degradation of FGNS. However, the char yield and temperature at a maximum mass loss (T_{max}) were increased by increasing the loadings of FGNS. A cone calorimeter was used to evaluate the fire safety of PU nanocomposites, where incorporation of FGNS filler with a mass ratio of 4% caused the decrease of HRR of the pure polymer from 1431 to 768 kW/m².

The author described that the P- and N- containing hex(4-carboxylphenoxy)cyclo triphosphazene helped with the rapid construction of strong remaining char—in addition to the barrier role of graphene work—in inhibiting the escape of flammable products and retarding fire's progress on a PU surface.

A layer-by-layer method is a versatile approach that involves the alternative exposure of a substrate to negatively and positively charged solutions, resulting in the adsorption of a monolayer. Such a method has been proposed as a promising tool for the insertion of new FR coatings into various polymers (57–60). These coatings have received massive attention because of their easy fabrication and green chemistry. They can be efficient alternatives for immediate FRs that can insure health, flammability, and environmental specifications.

Table 1. Thermogravimetric Analysis Data for FGNS PU Nanocomposites in the Air[a]

Sample	$T_{5\%}$ (°C)	T_{max} (°C)	Char Residue at 700 °C (wt %)
PU	313.4	584.9	0.84
PU 1.0	313.0	581.3	1.31
PU 2.0	310.5	593.3	1.55
PU 4.0	309.2	627.1	2.41

[a] Reproduced with permission from reference (56). Copyright 2017 Elsevier.

Pan et al. (61) utilized a layer-by-layer technique for forming a coating from β-FeOOH nanorods and GO on flexible PUs to decrease their flammability properties, as shown in Scheme 2. The foam was immersed in the first cationic solution from the branched polyethylenimine (PEI), then in anionic sodium alginate (ALG)–graphene solution, and finally in a cationic suspension from the iron hydroxide nanorods. The obtained foam was dried and coated five times following the same protocol.

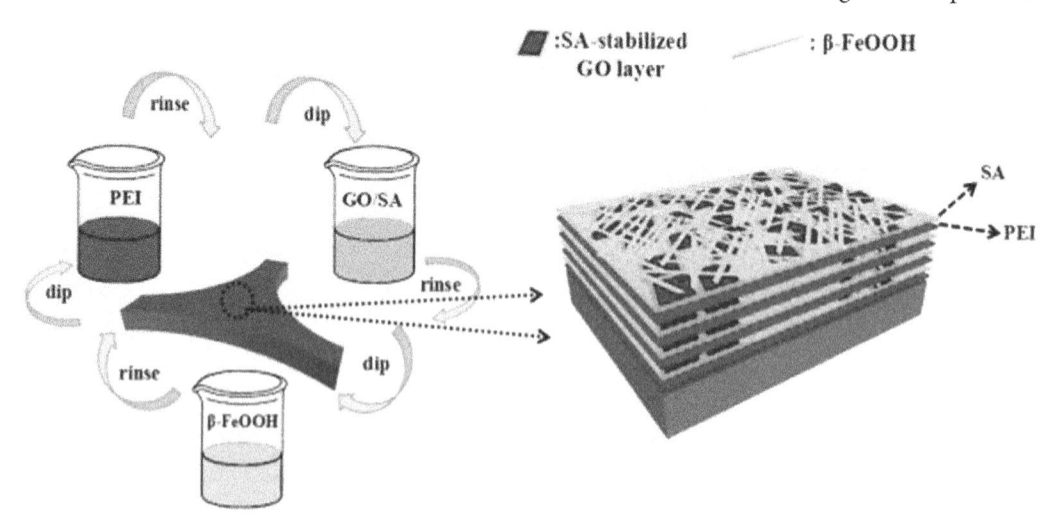

Scheme 2. Layer-by-layer assembly technique for preparing GO/β-FeOOH-coated PU foam. Reproduced with permission from reference (61). Copyright 2016 Elsevier.

The authors found that peak HRR (PHRR) for the bare PUs diminished from 738 to 370 kW/m². This achieved a decrease of 49.5% when self-assembling of GO and β-FeOOH nanorods happened to delay thermal decomposition by maintaining a physical barrier from the as-fabricated layers. This slowed down the conduction of the pyrolysis products to the gas-phase region.

Expandable Graphite as a PU Nanocomposite

Expandable graphite (EG), another 2D nanomaterial, is widely known as an intumescent system for reducing the fire propagation of PU, especially in foams. It is broadly utilized in vehicles and airplanes as seat material. More specifically, it can be used for thermal and sound insulation after required chemical treatment. EG structure is made up of flaky graphite sheets consisting of stacked layers with hexagonal oriented carbons in sp² hybridization systems, It is chemically intercalated with acids such as H_2SO_4, HNO_3, and CH_3COOH, as shown in Figure 6 (4, 28, 62).

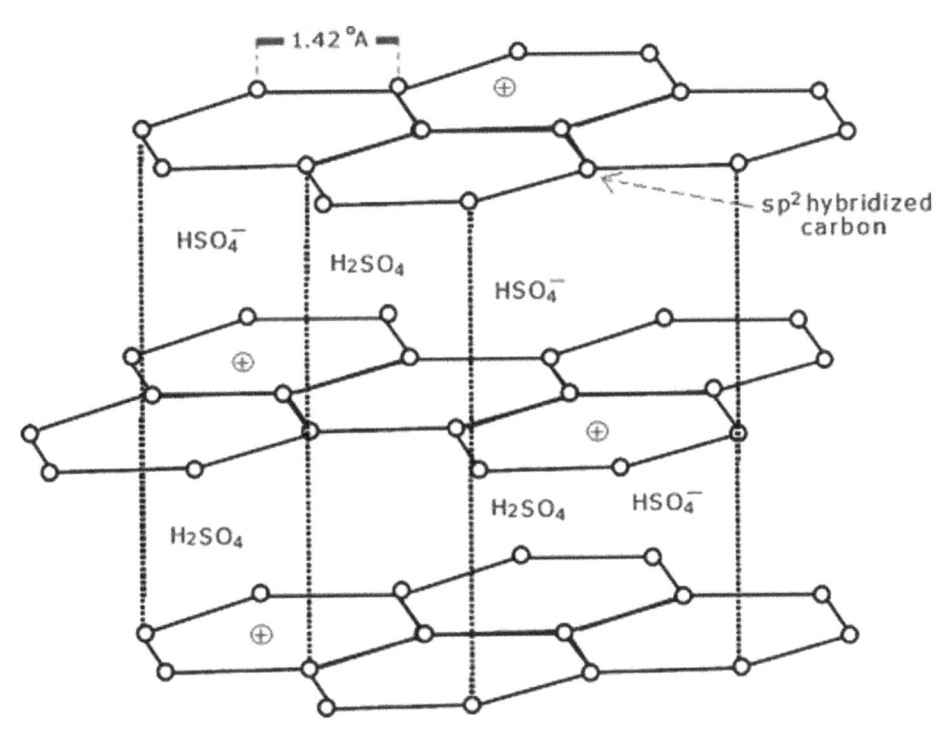

Figure 6. Structure of EG intercalated by H_2SO_4. Reproduced with permission from reference (4). Copyright 2009 Elsevier.

When the intercalated EG was heated, it exfoliated or expanded along the c-axis of the lattice by 100 times. EG does not contain any halogen atoms and it retards flames only in the condensed phase by creating an insulating barrier for suppressing smoke density and heat transfer. The exposure to the flame heat causes the formation of a wormlike thermal layer of a low density on the PU surface. The blowing reaction in EG can be conducted by the thermal degradation of sulfuric acid intercalated graphite. This yields gases such as SO_2 and CO_2, which along with the heat initiate a pressure that finally maximizes the distance between the basal planes of graphite (62, 63).

Wang et al. (64) attempted to coat the silicone resin and EG on PU foam using brush painting for acquiring high flame inhibition with extreme maintenance of the mechanical properties, as depicted in Figure 7.

The add-on for the coated layer was found to be 8.5 mg/cm² with a more mended fire retardancy behavior, where UL-94 tests revealed V-0 classification and LOI was maximized from 18 to 32.3%. In the cone calorimeter, the peaks of smoke release rate and HRR were diminished to 59 and 55% respectively concerning the pure foam. The ignition of coated foam also could only happen in the case of exposure for 10 min to the butane burner because of the tightly adhered layer from silicone resin/EG that could transfer under heat into a strong char layer from nanosilica-decorated EG.

Another study involves the incorporation of EG and 1,4-bis(diethylmethylenephosphonate) (DMP) piperazine in the precursors of PU, which are polymethylene polyphenyl isocyanate and polyether polyol, to obtain FR rigid PU foams. When EG has a mass ratio of 10% and 1,4-bis(DMP) has one of 15%, the PU achieves UL-94 V-0 and 25.7% LOI, with an increase in the char yield from 0.35 to 30.52% at 800 °C. Such data confirm the formation of a compact, intact, and thick char layer containing enormous amounts of nitrogen and phosphorous.

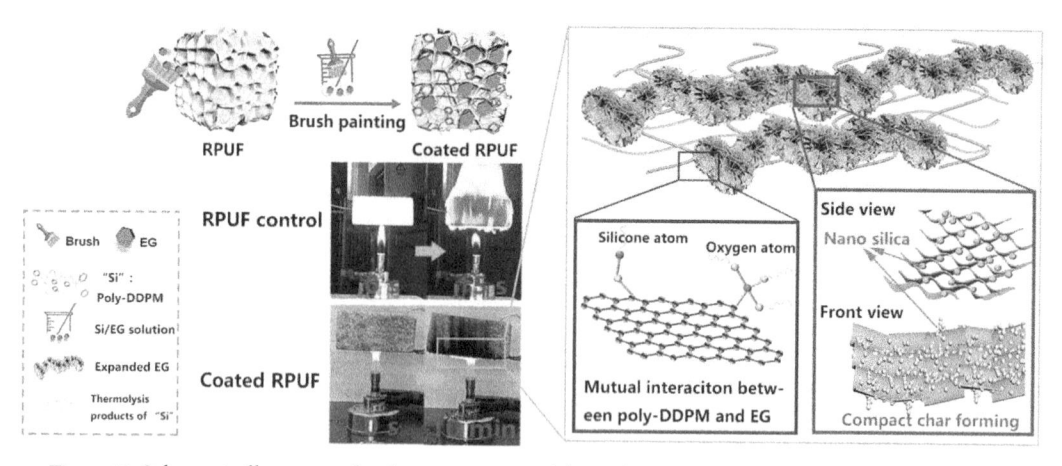

Figure 7. Schematic illustration for the preparation of the Si/EG-coated PU sample. Reproduced with permission from reference (64). Copyright 2020 Elsevier.

The performance of EG as an FR can be controlled by the amount, size, and density of used EG (65, 66). For example, the investigation of flame behavior of EG/PU foams with a density value of 0.035 g/cm³ was performed by Modesti et al. (67). They proposed that a raise in EG content could create a real improvement for the fire reaction. Additional research performed by Duquesne et al. revealed that using EG with a high density of 0.46–0.50 g/cm³ could outfit an FR with excellent properties (68). It was supposed that the utilization of PUs with EG could minify the evolution of harmful gases such as HCN or CO (69). Other research groups have explored the idea that EG with a low size diameter could display less volume expansion, suggesting an FR of lower efficiency (4, 65).

MoS$_2$ as a PU Nanocomposite

MoS$_2$ is a novel analog to graphite with an interesting 2D sheet morphology. The crystal unit contains three stacked layers from S–MO–S attached using van der Waals forces. MoS$_2$ with ultrathin nanosheet has prodigious features and receives tremendous interest for prospective applications such as lubrication, catalysis, and electronics. It is considered a nucleus for intercalation chemistry, where it is serving the host materials (70–72).

MoS$_2$ can be also exploited as a filler for inhibiting polymer flammability, as a consequence of its lower thermal conductivity and ability to form a compact insulation layer, aside from the capability to reduce the smoke production raised from the presence of Mo species (73). Previously, Zhou et al. introduced exfoliated MoS$_2$ for polystyrene to obtain greatly improved thermal stability, smoke suppression, and fire retardancy properties (74).

Exfoliated MoS$_2$ can face the problem of bad dispersion in a polymer matrix owing to the large surface area and van der Waals forces negatively affecting fire behavior. As a solution, the MoS$_2$ can be uniformly dispersed with a prosperous flame retardancy effect when it is modified at the surface with LDHs, metal oxide, graphene, and polyhedral oligomeric silsesquioxane (75–78). Few reports have been found for utilizing MoS$_2$ as an FR for PU foams.

Zhi et al. (79) demonstrated grafting of MoS$_2$ at the surface with 9,10-dihydro-9-oxa-10-phosphaphenanthrene-10-oxide (DOPO), utilizing allyl mercaptan as a smart FR for PU foam, as shown in Scheme 3. The mechanical properties of PUs were not changed largely when MoS$_2$–DOPO was incorporated, but the char yield was considerably enhanced. The PHRR, total heat release (THR), and maximum smoke density for PU composites with MoS$_2$–DOPO obtained from a cone calorimeter and smoke density chamber revealed a decrease of 41.3, 27.7, and 40.5%, respectively, compared with the bare foam. The authors gave three probable mechanisms for such interesting data. First, DOPO molecules that incapsulated MoS$_2$ could achieve compatibility with the polymer through a firm interface interaction, favorably exerting a synergistic influence in the flame retardancy recorded for DOPO and MoS$_2$. The second cause was the ability of 2D MoS$_2$ nanosheets to adsorb aromatic compounds and hydrocarbons through the weak van der Waals binding, achieving efficacious chemisorption and physisorption. Additionally, MoS$_2$ could produce a physical layer, assisting the retardation in liberating the combustible gases. Such a barrier and absorption dual functionality were important for aggregating smoke particles for forming char residues on the MoS$_2$ surface. Also, there was oxidation of MoS$_2$ to MoO$_3$ nanoparticles via the existing energetic oxygen species in the flame region. Such nanoparticles could ameliorate the thermal stability of the formed barrier layer. Third, the FR of DOPO could act simultaneously in the condensed and gas phases, where DOPO decomposed to release PO• radicals to capture the energetic hydrogen and hydroxyl-free radicals in the flame region. This stopped their chain reactions and thus decreased the supply of heat and obtained incomplete combustion. In the condensed phase, DOPO could catalyze char formation by degradation to yield oxygen-containing phosphorus acids, This established the dehydration and charring reactions of the hydroxyl-based compounds for producing a phosphorus-containing carbonaceous layer (80–83). Another author studied the preparation of FR coating based on MoS$_2$ and chitosan (CS) for PUs using the layer-by-layer approach. The foam-containing MoS$_2$ caused burning without melt dripping. After the combustion test, the coated sample still remained in its initial shape, whereas the bare PU was entirely burned. A cone calorimeter showed that a PU with a mass loading of 8.5 wt % attained a remarkable decrease in the peak of HRR by 70%, in smoke production rate by 62.4%, and in total smoke released by 33.3% (84).

Carbides/Carbonitrides as PU Nanocomposites

Graphene was primarily discovered in the laboratory in 2004. A variety of 2D nanomaterials, such as BN and transition-metal dichalcogenides, have been inspected thoroughly because of their unrivaled physicochemical properties resulting from their unique ultrathin structure. Transition-metal carbides/carbonitrides (MXenes) have been included as novel potential members of the 2D nanomaterials' family. They are essentially prepared from the precursors of MAX phase by the selective etching of layered group 3A or 4A elements, where X refers to carbon or nitrogen and M is an early transition metal. Interestingly, a family of more than 70 structures with hexagonal, layered transition-metal nitrides and carbides can be harvested. To negatively charge the surface of MXenes, fluoride-based acids can be utilized, where hydrofluoric acid etching can at most obtain fluoride-based function groups on the MXenes surface, while LiF–HCl can produce predominately oxygen-containing groups. MXenes have shown a potent capability in a wide range of advanced areas, including catalysts, electromagnetic shielding, and energy storage devices. This is owing to their inherent flexibility, mechanical strength, and high conductivity (85–88). While MXenes are participating in these areas, there is a shortage in reports of exploiting these nanosheets as reinforcing agents for retarding smoke production, retarding flame growth, and mechanical properties of a polymer.

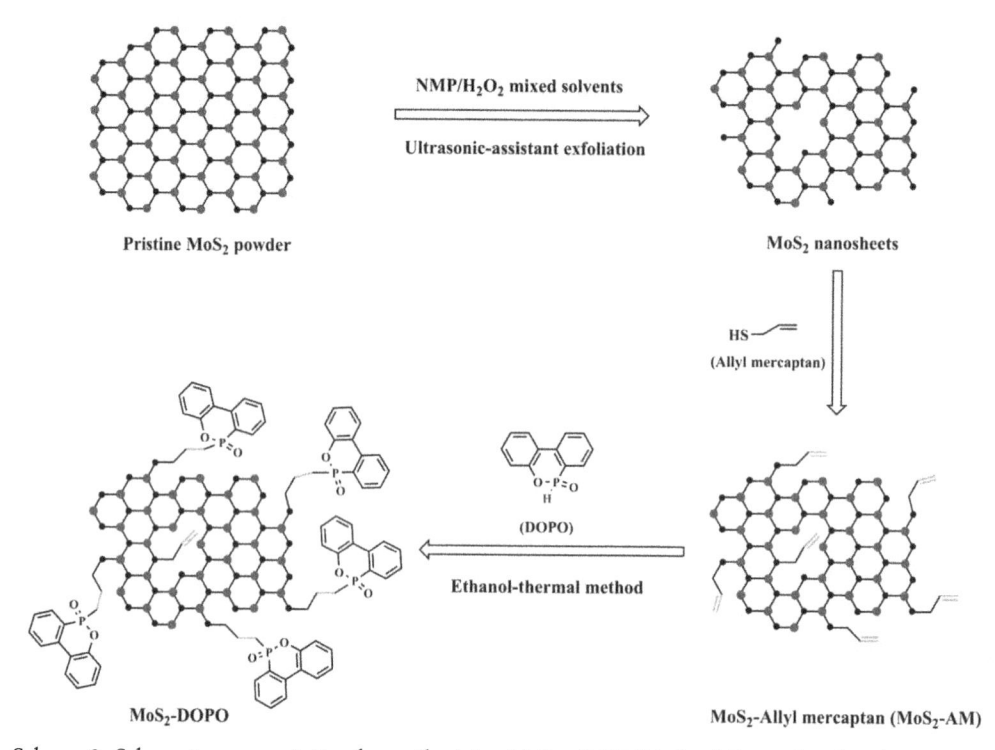

Scheme 3. Schematic representation for synthesizing MoS$_2$–DOPO hybrid. Reproduced with permission from reference (79). Copyright 2020 American Chemical Society.

Like graphene, the adjacent sheets of MXenes are unavoidably aggregated by van der Waals forces during material processing. Drying leads to the accumulation of 2D materials and deteriorates the mechanical and flammability properties. Consequently, surface modification with organic molecules can solve the aforementioned issue, where the distribution of 2D sheets in polymer chains can form a coherent insulator layer on a polymer surface during combustion. This prolongs the time of polymer degradation and decreases the evolution of flammable gases to the flame zone (89).

He et al. (85) followed the same strategy by employing the wet ball-milling approach for the functionalization and exfoliation of MXenes with a negative charge with long chains from poly(diallyldimethylammonium chloride) (PDDAC) for preventing recombination of the peeled-off sheets. This decreased the collision obtained from the steel balls and postponed the MXenes' oxidation in the air (see Figure 8a). More substantially, when PDDAC-coated MXene was incorporated in PUs with a mass ratio of only of 3%, the total smoke production and PHRR were minimized by 47 and 50%, respectively. The tensile strength at break and thermal conductivity were maximized in sequence by 31.2 and 88.6%. As shown in Figure 8b, the flame retardancy of PDDAC-modified Ti$_3$C$_2$ sheets was performed in two cases. In the beginning, the well-distributed nanosheets in the matrix of PU had a vital role in establishing a barrier layer for the suppression of mass and heat transfer. When the temperature was going up, the heat absorption caused the transformation of Ti$_3$C$_2$ into TiO$_2$ and carbon. The existence of well-dispersed TiO$_2$ into the carbon residue might have altered the redox pathway that happened in the fire, catalyzing the production of highly graphitized carbon char to suppress the release of flammable gases.

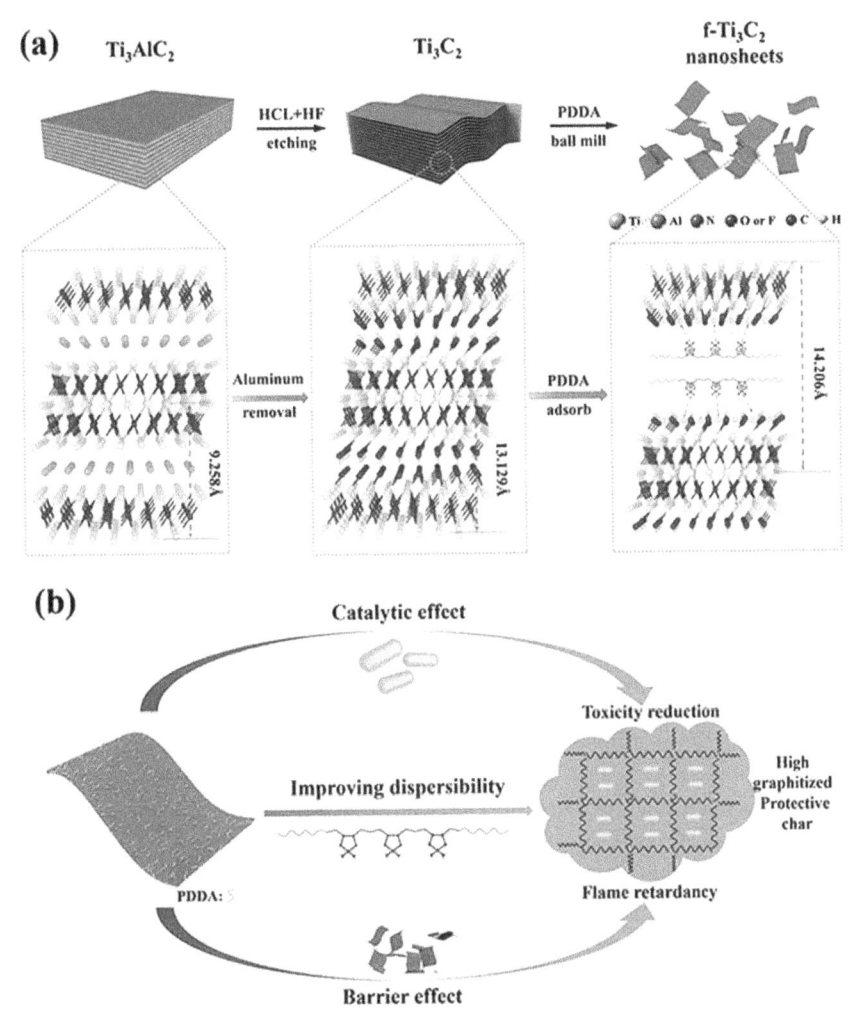

Figure 8. (a) A graphical representation for synthesizing functionalized Ti_3C_2, and (b) its mechanism for retarding flames on the surface of PU composites. Reproduced with permission from reference (85). Copyright 2019 Elsevier.

Another research group coated MXene with tetrabutyl phosphine chloride and cetyltrimethyl ammonium bromide. This was utilized as a filler for improving fire safety properties and the emitted toxic gases from the TPU because of the aforementioned barrier and catalytic role of MXenes (90).

An attractive study was conducted by Liu et al. (91) where the mutual intercalation and assembly of rGO on the sheets of titanium carbide through hydrogen bonding was done hydrothermally at 180 °C for 24 h to find out the aggregation obstacle of 2D nanomaterials. The as-assembled filler was presenting with homogenous dispersion in a PU elastomer as a result of the good compatibility and interface adhesion reflecting good thermal stability. There was also a remarkable decrease in the smoke and carbon monoxide gas released from the polymer because of the catalytic charring and heat absorption behaviors of the filler.

Nanoclays as PUs Nanocomposites

The development of polymeric material–containing FRs has gained wide attention from both academic and industrial global researchers. This is because they exhibit outstanding improvement in flame retardancy of polymer by inhibiting ignition and enhancing polymer toughness. In addition, they slow flame spread compared with unmodified polymer (92–94). Nanoclays as naturally occurring minerals are vastly applied in the form of phyllosilicate or nanosheet structures with 1D thicknesses and surfaces of approximately 1 nm and 50–150 nm, respectively. Nanoclays are classified into different categories such as montmorillonite (MMT) and kaolinite, based on their chemical composition and the nanoparticles' morphology.

MMT as a PU Nanocomposite

MMT is a natural lamellate layered mineral composed of sandwiched octahedral aluminum hydroxide as an inner layer between two silicate sheets. As MMT has a high hydrophilic nature, it absorbs a great amount of water that increases significantly in volume (95, 96). MMT nanoclay has been utilized in various applications for flame retardancy, drug delivery, wastewater treatment, and energy storage because of low-cost material, less shrinkage, higher thermal stability, easy availability, natural abundance, and cationic exchangeability. As a result, MMT silicate layer nanoclay has been utilized to improve both the mechanical and combustion performance of PU through achieving compatibility between the inorganic MMT layers and the polymer, depending on the conditions of processing and their bonding. The layered silicate in low amounts disperses through polymeric chains by exfoliation and or intercalation states (92). In terms of nanoclay dispersion into the polymer matrix, there are three mentioned ways: exfoliation, intercalation, and microcomposite for preparing MMT–polymer nanocomposites.

A great deal of literature has reported the exfoliation capability of MMT clay (97, 98), which can be done through three prepared techniques including solution blending, in situ intercalative polymerization, and melt compounding (99–101). The last two techniques are being applied to hydrophilic polymers because of their ability to intercalate the polymer and MMT silicate layers (102), although exfoliation of clay has not been achieved reproducibly through a hydrophobic polymer, as PUs have faced a problem. Preexfoliation of an aluminum silicate clay surface, which is prepared with a blending solution technique, has attracted the most interest, as it is an effective way to ensure MMT exfoliation within different substrates; avoid agglomeration of the MMT layers; improve the MMT nanolayers' dispersion; and increase the surface active sites onto MMT layers through surface modification to be widely applied (103).

Liu et al. (104) reported the enhancement of PU's flame retardancy by the addition of exfoliated and modified MMT sheets with phosphorylated CS (PMT) under an ultrasonic technique, as illustrated in Figure 9a. They then functionalized the structure of TPU with organically modified MMT using extrusion–pelletization and hot compression molding in the presence of aluminum hypophosphite (AHP). The flammability resistance of PU nanocomposites was first studied by measuring LOI values and UL-94 rate. The findings demonstrated that LOI values increased significantly from 20.8% for unmodified PU to a peak of 28.4% in the case of PU/PMT/AHP, with a ratio of 90/1/9 and a UL-94 rate reaching V-0, as shown in Figure 9b. Additionally, the PHRR and total smoke rate (TSR) of PUs sharply plunged from 1090 to 195 kW/m^2 and from 831 to 91 m^2/m^2, respectively when the TPU/PMT/AHP nanocomposite was utilized.

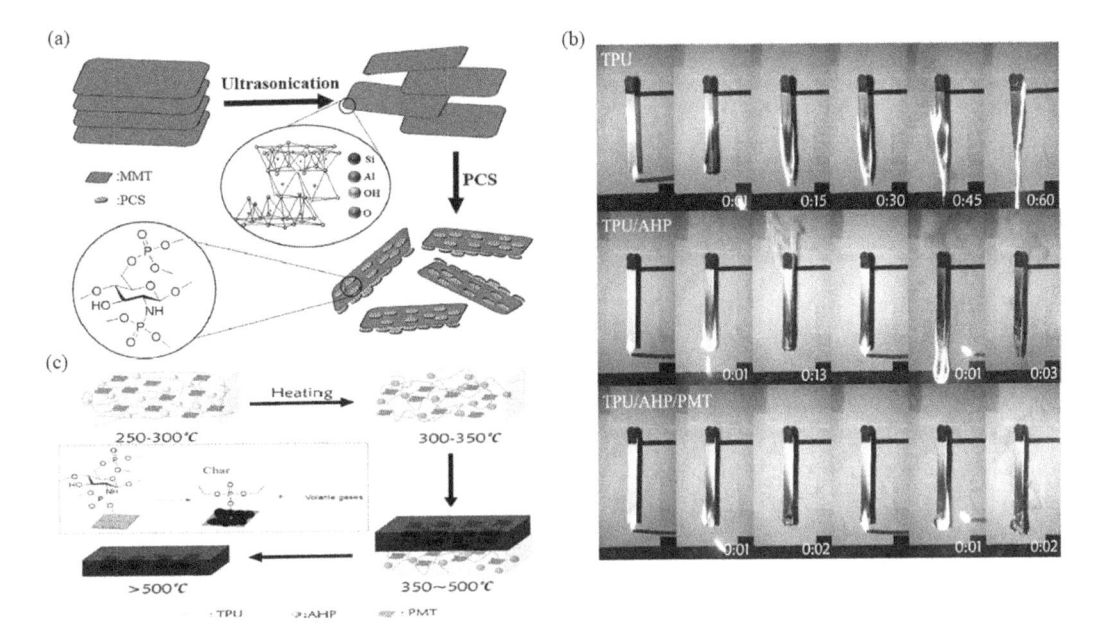

Figure 9. (a) Schematic preparation of phosphorylated CS-modified MMT sheets, (b) digital camera images for the combustion process of TPU nanocomposites using the (c) the proposed mechanism for flame retardancy of TPU nanocomposite. Reproduced with permission from reference (104). Copyright 2019 Elsevier.

Kaolin as a PU Nanocomposite

In another 2D silicate nanoclay, kaolin (KA) is also a natural white layered nanoclay with a chemical formula of $Al_2Si_2O_5(OH)_4$, an asymmetric negatively plated structure, and a superposition with an equal ratio between the tetrahedral and octahedral sheets (*105, 106*). Because of its outstanding properties—such as natural abundance, low cost, shielding effect on polymer, good adsorption behavior, thermal stability, and good whiteness—it can be exploited as a filler to reduce the fire growth behavior of polymers, especially PUs. Seifi et al. (*107*) boosted the thermal stability of KA through the intercalated modified KA with urea through a method of mixing and ball milling together at room temperature.

Thermal stability and flame retardancy of PUs also were enhanced by using KA and castor oil in a study by Agrawal and his coworkers (*108*) where the temperature of 5% mass loss was increased to 260 °C from 192 °C with unmodified PUs, giving a char residue percentage of around 9%. The cone calorimeter technique revealed a sharp decrease in PHRR, THR, and TSR from 118 kW/m², 29.8 MJ/m², and 347 m²/m², respectively for unmodified PU to 98 kW/m², 16.4 MJ/m², and 129 m²/m², respectively for modified PU.

LDHs as PU Nanocomposites

LDHs structurally consist of positively charged brucite layers with intercalated regions including compensated anionic charges. Generally, the chemical formula of LDHs is $(M^{2+}_{1-x} M^{3+}_{x} (OH)_2/(A^{n-}_{x/n} \cdot H_2O)$. In this formula, M^{2+} represents divalent metal ions (including Mg^{2+}, Ca^{2+}, and Fe^{2+}); M^{3+} is the trivalent one (as Al^{3+}, Fe^{3+}, and Co^{3+}); A^n represents intercalated anionic compounds (such as bromides, nitrates, chlorides, and carbonates); and M is the molar ratio between trivalent and divalent cations. Specifically, upon exposure to the flame, the thermal decomposition of LDHs to evolve into water vapor causes the endothermic absorption of heat and forms a carbon char containing thermally stable inorganic minerals. They can be utilized as efficient FRs for polymers (32, 109, 110).

LDHs can be modified with natural polymers to improve the properties of PUs. For example, Liu et al. (111) enhanced the flame resistance of PUs by submersion in three solutions from CS, ALG, and LDHs such as MgAl LDHs and NiAl LDHs. The authors found that NiAl-layered hydroxides with a positive charge could absorb the negative CS/ALG polyelectrolyte more than MgAl layers. The as-coated PUs were also observed to have lower PHRR, THR, and TSR compared with uncoated ones, owing to the LDHs' physical barrier shield.

Particularly, the coating with NiAl LDHs shows more effective flame retardancy given by the cone calorimeter data than that with MgAl LDHs.

In another trial, rGO was decorated with MgAl LDHs using a coprecipitation method and was modified with heptaheptamolybdate via the ion exchange method employed by Xu et al. (112). The former filler was utilized as a smart FR and smoke suppression agent for PUs. The PUs composites exhibited a drastic decline in both PHRR and TSR by about 36.4 and 18.2%, respectively, because of the formation of LDHs/rGO/Mo physical barrier on the PU surface, along with MoO_3, during the decomposition of nanocomposite onto the surface of PU. This has promoted smoke suppression and catalytic carbonization effects.

Black Phosphorous as a PU Nanocomposite

There is a large interest in 2D black phosphorous (BP) from both industrial and academic perspectives. Unlike graphene of zero bandgap, excellent carrier mobility, direct bandgap, and a higher on/off ratio have put BP of few-layered structures as a successive substrate in optoelectronics and chemicals/biosensing (113, 114). However, BP is always suffering from poor stability in ambient conditions (oxygen and moisture). The mechanism of BP degradation proceeds via three steps. The absorption of oxygen molecules on a BP surface comes first. This results in the transformation into O_2^- by the combination with free electrons raised from the illumination of the BP sheet. In the second step, the dissociation of O_2^- creates the bond of P=O. Finally, water molecules are able to catch the oxygen of this bond through a hydrogen bond, leading to the removal of phosphorous and breaking the top layer on the BP surface (115). To overcome this, a mechanism of protection is achieved by passivating the reactivity of a lone pair electron and the isolation coating. The structure of BP contains entirely phosphorous elements that could retard a fire reaction and improve flame retardancy more than other reported layered nanostructures, especially for PUs (116). In an attractive work, tannin (TA), a natural antioxidant, was exploited for the functionalization of BP sheets for yielding improved ambient stability and suppression of toxic gases for its nanocomposites with PUs. From cone calorimeter data, the incorporation of modified BP with 2 wt % into the matrix of PUs causes a sharp decrease in PHRR, THR, carbon dioxide concentration, and total amount of carbon mono gases by 56.5, 43, 57.3, and 55.1%, respectively. Mechanical properties are totally enhanced (116). The authors hypothesized that good interfacial bonding between TA and BP creates the good dispersion of BP sheets in PUs for forming a good barrier layer on a polymer surface. This therefore hinders the delivery of flammable gases. Also, the thermal decomposition of BP sheets can give quenching agents from phosphorous-based species radicals that react with carbon radicals to inhibit the combustion cycle. TA can decompose thermally and form a char layer by reacting with radicals coming from a polymer and increasing graphitization content to build a strong carbonaceous layer and protect the polymer from the flame.

BN as a PU Nanocomposite

In addition, 2D hexagonal BN (h-BN) nanosheets have gained attention from researchers around the world because of their properties such as high thermal stability, high thermal conductivity, and excellent chemical inertness. These features create great potential for h-BN in different applications, especially in the polymer matrix composite (117–119). Therefore, h-BN can enhance the thermal conductivity of different polymers. For example, Lin et al. (120) reported the improvement of the thermal conductivity of epoxy resin by 113% in the presence of exfoliated h-BN nanosheets between resin chains. Also, Yu et al. (121) prepared h-BN nanosheets/TPU composites through a ball-milling technique in N,N-dimethylformamide. The authors reported that the thermal conductivity of PUs increased sharply by up to 50.3 $Wm^{-1}K^{-1}$. Moreover, h-BN has been incorporated inside polymer chains for creating a rigid barrier to prevent the transfer of decomposed fragments; this boosts the fire retardancy of the polymer because of its significant thermal stability (122, 123).

Wang and his coworkers (124) modified the TPU structure with multifunctional h-BN nanosheets throughout the wrappings of phytic acid–doped polypyrrole, which was followed by adsorption of copper ions. The results exhibited a considerable decrease in the emission of toxic volatile gases (CO and HCN). The authors showed that PU/multifunctional h-BN nanosheet nanocomposites improved the fire resistance of PU, so cone calorimeter data have illustrated a sharp decrease in PHRR and TSR by about 35.6 and 5.0%, respectively (124).

Yin S. et al. (125) conducted synergistic flame retardancy studies for PUs by preparing high-performance FR composites from black phosphorene and h-BN nanosheets. They demonstrated an increase of LOI value up to 33.8% compared with 21.7% for virgin PU. Additionally, the fire-resistance data displayed a significant decline in both PHRR and THR by 50.94 and 23.92%, respectively, using only 0.4 wt %. The data also showed that the char residue percentage considerably increased by approximately 10.24%. Various reported FR systems containing 2D nanomaterials and their tested performances are listed in Table 2.

Table 2. The Flammability Properties of PU Nanocomposites Containing 2D Nanomaterials

PU Composites	Organic Modifier	LOI (%)	PHRR (kW/m²)	THR (MJ/m²)	TSR (m²/m²)	Char (%)	Ref.
PU/MMT/Bi₂O₃	Bi₂O₃	Increased to 23.5	Decreased from 1200 to 635	Declined from 153.2 to 80.5	-	-	(126)
PU/MMT/DETC[a]	DETC	-	Declined between 523 and 391	From 90 to 84	-	Increased by 3%	(127)
Rigid PU foams/MMT–AP[b] and DMP	AP and DMP	28.5	64% reduction	-	-	Boosted by 24%	(128)
PU/MMT	Melamine polyphosphate	Increased to 27.5	Reduced from 923 to 243	-	-	Increased by 9.47%	(129)
Flexible PU/ MMT-starch	Starch	-	54% reduction	-	Reduced by 39%	-	(130)
PU foam/PAA[c] and PDDAC–MMT	PAA and PDDAC	-	Reduced from 300 to 150	Decreased by 0.3	Reduced by 34%	Increased to 25%	(131)
PU/KA/CS/PAA/APP	CS, PAA, ammonium polyphosphate	-	Dipped from 488.8 to 162.4	Decreased by 7.5%	Decreased from 548.3 to 222.9	Enhanced to 91.3%	(132)
PU/Zn-Al LDHs/CNTs[d]	Zn-Al LDHs/CNTs		−13.5%	−4%	16.3%	(133)	
Rigid PU/MgAl LDHs	MgAl LDHs	29.5				27.84%	(134)
PU/lanthanum/ MgAl LDHs/GO	Lanthanum/MgAl LDHs/GO	23.2	−33.1%	−0.9	−51%	9.7	(135)
Flexible PU/h-BN/PEI/ALG	h-BN/PEI/ALG		−50.1%	−0.1	+1.6%	23.2	(136)
TPU/h-BN/cetyl–trimethylammonium bromide	h-BN/ cetyl–trimethylammonium bromide		−57.5%	−17.8%		8.3	(137)
PU/h-BN/SiO₂/phytic acid	h-BN/SiO₂/phytic acid	22.5	−13.9%	−22.1%	−6.1%	6.11	(138)
PU/h-BN/CuMoO₄	h-BN/CuMoO₄		−73.6%	−52.4%	−28.2%	24.4	(139)

[a] DETC, 2-(2-(5,5-dimethyl-1,3,2-dioxaphosphinyl-2-ylamino)-N,N,N-triethyl-2-oxoethanaminium chloride. [b] AP, ammonium phosphate. [c] PAA, polyacrylic acid. [d] CNT, carbon nanotube.

Summary and Perspective

In general, the unique features for nanomaterials of layered sheets—such as GO, EG, MMT, LDHs, BP, BN, and MoS_2—help with applications in many distinct fields, especially in the flame retardancy of polymers. The addition of these materials in PUs with low mass loading can assist in providing remarkable flame retardancy behavior. The fabrication and design of different FR systems can be obtained based on adjusting the synergism and catalysis influences. The understanding of the polymer nanocomposite field has allowed the development of modern products with specified features and functions. Straightforward research on the utilization of 2D layered nanomaterials as reinforcing agents for retarding the propagation of fire is required, regarding several paramount issues: (1) the compatibility between the hydrophobic polymer and the hydrophilic 2D FRs; (2) the synergism effect between 2D FRs and conventional FRs; and (3) the structural engineering of 2D nanomaterials for enhancing their properties. The mechanism of flame retardation is varied based on the applied 2D materials in PUs, such as BP sheets that can work in both condensed and gas phases. Considering the wide range of fields for FRs, much research is currently directed toward the design of green and multifunctional ones. Without a doubt, polymer nanocomposites based on 2D layered nanomaterials will find broad areas of use.

References

1. Ghasemlou, M.; Daver, F.; Ivanova, E. P.; Adhikari, B. Polyurethanes from seed oil-based polyols: A review of synthesis, mechanical and thermal properties. *Industrial Crops and Products* **2019**, *142*, 111841.
2. Nohra, B.; Candy, L.; Blanco, J.-F.; Guerin, C.; Raoul, Y.; Mouloungui, Z. From Petrochemical Polyurethanes to Biobased Polyhydroxyurethanes. *Macromolecules* **2013**, *46* (10), 3771–3792.
3. Zia, K. M.; Anjum, S.; Zuber, M.; Mujahid, M.; Jamil, T. Synthesis and molecular characterization of chitosan based polyurethane elastomers using aromatic diisocyanate. *International Journal of Biological Macromolecules* **2014**, *66*, 26–32.
4. Chattopadhyay, D. K.; Webster, D. C. Thermal stability and flame retardancy of polyurethanes. *Progress in Polymer Science* **2009**, *34* (10), 1068–1133.
5. Woods, G. *Flexible Polyurethane Foams: Chemistry and Technology*; Applied Sci. Publishers: GBR; London; New Jersey, 1982; 334 p.
6. Van Maris, R.; Tamano, Y.; Yoshimura, H.; Gay, K. M. Polyurethane Catalysis by Tertiary Amines. *Journal of Cellular Plastics* **2005**, *41* (4), 305–322.
7. Silva, A. L.; Bordado, J. C. Recent Developments in Polyurethane Catalysis: Catalytic Mechanisms Review. *Catalysis Reviews* **2004**, *46* (1), 31–51.
8. Dworakowska, S.; Bogdał, D.; Zaccheria, F.; Ravasio, N. The role of catalysis in the synthesis of polyurethane foams based on renewable raw materials. *Catalysis Today* **2014**, *223*, 148–156.
9. Chen-Tsai, C. H. Y.; Thomas, E. L.; MacKnight, W. J.; Schneider, N. S. Structure and morphology of segmented polyurethanes: 3. Electron microscopy and small angle X-ray scattering studies of amorphous random segmented polyurethanes. *Polymer* **1986**, *27* (5), 659–666.
10. Qi, H. J.; Boyce, M. C. Stress–strain behavior of thermoplastic polyurethanes. *Mechanics of Materials* **2005**, *37* (8), 817–839.

11. Petrović, Z. S.; Ferguson, J. Polyurethane elastomers. *Progress in Polymer Science* **1991**, *16* (5), 695–836.
12. Estes, G. M.; Seymour, R. W.; Cooper, S. L. Infrared Studies of Segmented Polyurethane Elastomers. II. Infrared Dichroism. *Macromolecules* **1971**, *4* (4), 452–457.
13. Nie, S.; Peng, C.; Yuan, S.; Zhang, M. Thermal and flame retardant properties of novel intumescent flame retardant polypropylene composites. *Journal of Thermal Analysis and Calorimetry* **2013**, *113* (2), 865–871.
14. Hassan, M.; Nour, M.; Abdelmonem, Y.; Makhlouf, G.; Abdelkhalik, A. Synergistic effect of chitosan-based flame retardant and modified clay on the flammability properties of LLDPE. *Polymer Degradation and Stability* **2016**, *133*, 8–15.
15. Xue, C.-H.; Wu, Y.; Guo, X.-J.; Liu, B.-Y.; Wang, H.-D.; Jia, S.-T. Superhydrophobic, flame-retardant and conductive cotton fabrics via layer-by-layer assembly of carbon nanotubes for flexible sensing electronics. *Cellulose* **2020**, *27* (6), 3455–3468.
16. Goda, E. S.; Abu Elella, M. H.; Hong, S. E.; Pandit, B.; Yoon, K. R.; Gamal, H. Smart flame retardant coating containing carboxymethyl chitosan nanoparticles decorated graphene for obtaining multifunctional textiles. *Cellulose* **2021**, *28*, 5085–5105.
17. Abu Elella, M. H.; ElHafeez, E. A.; Goda, E. S.; Lee, S.; Yoon, K. R. Smart bactericidal filter containing biodegradable polymers for crystal violet dye adsorption. *Cellulose* **2019**, *26* (17), 9179–9206.
18. Abu Elella, M. H.; Goda, E. S.; Yoon, K. R.; Hong, S. E.; Morsy, M. S.; Sadak, R. A.; Gamal, H. Novel vapor polymerization for integrating flame retardant textile with multifunctional properties. *Composites Communications* **2021**, 100614.
19. McKenna, S. T.; Hull, T. R. The fire toxicity of polyurethane foams. *Fire Science Reviews* **2016**, *5* (1), 3.
20. Wang, J.; Wang, H.; Geng, G. Flame-retardant superhydrophobic coating derived from fly ash on polymeric foam for efficient oil/corrosive water and emulsion separation. *Journal of Colloid and Interface Science* **2018**, *525*, 11–20.
21. Goda, E. S.; Yoon, K. R.; El-sayed, S. H.; Hong, S. E. Halloysite nanotubes as smart flame retardant and economic reinforcing materials: A review. *Thermochimica Acta* **2018**, *669*, 173–184.
22. Daimatsu, K.; Sugimoto, H.; Kato, Y.; Nakanishi, E.; Inomata, K.; Amekawa, Y.; Takemura, K. Preparation and physical properties of flame retardant acrylic resin containing nano-sized aluminum hydroxide. *Polymer Degradation and Stability* **2007**, *92* (8), 1433–1438.
23. Yan, L.; Xu, Z.; Zhang, J. Influence of nanoparticle geometry on the thermal stability and flame retardancy of high-impact polystyrene nanocomposites. *Journal of Thermal Analysis and Calorimetry* **2017**, *130* (3), 1987–1996.
24. Dittrich, B.; Wartig, K.-A.; Hofmann, D.; Mülhaupt, R.; Schartel, B. The influence of layered, spherical, and tubular carbon nanomaterials' concentration on the flame retardancy of polypropylene. *Polymer Composites* **2015**, *36* (7), 1230–1241.
25. Elbasuney, S. Surface engineering of layered double hydroxide (LDH) nanoparticles for polymer flame retardancy. *Powder Technology* **2015**, *277*, 63–73.

26. Zhao, M.; Yi, D.; Yang, R. Enhanced mechanical properties and fire retardancy of polyamide 6 nanocomposites based on interdigitated crystalline montmorillonite–melamine cyanurate. *Journal of Applied Polymer Science* **2018**, *135* (13), 46039.
27. Kai, M. F.; Zhang, L. W.; Liew, K. M. Graphene and graphene oxide in calcium silicate hydrates: Chemical reactions, mechanical behavior and interfacial sliding. *Carbon* **2019**, *146*, 181–193.
28. Guler, T.; Tayfun, U.; Bayramli, E.; Dogan, M. Effect of expandable graphite on flame retardant, thermal and mechanical properties of thermoplastic polyurethane composites filled with huntite&hydromagnesite mineral. *Thermochimica Acta* **2017**, *647*, 70–80.
29. Zhang, Y.; Zhu, W.; Li, J.; Zhu, Y.; Wang, A.; Lu, X.; Li, W.; Shi, Y. Effects of ionic hydration and hydrogen bonding on flow resistance of ionic aqueous solutions confined in molybdenum disulfide nanoslits: Insights from molecular dynamics simulations. *Fluid Phase Equilibria* **2019**, *489*, 23–29.
30. Schartel, B.; Wilkie, C. A.; Camino, G. Recommendations on the scientific approach to polymer flame retardancy: Part 2—Concepts. *Journal of Fire Sciences* **2017**, *35* (1), 3–20.
31. Vahabi, H.; Laoutid, F.; Mehrpouya, M.; Saeb, M. R.; Dubois, P. Flame retardant polymer materials: An update and the future for 3D printing developments. *Materials Science and Engineering: R: Reports* **2021**, *144*, 100604.
32. Yue, X.; Li, C.; Ni, Y.; Xu, Y.; Wang, J. Flame retardant nanocomposites based on 2D layered nanomaterials: a review. *Journal of Materials Science* **2019**, *54* (20), 13070–13105.
33. Alexandre, M.; Dubois, P. Polymer-layered silicate nanocomposites: preparation, properties and uses of a new class of materials. *Materials Science and Engineering: R: Reports* **2000**, *28* (1), 1–63.
34. Goda, E. S.; Gab-Allah, M. A.; Singu, B. S.; Yoon, K. R. Halloysite nanotubes based electrochemical sensors: A review. *Microchemical Journal* **2019**, *147*, 1083–1096.
35. Goda, E. S.; Singu, B. S.; Hong, S. E.; Yoon, K. R. Good dispersion of poly(δ-gluconolactone)-grafted graphene in poly(vinyl alcohol) for significantly enhanced mechanical strength. *Materials Chemistry and Physics* **2020**, *254*, 123465.
36. Singu, B. S.; Goda, E. S.; Yoon, K. R. Carbon Nanotube–Manganese oxide nanorods hybrid composites for high-performance supercapacitor materials. *Journal of Industrial and Engineering Chemistry* **2021**, *97*, 239–249.
37. Gab-Allah, M. A.; Goda, E. S.; Shehata, A. B.; Gamal, H. Critical Review on the Analytical Methods for the Determination of Sulfur and Trace Elements in Crude Oil. *Critical Reviews in Analytical Chemistry* **2020**, *50* (2), 161–178.
38. Goda, E. S.; Hong, S. E.; Yoon, K. R. Facile synthesis of Cu-PBA nanocubes/graphene oxide composite as binder-free electrodes for supercapacitor. *Journal of Alloys and Compounds* **2020**, 157868.
39. Pandit, B.; Rondiya, S. R.; Dzade, N. Y.; Shaikh, S. F.; Kumar, N.; Goda, E. S.; Al-Kahtani, A. A.; Mane, R. S.; Mathur, S.; Salunkhe, R. R. High Stability and Long Cycle Life of Rechargeable Sodium-Ion Battery Using Manganese Oxide Cathode: A Combined Density Functional Theory (DFT) and Experimental Study. *ACS Applied Materials & Interfaces* **2021**, *13* (9), 11433–11441.

40. Abu Elella, M. H.; Goda, E. S.; Abdallah, H. M.; Shalan, A. E.; Gamal, H.; Yoon, K. R. Innovative bactericidal adsorbents containing modified xanthan gum/montmorillonite nanocomposites for wastewater treatment. *International Journal of Biological Macromolecules* **2021**, *167*, 1113–1125.
41. Goda, E. S.; Lee, S.; Sohail, M.; Yoon, K. R. Prussian blue and its analogues as advanced supercapacitor electrodes. *Journal of Energy Chemistry* **2020**, *50*, 206–229.
42. Lee, J.; Goda, E. S.; Choi, J.; Park, J.; Lee, S. Synthesis and characterization of elution behavior of nonspherical gold nanoparticles in asymmetrical flow field-flow fractionation (AsFlFFF). *Journal of Nanoparticle Research* **2020**, *22* (9), 256.
43. Abu Elella, M. H.; Goda, E. S.; Gab-Allah, M. A.; Hong, S. E.; Pandit, B.; Lee, S.; Gamal, H.; Rehman, A. u.; Yoon, K. R. Xanthan gum-derived materials for applications in environment and eco-friendly materials: A review. *Journal of Environmental Chemical Engineering* **2021**, *9* (1), 104702.
44. Bonaccorso, F.; Colombo, L.; Yu, G.; Stoller, M.; Tozzini, V.; Ferrari, A. C.; Ruoff, R. S.; Pellegrini, V. Graphene, related two-dimensional crystals, and hybrid systems for energy conversion and storage. *Science* **2015**, *347* (6217), 1246501.
45. Wu, Z.; Xu, C.; Ma, C.; Liu, Z.; Cheng, H.-M.; Ren, W. Synergistic Effect of Aligned Graphene Nanosheets in Graphene Foam for High-Performance Thermally Conductive Composites. *Advanced Materials* **2019**, *31* (19), 1900199.
46. Mortazavi, B.; Shahrokhi, M.; Raeisi, M.; Zhuang, X.; Pereira, L. F. C.; Rabczuk, T. Outstanding strength, optical characteristics and thermal conductivity of graphene-like BC3 and BC6N semiconductors. *Carbon* **2019**, *149*, 733–742.
47. Mohanraj, J.; Durgalakshmi, D.; Rakkesh, R. A.; Balakumar, S.; Rajendran, S.; Karimi-Maleh, H. Facile synthesis of paper based graphene electrodes for point of care devices: A double stranded DNA (dsDNA) biosensor. *Journal of Colloid and Interface Science* **2020**, *566*, 463–472.
48. Ramanathan, T.; Abdala, A. A.; Stankovich, S.; Dikin, D. A.; Herrera-Alonso, M.; Piner, R. D.; Adamson, D. H.; Schniepp, H. C.; Chen, X.; Ruoff, R. S.; Nguyen, S. T.; Aksay, I. A.; Prud'Homme, R. K.; Brinson, L. C. Functionalized graphene sheets for polymer nanocomposites. *Nature Nanotechnology* **2008**, *3* (6), 327–331.
49. Eda, G.; Fanchini, G.; Chhowalla, M. Large-area ultrathin films of reduced graphene oxide as a transparent and flexible electronic material. *Nature Nanotechnology* **2008**, *3* (5), 270–274.
50. Erickson, K.; Erni, R.; Lee, Z.; Alem, N.; Gannett, W.; Zettl, A. Determination of the Local Chemical Structure of Graphene Oxide and Reduced Graphene Oxide. *Advanced Materials* **2010**, *22* (40), 4467–4472.
51. Du, W.; Zhang, Z.; Su, H.; Lin, H.; Li, Z. Urethane-Functionalized Graphene Oxide for Improving Compatibility and Thermal Conductivity of Waterborne Polyurethane Composites. *Industrial & Engineering Chemistry Research* **2018**, *57* (21), 7146–7155.
52. Du, W.; Jin, Y.; Lai, S.; Shi, L.; Shen, Y.; Pan, J. Urethane-silica functionalized graphene oxide for enhancing mechanical property and fire safety of waterborne polyurethane composites. *Applied Surface Science* **2019**, *492*, 298–308.
53. Cao, Y.; Feng, J.; Wu, P. Preparation of organically dispersible graphene nanosheet powders through a lyophilization method and their poly(lactic acid) composites. *Carbon* **2010**, *48* (13), 3834–3839.

54. Guo, K.-Y.; Wu, Q.; Mao, M.; Chen, H.; Zhang, G.-D.; Zhao, L.; Gao, J.-F.; Song, P.; Tang, L.-C. Water-based hybrid coatings toward mechanically flexible, super-hydrophobic and flame-retardant polyurethane foam nanocomposites with high-efficiency and reliable fire alarm response. *Composites Part B: Engineering* **2020**, *193*, 108017.
55. Jamsaz, A.; Goharshadi, E. K. Flame retardant, superhydrophobic, and superoleophilic reduced graphene oxide/orthoaminophenol polyurethane sponge for efficient oil/water separation. *Journal of Molecular Liquids* **2020**, *307*, 112979.
56. Cai, W.; Feng, X.; Wang, B.; Hu, W.; Yuan, B.; Hong, N.; Hu, Y. A novel strategy to simultaneously electrochemically prepare and functionalize graphene with a multifunctional flame retardant. *Chemical Engineering Journal* **2017**, *316*, 514–524.
57. Alongi, J.; Di Blasio, A.; Carosio, F.; Malucelli, G. UV-cured hybrid organic–inorganic Layer by Layer assemblies: Effect on the flame retardancy of polycarbonate films. *Polymer Degradation and Stability* **2014**, *107*, 74–81.
58. Li, Y.-C.; Kim, Y. S.; Shields, J.; Davis, R. Controlling polyurethane foam flammability and mechanical behaviour by tailoring the composition of clay-based multilayer nanocoatings. *Journal of Materials Chemistry A* **2013**, *1* (41), 12987–12997.
59. Zhang, T.; Yan, H.; Peng, M.; Wang, L.; Ding, H.; Fang, Z. Construction of flame retardant nanocoating on ramie fabric via layer-by-layer assembly of carbon nanotube and ammonium polyphosphate. *Nanoscale* **2013**, *5* (7), 3013–3021.
60. Li, Y.-C.; Mannen, S.; Morgan, A. B.; Chang, S.; Yang, Y.-H.; Condon, B.; Grunlan, J. C. Intumescent All-Polymer Multilayer Nanocoating Capable of Extinguishing Flame on Fabric. *Advanced Materials* **2011**, *23* (34), 3926–3931.
61. Pan, H.; Lu, Y.; Song, L.; Zhang, X.; Hu, Y. Construction of layer-by-layer coating based on graphene oxide/β-FeOOH nanorods and its synergistic effect on improving flame retardancy of flexible polyurethane foam. *Composites Science and Technology* **2016**, *129*, 116–122.
62. Duquesne, S.; Bras, M. L.; Bourbigot, S.; Delobel, R.; Vezin, H.; Camino, G.; Eling, B.; Lindsay, C.; Roels, T. Expandable graphite: A fire retardant additive for polyurethane coatings. *Fire and Materials* **2003**, *27* (3), 103–117.
63. Duquesne, S.; Delobel, R.; Le Bras, M.; Camino, G. A comparative study of the mechanism of action of ammonium polyphosphate and expandable graphite in polyurethane. *Polymer Degradation and Stability* **2002**, *77* (2), 333–344.
64. Wang, S.; Wang, X.; Wang, X.; Li, H.; Sun, J.; Sun, W.; Yao, Y.; Gu, X.; Zhang, S. Surface coated rigid polyurethane foam with durable flame retardancy and improved mechanical property. *Chemical Engineering Journal* **2020**, *385*, 123755.
65. Bian, X.-C.; Tang, J.-H.; Li, Z.-M.; Lu, Z.-Y.; Lu, A. Dependence of flame-retardant properties on density of expandable graphite filled rigid polyurethane foam. *Journal of Applied Polymer Science* **2007**, *104* (5), 3347–3355.
66. Modesti, M.; Lorenzetti, A.; Simioni, F.; Camino, G. Expandable graphite as an intumescent flame retardant in polyisocyanurate–polyurethane foams. *Polymer Degradation and Stability* **2002**, *77* (2), 195–202.
67. Modesti, M.; Lorenzetti, A. Improvement on fire behaviour of water blown PIR–PUR foams: use of an halogen-free flame retardant. *European Polymer Journal* **2003**, *39* (2), 263–268.

68. Duquesne, S.; Le Bras, M.; Bourbigot, S.; Delobel, R.; Camino, G.; Eling, B.; Lindsay, C.; Roels, T. Thermal degradation of polyurethane and polyurethane/expandable graphite coatings. *Polymer Degradation and Stability* **2001**, *74* (3), 493–499.

69. Duquesne, S.; Le Bras, M.; Bourbigot, S.; Delobel, R.; Poutch, F.; Camino, G.; Eling, B.; Lindsay, C.; Roels, T. Analysis of Fire Gases Released from Polyurethane and Fire-Retarded Polyurethane Coatings. *Journal of Fire Sciences* **2000**, *18* (6), 456–482.

70. Yu, L.; Zhang, P.; Du, Z. Tribological behavior and structural change of the LB film of MoS2 nanoparticles coated with dialkyldithiophosphate. *Surface and Coatings Technology* **2000**, *130* (1), 110–115.

71. Breysse, M.; Geantet, C.; Afanasiev, P.; Blanchard, J.; Vrinat, M. Recent studies on the preparation, activation and design of active phases and supports of hydrotreating catalysts. *Catalysis Today* **2008**, *130* (1), 3–13.

72. Lukowski, M. A.; Daniel, A. S.; Meng, F.; Forticaux, A.; Li, L.; Jin, S. Enhanced Hydrogen Evolution Catalysis from Chemically Exfoliated Metallic MoS2 Nanosheets. *Journal of the American Chemical Society* **2013**, *135* (28), 10274–10277.

73. Jiang, J.-W. Graphene versus MoS2: A short review. *Frontiers of Physics* **2015**, *10* (3), 287–302.

74. Zhou, K.; Yang, W.; Tang, G.; Wang, B.; Jiang, S.; Hu, Y.; Gui, Z. Comparative study on the thermal stability, flame retardancy and smoke suppression properties of polystyrene composites containing molybdenum disulfide and graphene. *RSC Advances* **2013**, *3* (47), 25030–25040.

75. Zhou, K.; Gao, R.; Qian, X. Self-assembly of exfoliated molybdenum disulfide (MoS2) nanosheets and layered double hydroxide (LDH): Towards reducing fire hazards of epoxy. *Journal of Hazardous Materials* **2017**, *338*, 343–355.

76. Wenelska, K.; Mijowska, E. Preparation, thermal conductivity, and thermal stability of flame retardant polyethylene with exfoliated MoS2/MxOy. *New Journal of Chemistry* **2017**, *41* (22), 13287–13292.

77. Jiang, S.-D.; Tang, G.; Bai, Z.-M.; Wang, Y.-Y.; Hu, Y.; Song, L. Surface functionalization of MoS2 with POSS for enhancing thermal, flame-retardant and mechanical properties in PVA composites. *RSC Advances* **2014**, *4* (7), 3253–3262.

78. Wang, D.; Xing, W.; Song, L.; Hu, Y. Space-Confined Growth of Defect-Rich Molybdenum Disulfide Nanosheets Within Graphene: Application in The Removal of Smoke Particles and Toxic Volatiles. *ACS Applied Materials & Interfaces* **2016**, *8* (50), 34735–34743.

79. Zhi, M.; Liu, Q.; Zhao, Y.; Gao, S.; Zhang, Z.; He, Y. Novel MoS2–DOPO Hybrid for Effective Enhancements on Flame Retardancy and Smoke Suppression of Flexible Polyurethane Foams. *ACS Omega* **2020**, *5* (6), 2734–2746.

80. Cho, S.-Y.; Kim, S. J.; Lee, Y.; Kim, J.-S.; Jung, W.-B.; Yoo, H.-W.; Kim, J.; Jung, H.-T. Highly Enhanced Gas Adsorption Properties in Vertically Aligned MoS2 Layers. *ACS Nano* **2015**, *9* (9), 9314–9321.

81. Wang, D.; Song, L.; Zhou, K.; Yu, X.; Hu, Y.; Wang, J. Anomalous nano-barrier effects of ultrathin molybdenum disulfide nanosheets for improving the flame retardance of polymer nanocomposites. *Journal of Materials Chemistry A* **2015**, *3* (27), 14307–14317.

82. Feng, X.; Xing, W.; Song, L.; Hu, Y. In situ synthesis of a MoS2/CoOOH hybrid by a facile wet chemical method and the catalytic oxidation of CO in epoxy resin during decomposition. *Journal of Materials Chemistry A* **2014**, *2* (33), 13299–13308.

83. Salmeia, K. A.; Fage, J.; Liang, S.; Gaan, S. An Overview of Mode of Action and Analytical Methods for Evaluation of Gas Phase Activities of Flame Retardants. *Polymers* **2015**, *7* (3), 504–526.
84. Pan, H.; Shen, Q.; Zhang, Z.; Yu, B.; Lu, Y. MoS2-filled coating on flexible polyurethane foam via layer-by-layer assembly technique: flame-retardant and smoke suppression properties. *Journal of Materials Science* **2018**, *53* (12), 9340–9349.
85. He, L.; Wang, J.; Wang, B.; Wang, X.; Zhou, X.; Cai, W.; Mu, X.; Hou, Y.; Hu, Y.; Song, L. Large-scale production of simultaneously exfoliated and Functionalized Mxenes as promising flame retardant for polyurethane. *Composites Part B: Engineering* **2019**, *179*, 107486.
86. Naguib, M.; Mochalin, V. N.; Barsoum, M. W.; Gogotsi, Y. 25th Anniversary Article: MXenes: A New Family of Two-Dimensional Materials. *Advanced Materials* **2014**, *26* (7), 992–1005.
87. Naguib, M.; Kurtoglu, M.; Presser, V.; Lu, J.; Niu, J.; Heon, M.; Hultman, L.; Gogotsi, Y.; Barsoum, M. W. Two-Dimensional Nanocrystals Produced by Exfoliation of Ti3AlC2. *Advanced Materials* **2011**, *23* (37), 4248–4253.
88. Er, D.; Li, J.; Naguib, M.; Gogotsi, Y.; Shenoy, V. B. Ti3C2 MXene as a High Capacity Electrode Material for Metal (Li, Na, K, Ca) Ion Batteries. *ACS Applied Materials & Interfaces* **2014**, *6* (14), 11173–11179.
89. Wang, X.; Wang, L.; He, Y.; Wu, M.; Zhou, A. The effect of two-dimensional d-Ti3C2 on the mechanical and thermal conductivity properties of thermoplastic polyurethane composites. *Polymer Composites* **2020**, *41* (1), 350–359.
90. Yu, B.; Tawiah, B.; Wang, L.-Q.; Yin Yuen, A. C.; Zhang, Z.-C.; Shen, L.-L.; Lin, B.; Fei, B.; Yang, W.; Li, A.; Zhu, S.-E.; Hu, E.-Z.; Lu, H.-D.; Yeoh, G. H. Interface decoration of exfoliated MXene ultra-thin nanosheets for fire and smoke suppressions of thermoplastic polyurethane elastomer. *Journal of Hazardous Materials* **2019**, *374*, 110–119.
91. Liu, C.; Wu, W.; Shi, Y.; Yang, F.; Liu, M.; Chen, Z.; Yu, B.; Feng, Y. Creating MXene/reduced graphene oxide hybrid towards highly fire safe thermoplastic polyurethane nanocomposites. *Composites Part B: Engineering* **2020**, *203*, 108486.
92. Lu, H.; Song, L.; Hu, Y. A review on flame retardant technology in China. Part II: flame retardant polymeric nanocomposites and coatings. *Polymers for Advanced Technologies* **2011**, *22* (4), 379–394.
93. Karatrantos, A.; Clarke, N.; Kröger, M. Modeling of polymer structure and conformations in polymer nanocomposites from atomistic to mesoscale: A Review. *Polymer reviews* **2016**, *56* (3), 385–428.
94. Zhang, S.; Yan, Y.; Wang, W.; Gu, X.; Li, H.; Li, J.; Sun, J. Intercalation of phosphotungstic acid into layered double hydroxides by reconstruction method and its application in intumescent flame retardant poly (lactic acid) composites. *Polymer Degradation and Stability* **2018**, *147*, 142–150.
95. Montero, B.; Bellas, R.; Ramírez, C.; Rico, M.; Bouza, R. Flame retardancy and thermal stability of organic–inorganic hybrid resins based on polyhedral oligomeric silsesquioxanes and montmorillonite clay. *Composites Part B: Engineering* **2014**, *63*, 67–76.
96. Nakato, T.; Miyamoto, N. Liquid crystalline behavior and related properties of colloidal systems of inorganic oxide nanosheets. *Materials* **2009**, *2* (4), 1734–1761.

97. Kong, Q.; Wu, T.; Zhang, H.; Zhang, Y.; Zhang, M.; Si, T.; Yang, L.; Zhang, J. Improving flame retardancy of IFR/PP composites through the synergistic effect of organic montmorillonite intercalation cobalt hydroxides modified by acidified chitosan. *Applied Clay Science* **2017**, *146*, 230–237.
98. Cheng, H.; Li, J.; Wang, H.; Jiao, S.; Ma, Q. Adsorption kinetics of chitosan/montmorillonite intercalation composites for Reactive Red dye and desorption properties. *Environmental Chemistry* **2014**, *33* (1), 115–122.
99. Zeng, Q.; Wang, D.; Yu, A.; Lu, G. Synthesis of polymer–montmorillonite nanocomposites by in situ intercalative polymerization. *Nanotechnology* **2002**, *13* (5), 549.
100. Dal Castel, C.; Pelegrini, T., Jr; Barbosa, R.; Liberman, S.; Mauler, R. Properties of silane grafted polypropylene/montmorillonite nanocomposites. *Composites Part A: Applied Science and Manufacturing* **2010**, *41* (2), 185–191.
101. Zhu, J.; Tian, M.; Hou, J.; Wang, J.; Lin, J.; Zhang, Y.; Liu, J.; Van der Bruggen, B. Surface zwitterionic functionalized graphene oxide for a novel loose nanofiltration membrane. *Journal of Materials Chemistry A* **2016**, *4* (5), 1980–1990.
102. Yi, D.; Yang, H.; Zhao, M.; Huang, L.; Camino, G.; Frache, A.; Yang, R. A novel, low surface charge density, anionically modified montmorillonite for polymer nanocomposites. *RSC advances* **2017**, *7* (10), 5980–5988.
103. Saba, N.; Jawaid, M.; Asim, M. Recent advances in nanoclay/natural fibers hybrid composites. *Nanoclay reinforced polymer composites* **2016**, 1–28.
104. Liu, X.; Guo, J.; Tang, W.; Li, H.; Gu, X.; Sun, J.; Zhang, S. Enhancing the flame retardancy of thermoplastic polyurethane by introducing montmorillonite nanosheets modified with phosphorylated chitosan. *Composites Part A: Applied Science and Manufacturing* **2019**, *119*, 291–298.
105. Tang, W.; Song, L.; Zhang, S.; Li, H.; Sun, J.; Gu, X. Preparation of thiourea-intercalated kaolinite and its influence on thermostability and flammability of polypropylene composite. *Journal of Materials Science* **2017**, *52* (1), 208–217.
106. Tang, W.; Zhang, S.; Sun, J.; Gu, X. Flame retardancy and thermal stability of polypropylene composite containing ammonium sulfamate intercalated kaolinite. *Industrial & Engineering Chemistry Research* **2016**, *55* (28), 7669–7678.
107. Seifi, S.; Diatta-Dieme, M. T.; Blanchart, P.; Lecomte-Nana, G. L.; Kobor, D.; Petit, S. Kaolin intercalated by urea. Ceramic applications. *Construction and Building Materials* **2016**, *113*, 579–585.
108. Agrawal, A.; Kaur, R.; Walia, R. S. Investigation on flammability of rigid polyurethane foam-mineral fillers composite. *Fire and Materials* **2019**, *43* (8), 917–927.
109. Becker, C. M.; Dick, T. A.; Wypych, F.; Schrekker, H. S.; Amico, S. C. Synergetic effect of LDH and glass fiber on the properties of two-and three-component epoxy composites. *Polymer testing* **2012**, *31* (6), 741–747.
110. Wu, Y.; Tang, M.; Wang, N.; Qin, J.; Chen, X.; Zhang, K. Preparation and Investigation on Morphology, Thermal Stability and Crystallization Behavior of HDPE/EVA/Organo-Modified Layered Double Hydroxide Nanocomposites. *Polymer Composites* **2018**, *39* (S3), E1849–E1857.

111. Liu, L.; Wang, W.; Hu, Y. Layered double hydroxide-decorated flexible polyurethane foam: significantly improved toxic effluent elimination. *RSC advances* **2015**, *5* (118), 97458–97466.
112. Xu, W.; Zhang, B.; Xu, B.; Li, A. The flame retardancy and smoke suppression effect of heptaheptamolybdate modified reduced graphene oxide/layered double hydroxide hybrids on polyurethane elastomer. *Composites Part A: Applied Science and Manufacturing* **2016**, *91*, 30–40.
113. Tan, C.; Cao, X.; Wu, X.-J.; He, Q.; Yang, J.; Zhang, X.; Chen, J.; Zhao, W.; Han, S.; Nam, G.-H.; Sindoro, M.; Zhang, H. Recent Advances in Ultrathin Two-Dimensional Nanomaterials. *Chemical Reviews* **2017**, *117* (9), 6225–6331.
114. Wu, L.; Xie, Z.; Lu, L.; Zhao, J.; Wang, Y.; Jiang, X.; Ge, Y.; Zhang, F.; Lu, S.; Guo, Z.; Liu, J.; Xiang, Y.; Xu, S.; Li, J.; Fan, D.; Zhang, H. Few-Layer Tin Sulfide: A Promising Black-Phosphorus-Analogue 2D Material with Exceptionally Large Nonlinear Optical Response, High Stability, and Applications in All-Optical Switching and Wavelength Conversion. *Advanced Optical Materials* **2018**, *6* (2), 1700985.
115. Zhou, Q.; Chen, Q.; Tong, Y.; Wang, J. Light-Induced Ambient Degradation of Few-Layer Black Phosphorus: Mechanism and Protection. *Angewandte Chemie International Edition* **2016**, *55* (38), 11437–11441.
116. Cai, W.; Cai, T.; He, L.; Chu, F.; Mu, X.; Han, L.; Hu, Y.; Wang, B.; Hu, W. Natural antioxidant functionalization for fabricating ambient-stable black phosphorus nanosheets toward enhancing flame retardancy and toxic gases suppression of polyurethane. *Journal of Hazardous Materials* **2020**, *387*, 121971.
117. Cui, X.; Ding, P.; Zhuang, N.; Shi, L.; Song, N.; Tang, S. Thermal conductive and mechanical properties of polymeric composites based on solution-exfoliated boron nitride and graphene nanosheets: a morphology-promoted synergistic effect. *ACS applied materials interfaces* **2015**, *7* (34), 19068–19075.
118. Gu, J.; Lv, Z.; Wu, Y.; Guo, Y.; Tian, L.; Qiu, H.; Li, W.; Zhang, Q. Dielectric thermally conductive boron nitride/polyimide composites with outstanding thermal stabilities via in-situ polymerization-electrospinning-hot press method. *Composites Part A: Applied Science and Manufacturing* **2017**, *94*, 209–216.
119. Wang, J.; Wu, Y.; Xue, Y.; Liu, D.; Wang, X.; Hu, X.; Bando, Y.; Lei, W. Super-compatible functional boron nitride nanosheets/polymer films with excellent mechanical properties and ultra-high thermal conductivity for thermal management. *Journal of Materials Chemistry C* **2018**, *6* (6), 1363–1369.
120. Lin, Z.; Mcnamara, A.; Liu, Y.; Moon, K.-s.; Wong, C.-P. Exfoliated hexagonal boron nitride-based polymer nanocomposite with enhanced thermal conductivity for electronic encapsulation. *Composites Science and Technology* **2014**, *90*, 123–128.
121. Yu, C.; Gong, W.; Tian, W.; Zhang, Q.; Xu, Y.; Lin, Z.; Hu, M.; Fan, X.; Yao, Y. Hot-pressing induced alignment of boron nitride in polyurethane for composite films with thermal conductivity over 50 Wm−1 K−1. *Composites Science and Technology* **2018**, *160*, 199–207.
122. Bhimanapati, G. R.; Kozuch, D.; Robinson, J. A. Large-scale synthesis and functionalization of hexagonal boron nitride nanosheets. *Nanoscale* **2014**, *6* (20), 11671–11675.
123. Lei, W.; Mochalin, V. N.; Liu, D.; Qin, S.; Gogotsi, Y.; Chen, Y. Boron nitride colloidal solutions, ultralight aerogels and freestanding membranes through one-step exfoliation and functionalization. *Nature communications* **2015**, *6* (1), 1–8.

124. Wang, J.; Zhang, D.; Zhang, Y.; Cai, W.; Yao, C.; Hu, Y.; Hu, W. Construction of multifunctional boron nitride nanosheet towards reducing toxic volatiles (CO and HCN) generation and fire hazard of thermoplastic polyurethane. *Journal of Hazardous Materials* **2019**, *362*, 482–494.

125. Yin, S.; Ren, X.; Lian, P.; Zhu, Y.; Mei, Y. Synergistic effects of black phosphorus/boron nitride nanosheets on enhancing the flame-retardant properties of waterborne polyurethane and its flame-retardant mechanism. *Polymers* **2020**, *12* (7), 1487.

126. Son, B.-G.; Hwang, T.-s.; Goo, D.-C. Fire-retardation properties of polyurethane nanocomposite by filling inorganic nano flame retardant. *Polymer Korea* **2007**, *31* (5), 404–409.

127. Huang, G.; Gao, J.; Li, Y.; Han, L.; Wang, X. Functionalizing nano-montmorillonites by modified with intumescent flame retardant: Preparation and application in polyurethane. *Polymer Degradation and Stability* **2010**, *95* (2), 245–253.

128. Xu, W.; Wang, G.; Zheng, X. Research on highly flame-retardant rigid PU foams by combination of nanostructured additives and phosphorus flame retardants. *Polymer Degradation and Stability* **2015**, *111*, 142–150.

129. Song, L.; Hu, Y.; Tang, Y.; Zhang, R.; Chen, Z.; Fan, W. Study on the properties of flame retardant polyurethane/organoclay nanocomposite. *Polymer Degradation and Stability* **2005**, *87* (1), 111–116.

130. Zhang, C.; Milhorn, A.; Haile, M.; Mai, G.; Grunlan, J. C. Nanocoating of starch and clay that reduces the flammability of polyurethane foam. *Green Materials* **2017**, *5* (4), 182–186.

131. Carosio, F.; Fina, A. Three organic/inorganic nanolayers on flexible foam allow retaining superior flame retardancy performance upon mechanical compression cycles. *Frontiers in Materials* **2019**, *6*, 20.

132. Liu, X.; Qin, S.; Li, H.; Sun, J.; Gu, X.; Zhang, S.; Grunlan, J. C. Combination Intumescent and Kaolin-Filled Multilayer Nanocoatings that Reduce Polyurethane Flammability. *Macromolecular Materials and Engineering* **2019**, *304* (2), 1800531.

133. Xie, H.; Ye, Q.; Si, J.; Yang, W.; Lu, H.; Zhang, Q. Synthesis of a carbon nanotubes/ZnAl-layered double hydroxide composite as a novel flame retardant for flexible polyurethane foams. *Polymers for Advanced Technologies* **2016**, *27* (5), 651–656.

134. Peng, H. K.; Wang, X.; Li, T. T.; Lou, C. W.; Wang, Y.; Lin, J. H. Mechanical properties, thermal stability, sound absorption, and flame retardancy of rigid PU foam composites containing a fire-retarding agent: Effect of magnesium hydroxide and aluminum hydroxide. *Polymers for Advanced Technologies* **2019**, *30* (8), 2045–2055.

135. Qian, Y.; Qiao, P.; Li, L.; Han, H.; Zhang, H.; Chang, G. Hydrothermal synthesis of lanthanum-doped MgAl-layered double hydroxide/graphene oxide hybrid and its application as flame retardant for thermoplastic polyurethane. *Advances in Polymer Technology* **2020**, *2020*, 1–10.

136. Qiu, X.; Li, Z.; Li, X.; Yu, L.; Zhang, Z. Construction and flame-retardant performance of layer-by-layer assembled hexagonal boron nitride coatings on flexible polyurethane foams. *Journal of Applied Polymer Science* **2019**, *136* (29), 47839.

137. Cai, W.; Mu, X.; Pan, Y.; Guo, W.; Wang, J.; Yuan, B.; Feng, X.; Tai, Q.; Hu, Y. Facile fabrication of organically modified boron nitride nanosheets and its effect on the thermal

stability, flame retardant, and mechanical properties of thermoplastic polyurethane. *Polymers for Advanced Technologies* **2018**, *29* (9), 2545–2552.

138. Cai, W.; Wang, B.; Liu, L.; Zhou, X.; Chu, F.; Zhan, J.; Hu, Y.; Kan, Y.; Wang, X. An operable platform towards functionalization of chemically inert boron nitride nanosheets for flame retardancy and toxic gas suppression of thermoplastic polyurethane. *Composites Part B: Engineering* **2019**, *178*, 107462.

139. Xu, W.; Li, A.; Liu, Y.; Chen, R.; Li, W. CuMoO 4@ hexagonal boron nitride hybrid: an ecofriendly flame retardant for polyurethane elastomer. *Journal of Materials Science* **2018**, *53* (16), 11265–11279.

Chapter 10

Flame Retardant Polyurethane Nanocomposites

Wen-Jie Yang,[1,2] Chun-Xiang Wei,[1] Hong-Dian Lu,[1] Wei Yang,[*,1] and Richard K. K. Yuen[*,2]

[1]School of Energy, Materials and Chemical Engineering, Hefei University, 99 Jinxiu Avenue, Hefei, Anhui 230601, P.R. China
[2]Department of Architecture and Civil Engineering, City University of Hong Kong, Tat Chee Avenue, Kowloon, Hong Kong, P.R. China
*Email: yangwei@hfuu.edu.cn
*Email: Richard.Yuen@cityu.edu.hk

The development of polyurethane (PU) has been greatly limited because of its flammable nature. Nanomaterials are widely used in various fields because of their unique nanoscale structure and high thermal stability, especially showing remarkable advantage in flame retardancy. The introduction of different nanomaterials, including zero-, one-, and two-dimensional nanoparticles, can effectively enhance the flame-retardant and mechanical properties of PU by using suitable preparation approaches. To further improve the overall properties, nanomaterials are commonly functionalized by various organics via chemical functionalization for achieving excellent dispersion and improved compatibility. In this chapter, the surface engineering strategies of nanomaterials and the fabrication of PU nanocomposites were described and the flame-retardant mechanism was discussed in detail.

© 2021 American Chemical Society

Introduction

Polyurethane (PU) is a kind of macromolecule containing carbamate groups in the molecular chain. It was invented in the 1930s and synthesized by German chemist Bayer (1). After nearly 90 years of technical development, PU has been widely used in household, construction, daily necessities, transportation, and home applications. According to the different application fields, PU can be divided into thermoplastic PU elastomer (TPU), rigid PU, PU fiber, and flexible PUF (FPUF) or rigid PUF (2, 3). TPU has a linear structure constructed by alternating hard segments and soft segments. It exhibits better thermal stability, chemical resistance, resilience, and mechanical properties than other polymers, which is extensively used in packaging, electronics, sound insulation, and filtration materials. Rigid PU shows the properties of light weight and sound and thermal insulation, which can be applied in construction, automobile, aviation industry, and thermal insulation structural materials (4).

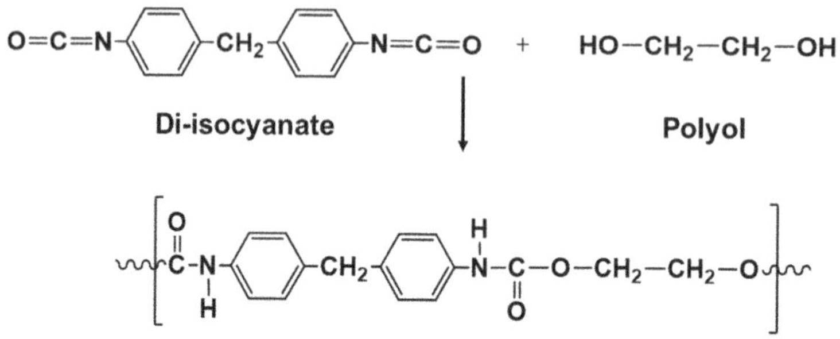

Figure 1. The synthesis route of PUs.

PU is generally synthesized via the reaction between isocyanate and hydroxyl groups from polyols, as shown in Figure 1 (2). The isocyanate monomers include methylene diphenyl diisocyanate, toluene diisocyanate (TDI), isophorone diisocyanate, and xylene diisocyanatea. The most commonly used polyols are polyethers, polyester polyols, polyethylene glycol, polypropylene glycol, and acrylic polyols. Despite the wide range of applications, the development of PU is restricted by its high flammability. The combustion of PU undergoes two stages, as illustrated in Figure 2. First, as the ambient temperature increases, the covalent bonds are stimulated and gradually decompose, resulting in the formation of combustible gas products (pyrolyzed molecular isocyanate and polyols). These combustible gases rise up and mix with oxygen in air to form the gas phase. When the flammable volatiles concentration in the gas phase approaches a certain level, they start to burn. As a result, a large amount of heat and gases that are harmful to humans are released. Furthermore, there are numerous nitrogen elements in PU chains that can generate toxic HCN gases during combustion. When PU burns, the degraded isocyanates diffuse upward with yellow smoke that decomposes into HCN and combustible volatiles. Burning HCN leads to the formation of nitrogen oxides that greatly damage the human respiratory system. Particularly, polyols will also burn to release a lot of heat, as well as CO and CO_2, to harm people's health (5–7).

As a result, to further extend the application ranges of PU, it is imperative to improve its flame-retardant (FR) properties. Incidentally, the most widely used FRs were halogen compounds that would produce toxic gases when they were burned or degraded. Therefore, the halogen-free, low-smoke, and nontoxic FRs have attracted more attention and become a hot topic. At present, FRs are mainly composed of organics, inorganics, and hybrids. Organics, such as phosphorus-nitrogen intumescent FRs (IFRs), can efficiently suppress the burning via rapid carbonization. However, some organic FRs such as chlorinated and brominated ones tend to release toxic gases during combustion. On the contrary, inorganic additives are relatively cheap, nontoxic, and nonpolluting. Furthermore, the addition of inorganics can enhance the mechanical properties of polymers. In consequence, inorganic FRs have attracted the keen interest of researchers.

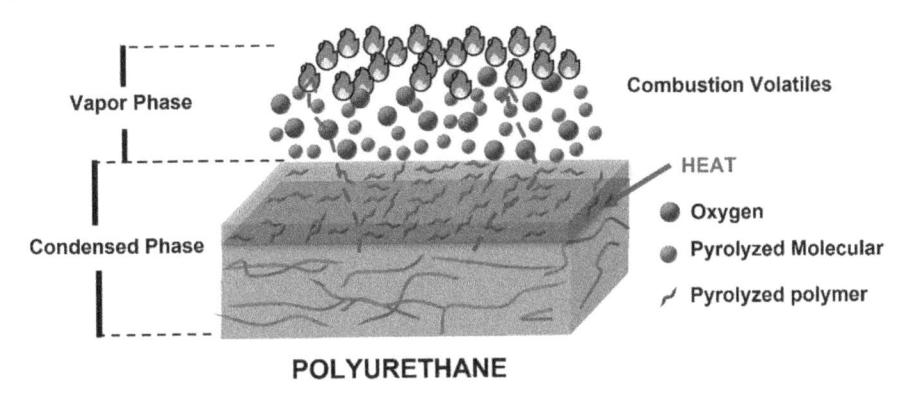

Figure 2. A typical combustion process of PU without flame retardants (FRs).

Among inorganics, nanomaterial-based FRs account for a large proportion. Nanomaterials refer to the materials that have at least one dimension ranging from 1 to 100 nm in a three-dimensional (3D) space or are composed of them as basic units. According to the distribution of nanometer sizes, nanomaterials can be divided into zero-dimensional (0D), one-dimensional (1D), and two-dimensional (2D) nanoparticles (Figure 3). Because of their uniform morphology and high thermal stability, they can catalyze the formation of char layers to isolate volatile and oxygen for protecting the polymer from burning. Until now, PU nanocomposites with improved mechanical and FR properties have been extensively reported. In this chapter, we first introduce the preparation of PU nanocomposites. The advantages and disadvantages of these preparation methods are discussed. The progress of PU nanocomposites on FR properties is then reviewed. Finally, the prospects of FR-PU nanocomposites are noted.

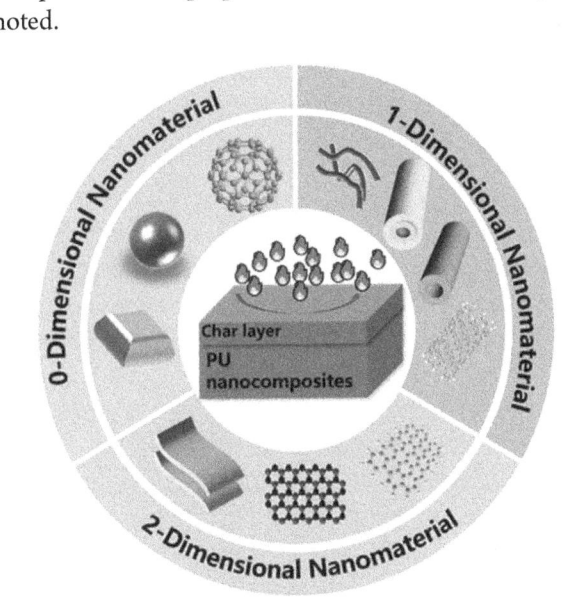

Figure 3. PU nanocomposites based on different nanomaterials.

Preparation of FR-PU Nanocomposites

According to different materials, there are various approaches for preparing polymer nanocomposites. Compared with other materials, nanoadditives tend to stack together and aggregate when incorporated with polymers because of their incompatibility, resulting in inferior mechanical and limited flame-resistant performance of the final nanocomposites. To obtain a better dispersion, it is necessary to choose appropriate preparation or pretreatment methods. At present, the strategies for fabricating polymer nanocomposites include melt blending, solvent exchange, and in situ polymerization.

Melt Blending

The melt blending method refers to mixing nanomaterials and PU in a molten state through high-speed stirring, in which the nanomaterials can be homogeneously dispersed under the action of heat and force. For example, titanium dioxide (TiO_2) nanoparticles were added into TPU by melt blending (8). Anatase TiO_2 can be uniformly dispersed in TPU, indicating that melt blending treatment contributed to maintaining the particle interspaced distance of anatase TiO_2. In addition, because of the high chemical activity and electron-hole separation performance, it not only enhanced the compatibility between TiO_2 nanoparticles and PU but also increased the degree of aggregation PU chains wrapped with anatase TiO_2, thereby improving the mechanical properties of the nanocomposites.

Melt blending is currently one of the most common methods for preparing composite materials because of its simple operation, low cost, and excellent commercial potential. However, some issues are frequently encountered during melt blending, namely the strong tendency for nanomaterials to aggregate because of incompatibility and lack of interfacial bonding with macromolecular chains of PU. Numerous studies have demonstrated that the dispersion of materials prepared via melt blending is not as good as the solvent exchange method. This is because high-speed stirring in the molten state makes it difficult to completely interrupt the molecular force between the nanoparticles. Especially in cases with high content of nanomaterials, the nanoparticles are easier to agglomerate, thus reducing the mechanical performance of polymers (9, 10).

Solvent Exchange

The solvent exchange method includes two steps of mixing and phase inversion. Polymers first need to be dissolved or dispersed in a specific solvent with nanoparticles. Through rapid stirring and external applied conditions (ultrasound and heating), nanomaterials and polymers can form a evenly stable phase in the dispersion. Taking advantage of evaporation or nonsolvent phase inversion method, the solvent in the mixture can be removed. For example, the modified MXene nanosheets were dispersed in a dimethylformamide solution with the subsequent addition of TPU to obtain a uniform suspension (11). The mixture was poured into deionized water with the aid of stirring for phase inversion, and the TPU nanocomposites were precipitated, in which MXene nanosheets were homogeneously dispersed.

The nanocomposites fabricated via the solvent exchange method exhibit excellent dispersion of nanoparticles in polymer matrix. However, this approach demands large amounts of organic solvents that hurt the environment and increase manufacturing costs during industrial production. Meanwhile, a certain amount of solvents that remain in polymers may influence the performance of nanocomposites. In addition, this method is not applicable when the nanoadditives cannot homogeneously dispersed in solvents or polymers. Therefore, the solution blending method is more suitable for preparing nanocomposites in laboratory instead of industrial scale.

In Situ Polymerization

For in situ polymerization, nanomaterials and precursors are diffused or dissolved in the solvent before polymerization is initiated. The monomer starts to polymerize under the action of a catalyst or initiator to construct the polymer chains, and the nanoparticles are meanwhile embedded into polymers for fabricating nanocomposites. It should be noted that nanomaterials are added into the polymers at a relatively lower molecular weight during the reaction, which makes it easier for them to form a uniform dispersion in the polymer matrix. For example, graphene was firstly mixed with polyether polyol and TDI. Under the action of amine and organic tin catalyst, polymerization was carried out in an aqueous solution (12). −NCO in TDI reacted with −OH in polyether polyol to generate carbamate under the action of catalysts. Furthermore, –NCO reacted with H_2O can generate CO_2 to obtain porous materials such as PU foam (PUF). The prepared graphene–PU nanocomposites showed excellent dispersion because of the hydrophilicity of graphene. In situ polymerization is considered to be a highly effective method to enhance dispersity of nanomaterials in polymers. However, the molecular weight of PU synthesized after polymerization is difficult to be controlled, and that will influence the overall performance of nanocomposites.

Surface Coating

Surface coating is mainly utilized in the preparation of FR PUF. PUF is a synthetic ultra-light material comprised of highly porous 3D architectures that exhibit versatile properties, including high specific surface area, low density, and low thermal conductivity, which has been widely applied in building insulation and packaging materials (13, 14). The direct addition of nanomaterials into PU matrix may be inappropriate for the fabrication of FR PUF because the presence of nanomaterials can affect the PUF structure—resulting in the reduction of elasticity. Therefore, manufacturing nanomaterials coating on the surface of PUF is an effective approach to improve the FR performance and simultaneously protect the foam structure (15, 16). The surface coating methods include layer-by-layer assembly, dip-coating, and plasma treatment. For example, FPUF was treated by HNO_3 to make the surface positively charged (17). Water-soluble phenolic resin (WSPR) and zeolitic imidazolate framework-67 (ZIF-67) were then coated on the surface of FPUF using layer-by-layer assembling. The water-soluble phenolic solution was used as a negatively charged polyelectrolyte solution, which can react with modified FPUF to generate a dense coating (Figure 4). ZIF-67 and WSPR connected via electrostatic attraction and π-π stacking between the interaction of groups.

The surface coating can endow PUF with a better FR effect as long as the coating is uniform and compact. The dispersion issue of nanomaterials in PU via the surface coating method can be neglected compared with other methods because this approach avoids the participation of nanomaterials during the preparation of PUF. However, the surface coating requires higher adhesion between nanomaterials and PUF. Otherwise, the layer can inevitably be exfoliated, which has an adverse effect on the final performance of nanocomposites.

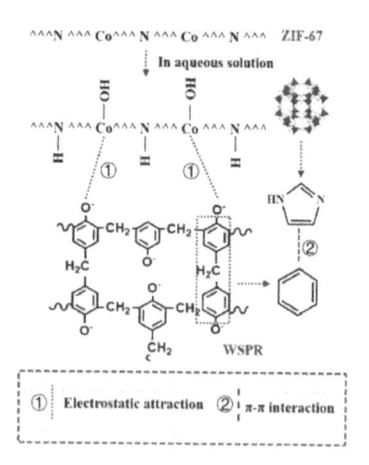

Figure 4. Assembly mechanism of ZIF-67, WSPR, and FPUF. Reproduced with permission from reference (17). Copyright 2020 Elsevier.

Various Nanomaterial-Based FR-PU Nanocomposites

0D Nanomaterial-Based PU Nanocomposites

0D materials refer to matter that are in the nanoscale range (1–100 nm) in three dimensions or are composed of them as basic units, roughly equivalent to the scale of 10–100 atoms that are closely packed together. Typical 0D materials include clusters, nanoparticles, and quantum dots. The 0D materials, including metal organic framework (MOF, (18, 19), metal oxides (20, 21), silica (22–24), clay (25), fullerene (26), have been reported as nanoadditives for the preparation of FR-PU nanocomposites. These inorganic materials have relatively good thermal stability that can effectively improve the FR properties of PU materials.

Silica was used to improve the flame retardancy of PU materials, which can act as a barrier to block heat transfer and prevent the transmission of oxygen and combustible organic volatiles produced by polymer cracking. In addition, silica catalyzed the formation of carbon on the surface resulting in the generation of char layers that further enhanced the FR properties of polymer (23). As illustrated in Figure 5, Brannum et al. embedded silica nanoparticles on the surface of PUF via sol-gel methods based on the Stöber process. Tetraethyl orthosilicate (TEOS) was utilized as the start monomer, followed by the hydrolysis and condensation processes (27). At the beginning of the combustion, the PU composite foam gradually burned and formed a char layer on the surface until it fully covered the outer layer of the foam, leading it to extinguish itself. Additionally, compared to pristine PUF, the silica nanoparticles coating prevented the modified PUF from dripping during burning. As a result, the peak heat release rate (PHRR) of the modified PU composites foam was further reduced (from 560 to 262 kW/m^2), showing good flame retardancy.

The metal oxides are another commonly used FR nanoparticles because of their good barrier effect for protecting polymers. It also enables to catalyze formation of char layers during combustion, thus improving FR properties. Ciesielczyk et al. prepared a aminoethylaminopropylIsobutyl polyhedral oligomeric silsesquioxane (N-POSS) coated MgO-SiO$_2$ binary oxide system via a sol-gel process that can be used as a FR for PUFs with a V-0 classification as a nonflammable material in vertical test (UL-94 V, (28). Benefitting from its excellent thermodynamic stability and compatibility, TiO$_2$ nanoparticles were added into PU by means of melt blending. TiO$_2$ can effectively promote the formation of compact carbon barrier to block heat and oxygen. Moreover, the dielectric constant of TiO$_2$ can be enhanced with the increase of temperature, leading it to hinder the free movement of flammable volatiles, and this gives a stronger adhesion effect to polymer chains. Furthermore, a more compact and smoother char surface restricted the penetration of oxygen and heat to enhance flame retardancy. Therefore, the PHRR of TPU incorporated with TiO$_2$ was 648 kW/m^2, which was much lower than that of pure TPU, which was 1510 kW/m^2 (29). In addition, to further improve the dispersion of TiO$_2$ in a polymer matrix, Dong et al. modified TiO$_2$ particles with 9,10-dihydro-9-oxa-10-phosphaphenanthrene-10-oxide (DOPO)-methacryloxy propyl trimethoxyl silane to reduce the agglomeration. As a traditional FR, KH570-DOPO can capture the free radicals generated by polymer cleavage in the gas phase to decrease the concentration of combustibles. When added into FPUF via in situ polymerization, the PHRR of FPUF was reduced from 657.0 to 519.2 W/g under a synergistic effect (30).

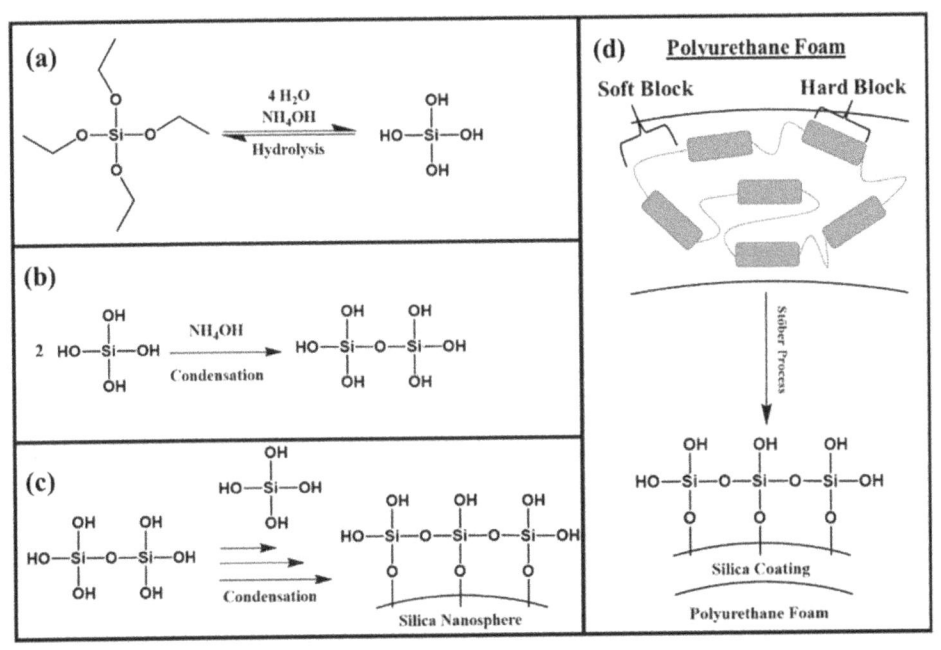

Figure 5. The general synthesis process of silica-based PU nanocomposites via Stöber Process. (a) Hydrolysis of the TEOS; (b) Condensation and dimerizing of the orthosilithic acid; (c) formation of silica nanospheres; and (d) PU structure and its subsequent silica treatment. Reproduced with permission from reference (27). Copyright 2019 American Chemical Society.

As a unique 0D nanomaterial, fullerene has attracted lots of attention as carbon-based nanomaterials. However, neat fullerene particles are easy to agglomerate, which limits its further application. Kanbur et al. added fullerene to a PU elastomer. To achieve a better dispersion, they modified the fullerene with HNO$_3$/H$_2$SO$_4$ to obtain hydrogen sulfated fullerene, thereby enhancing the interaction with TPU. Because of its excellent free radical capture function, fullerene can adhere to the polymer's condensed phase to reduce flammability (31, 32).

MOFs are manufactured by self-assembly of organic ligands and metal ions that have a 3D porous structure, usually with metal ions as the connection point in which the organic ligand supports the formation of a 3D extension of its space. MOFs can be prepared with different morphologies, whereas 0D MOFs are the most widely used nanoparticles in FR applications (33). Among MOFs, zeolitic imidazolate frameworks (ZIFs) possess high porosity and good thermal and chemical stability. Xu et al. prepared ZIF-8/α-ZrP hybrids by depositing ZIF-8 on the surface of α-ZrP for fabricating PU nanocomposites via in situ polymerization (8). During combustion, ZIF-8 can produce the metal oxides that catalyzed the combustibles decomposed from PU into thermally stable char residues. Furthermore, because of the physical barrier and catalytic carbonization effects of α-ZrP, it can block the rise of flammable gas and oxygen to prevent contact with the PU. In consequence, under the synergistic effect between α-ZrP and ZIF-8, PHRR and total heat release (THR) of PU nanocomposites were decreased by 70% and 46%, respectively. The WSPR/ZIF-67 FR coating, as illustrated previously, also can enhance the flame resistance of FPUF (17). In the vertical burning experiment, the WSPR/ZIF-67 coating can effectively prevent the cracking of the structure with improved self-extinguishing performance. The PHRR value was reduced by up to 67% compared to FPUF. Meanwhile, the smoke production rate and total smoke production were reduced by 63% and 78%, respectively, suggesting that the WSPR/ZIF-67 coating endowed FPUF with outstanding smoke suppression properties. This enhanced FR performance can be attributed to the compact, uniform WSPR/ZIF-67 coating. During combustion, a dense char layer was formed to protect FPUF and avoid further decomposition and collapse of the cell structure. The char layer effectively prevented the migration of cracked products and smoke. Additionally, the Co_3O_4 nanoparticles produced from ZIF-67 promoted the oxidation of CO to reduce its release (34).

1D Nanomaterial-Based PU Nanocomposites

1D nanomaterials are materials with 2D directions in the nanometer scale, such as nanotubes, nanofibers, and nanorods. These linear materials are often entangled with the polymer chains after they are added into polymers—leading to improved mechanical properties. Moreover, they can also be used as a barrier to block heat transmission for enhancing FR performance. Nevertheless, the agglomeration of 1D materials often occurs in the polymers when the concentration increases, which may destroy the mechanical properties with a limited FR effect.

β-FeOOH nanorods were utilized as FRs that can generate moisture during combustion to reduce the heat release. The inert gases decomposed from β-FeOOH to dilute the concentration of the surrounding combustible volatiles, as well as catalyzing polymers into char residues (35). Hu et al. carried out a series of experiments on the enhancement of flame resistance of PUF based on the β-FeOOH nanorods. β-FeOOH and montmorillonite (MMT) were alternately coated on the surface of FPUF with the formation of a sandwich-like topology structure based on charge interaction via layer-by-layer assembling (36). The synergetic FR performance of two materials was better than using a single one. −NCO containing volatiles and other toxic gases released from FPU foams were significantly reduced. It was attributed to the network formed by MMT layers and β-FeOOH nanorods, which inhibited the production of gases. The char layers also acted as a protective barrier for FPUF to enhance the flame resistance. In another work, the synergistic system composed of graphene and β-FeOOH showed great superiority for improving the flame retardancy of FPUF (37). This binary hybrid coating led to a significant reduction of PHRR (49.5%) with the inhibition on the appearance of the second PHRR.

Carbon-based 1D nanomaterials with superior thermal stability and electrical properties have been widely used to improve the FR properties of polymers, as well as other functions such as electromagnetic interference shielding. Carbon nanotubes (CNTs) were discovered by Dr. Lijima in Japan (38), and they are nanotube materials with extremely high structural integrity and large aspect ratio (100–1000). Their low density and other characteristics enable them to be applied in polymer reinforcement. CNTs have become widely researched 1D materials for improving the fire resistance properties of polymers because of their superior thermal stability and catalytic carbonization effect.

However, the strong van der Waals forces and π-π interactions between pure CNTs lead to the agglomeration of the material, thereby influencing their dispersion in the polymer matrix. These agglomerated particles reduce the interaction with polymers. Therefore, this research seeks to improve the dispersion of CNTs. For example, Xia et al. utilized ball milling to distribute CNTs in polyol solution with the aid of dispersion agent BYK 9077. PU nanocomposites were prepared via in situ polymerization. During the ball milling process, the moving ball converted the kinetic energy of the CNT agglomerated particles into mechanical pressure for breaking up these particles to obtain better dispersion (39). Another commonly used method is chemical functionalization. For example, CNTs were treated with mixed strong acids resulting in the grafting of carboxyl groups on the surface to enhance the interaction between CNTs and PU chains via hydrogen bonds (40).

Moreover, CNTs were also added into polymers together with other additives to achieve better FR properties on the basis of synergistic effects, including pyrene-modified branched polyethylenimine (BPEI) (41), ZnAl-layered double hydroxide (42), MMT (43), and aluminum hydroxide (44). Burgaz et al. introduced nanosilica and carboxylic acids modified CNT to rigid PU nanocomposites foams (45). Based on the hydrogen bonding interaction between CNT-COOH and nanosilica and PU chains, the dispersion was improved, and it significantly enhanced the thermal stability of PU. In Ji et al.'s research (46), CNTs combined with IFRs were used to improve the fire resistance of TPU. The synergy between CNTs and IFRs can effectively promote the formation of char layers, preventing melt dripping, as well as protecting the polymer suffered from heat. The LOI value of the nanocomposite was increased by 30%, and the PHRR was reduced by 92% compared to neat PU. Furthermore, because of the high electrical conductivity, the incorporation of CNTs endowed PU with outstanding electromagnetic interference shielding effectiveness values up to 20 dB, meeting the requirement of commercial electronic devices with the extension of the application range of PU. Ji et al. prepared the multilayer structure of CNT- and FR-based PU nanocomposites via layer-multiplying co-extrusion method. CNTs and FRs were deposited separately in the PU matrix (47). The electromagnetic interference shielding effectiveness could reach 32.7 dB because of this unique structure. Furthermore, the char layer formed during the combustion process also presented excellent protection of PU material, which passed the V-0 rating.

In addition, carbon nanofibers (CNFs) have attracted interest because of their excellent mechanical properties and thermal conductivity, which can be used to improve the FR properties of PU (48). Zhao et al. demonstrated that the mechanical properties of TPU were remarkably improved (tensile strength increased from 39% to 42%) by incorporating CNF and ammonium polyphosphate (APP), (49). As a result of the synergistic FR effect, total smoke release (TSR) was drastically reduced because of the fact that the addition of CNF can strengthen the viscosity of molten carbon monomer induced by the catalyzation of APP. Therefore, a compact char layer was formed. The PHRR was reduced to 129 kW/m^2, indicating excellent flame resistance. Kim et al. deposited CNF on the surface of PUF via layer-by-layer assembling (50). The 40% reduction of PHRR for CNF/PUF was obtained compared to neat PUF, showing the outstanding FR performance.

The other tubular structural 1D nanomaterials were also employed to improve the FR properties of PU such as poly(cyclotriphosphazene-co-4,4(-sulfonyldiphenol) nanotubes ((51) and polyphosphazene nanotube@mesoporous silica@bimetallic phosphide ternary nanostructures (52). Most of them were modified with other materials before the introduction to PU for achieving a superior dispersion and flame resistance. Halloysite nanotubes (HNTs) is one of the natural clays exhibiting high specific surface area, good thermal stability and mechanical strength. Plenty of Si-OH and Al-OH bonds located on the surface of HNTs, suggesting that it was easy to be modified for improve its functionality (53). During combustion, HNTs can produce moisture to reduce heat, which leads to the inhibition of burning. When HNTs were loaded on PUF by dip-coating, a protective barrier composed of melted PU and HNTs was formed to isolate the heat and oxygen. Because of its high hydrophilcity, HNT-modified PUF can be applied in the oil or water separation, which showed a better MB adsorption (from 0.02 to 0.15 mg/g) compared to pure PUF (54). HNTs can also be deposited on the surface of PUF using a layer-by-layer method, with BPEI or poly(acrylic acid) (PAA) as a stabilizer (55). During the burning tests, only one thin char layer remained on the PUF surface compared to pure PUF. It can be noted that HNTs served as a barrier to prevent burning heat from spreading into PUF and protected the structure of the PUF from collapse, resulting in the retardation of melt dripping.

2D Nanomaterial-Based PU Nanocomposites

2D nanomaterials have attracted increasing attention and have become a popular research topic. They are crystalline materials with a thickness of only one or a few atomic layers in a 2D plane, and there are van der Waals forces and π-π interactions between the layers. Because of their unique layered structure and special electrical and optical properties, they are widely used in the fields of optoelectronics, semiconductors, and thermal management. As FR additives, the 2D nanomaterials such as, MXene (11), black phosphorus (56), layered double hydroxide (57), and graphite nanoplatelets (58) have been reported. Their FR effect is more efficient than that of other nanomaterials.

Unique properties of 2D materials have been extensively studied and revealed in various fields. For example, when doped into the polymer, black phosphorus can effectively improve the FR properties of polymers. In the burning process, black phosphorus participated in the formation of char layer. The difference was that the char layer promoted by black phosphorus was more continuous and uniform, which can block heat and mass transfer of the combustion reaction (56, 59). Graphene oxide (GO)GO can also improve the FR properties of polymers. The addition of GO significantly reduced the gas permeability of the polymer composites because of the curved channels generated by GO (12, 60). Graphitic carbon nitride (g-C_3N_4) added into TPU can act as an effective physic barrier, resulting in a decrease in PHRR (61). Clay-based 2D materials such as silica or MMT nanosheets benefited from their sheet structure and thermodynamic stability. The physical barrier effect blocked the transmission of heat and improved the compactness of char layers (62, 63). Layered double hydroxide decomposed to oxides (i.e., water and CO_2), which could dilute the flammable volatile gas and heat in the gas phase, leading to the enhancement of flame retardancy (64).

Traditionally, most of the current 2D materials are in a block shape, and the layers are connected by a strong intermolecular force. The properties of 2D layered units are different from bulk materials. When the 2D materials are exfoliated to a single layer or few layers, the aspect ratio of the nanosheets is largely increased. In this way, the atomic utilization is greatly improved, and the special structure is more conducive to the improvement of FR performance (11, 65). For example, MXene, a new emerging 2D material, has a chemical structure of $M_{n+1}A_nX_n$, where M represents a transition metal, A means Al, Si, and Ga, and X stands for C or N. After etching aluminum atoms from Ti_3AlC_2 via an HF solution, the etched layer is peeled off by simple manual shaking or sonication. In our previous work, we used a mixture of LiF and HCl to etch Ti_3AlC_2. The etched Ti_3AlC_2 surface clearly showed a flaky structure, and Ti_3C_2 nanosheets were prepared by ultrasonic treatment (11).

The exfoliated MXene was modified with two cationic modifiers, cetyltrimethyl ammonium bromide and tetra butyl phosphine chloride. The decorated MXene nanosheets were uniformly dispersed in TPU, resulting in a significant improvement of FR properties. During burning, MXene nanosheets were wrapped by polyaromatic carbon, leading to the formation of dense char layers, which enhanced the heat insulation and slowed down flame propagation and spread. In another work, we deposited MXene nanosheets on the surface of FPUF via layer-by-layer under the assistance of chitosan. During combustion, Ti_3C_2 sheets were oxidized to generate TiO_2, which can reduce the oxygen concentration during combustion, as well as catalyze chitosan and PU into char layers. As the decomposed molecules continued to participate in the carbonization process, a uniform, dense protective layer formed on the surface of PUF to block the transmission of heat and gas, resulting in the reduction of PHRR (−57%) and TSR (−71%) ((65).

Generally, the poor compatibility between 2D materials with polymers limits the development of PU nanocomposites. After exfoliation, the single-layered 2D materials often have various hydrophilic functional groups on the surface. They are not compatible with hydrophobic PU, thus causing agglomeration. Furthermore, there are very strong van der Waals forces and π-π interactions between graphene sheets, leading to the agglomeration. This phenomenon will deteriorate the mechanical properties of polymers. Consequently, chemical functionalization is required to improve the compatibility. For example, Cai et al. used phosphonated polyethyleneimine to modify boron nitride nanosheets (BNNSs) based on Lewis acid-base interaction (66). Thanks to the hydrogen bond, the modified BNNSs were uniformly dispersed in TPU. The mechanical properties were significantly improved, in which tensile strength at break was increased by 39% compared to pure TPU. In the case of combustion, better dispersion leads to a better barrier effect, thus blocking the rise of oxygen and combustible materials. In addition, combustible degradation products will accumulate on the surface of the BNNSs. Simultaneously, phosphonated polyethyleneimine will react with these products to form a more compact char layer. Wang et al. utilized polypyrrole and phytic acid to decorate BNNSs followed by the adsorption of Cu^{2+} (67). After it was added into TPU, improved dispersion of BNNSs was achieved. During combustion, the formation of a continuous dense barrier network blocked the transmission of volatile toxic gases such as CO and HCN generated by the cracking of TPU.

To highlight the progress in FR-PU nanocomposites, a comparative performance evaluation of the reported PU filled with various nanomaterials was carried out (Table 1). It is clearly observed that the addition of different nanomaterials can enhance the fire safety properties of PU. To achieve better performance, it is necessary to combine with other FRs (e.g., IFRs, hydroxides, etc.) to generate a synergistic effect. The required incorporation content of these nanomaterials is much lower than that of traditional FRs, regardless of how they are prepared. Although the loading of 0D nanomaterials is the lowest, the corresponding FR effect is very limited. 1D nanomaterials that are usually combined with conventional FRs can maximize their FR effect during combustion in which the reduction of PHRR is more significant. Compared with 1D and 0D nanomaterials, 2D nanomaterials are more efficient in suppressing the THR and TSR of PU based on the excellent barrier effect. However, the synergistic effect between 2D nanomaterials and traditional FRs is limited. Therefore, it can be concluded that 2D nanomaterials show more remarkable advantages in improving the fire safety properties of PU, as well as preferable application potential.

Table 1. FR Properties of Selected PU Nanocomposites[a]

Category	Type of Nanomaterials	Content	PHRR	THR	TSR	Ref
0D nanoadditives/PU nanocomposites	SiO$_2$	5ML TEOS	−40%	−29%	—	(23)
	Silica nanoparticle	0.5 M TEOS	−50%	−11%	+9%	(27)
	TiO$_2$	1 wt %	−57%	−21%	−18%	(29)
	TiO$_2$-KH570-DOPO	10 wt %	−21%	−7%	—	(30)
	Fullerene	2 wt %	−31%	—	—	(32)
1D nanoadditives/PU nanocomposites	β-FeOOH, PEI, SA	8 trilayers	−62%	−5%	−21%	(35)
	PEI-Py/ PAAPEI-Py/ PAA + MWCNT	6 bilayer	−67%	+0.025%	−80%	(41)
	CH-CNT and MMT	2.81 wt %	−70%	−3%	—	(43)
	CF	15.00 wt % APP, 5.00 wt % CF	−91	−71%	−54%	(49)
	BPEI-HNT/PAA-HNT	34.2 wt %	−62%	+2%	60%	(55)
	PZS@M-SiO2@CoCuP	3 wt % of	−58%	19%	—	(52)
2D nanoadditives/PU nanocomposites	GO	3 wt %	−42	−38%	—	(68)
	BP-EC-Exf	3.0 wt %	−45%	−35%	—	(56)
	GNPs	3.9 vol %	−32%	−14%	—	(58)
	g-C$_3$N$_4$/carbon sphere/Cu	2.0 wt %	−36%	−14%	—	(61)
	PMT	9% AHP and 1% PMT	−82%	−62%	−89%	(62)
	MXene (Ti$_3$C$_2$)	6.9 wt %	−57%	−66%	−71%	(65)
	f-BN	5.0 wt %	−68%	−30%	—	(66)

[a] AHP: aluminum hypophosphite; BP-EC-Exf: cobaltous phytate-functionalized BP nanosheets; BPEI-HNT/PAA-HNT: branched polyethylenimine and poly(acrylic acid) stabilized halloysite nanotubes; CF: Carbon fibers; CH-CNT: chitosan-wrapped carbon nanotube; DOPO: 9,10-dihydro-9-oxa-10-phosphaphenanthrene-10-oxide; f-BN: phosphonated polyethyleneimine modified hexagonal boron nitride; GNPs: graphite nanoplatelets; KH570: 3-methacryloxypropyltrimethoxysilane; PEI: Polyethylenimine; PEI-Py/PAA + MWCNT: pyrene, anionic poly(acrylic acid), and multiwalled carbon nanotubes; PMT: phosphorylated chitosan modified montmorillonite nanosheets; PZS@M-SiO2@CoCuP: polyphosphazene nanotube@Mesoporous Silica@Bimetallic Phosphide; SA: sodium alginate.

Conclusion

In this chapter, we generally introduced the development of FR-PU nanocomposites, including the preparation strategies and the FR mechanism. The addition of nanomaterials endow PU with improved flame resistance. The dispersion of nanomaterials in PU is of great importance for further enhancing its fire safety and mechanical properties. Therefore, it is necessary to modify nanomaterials via commonly used chemical functionalizations for enhancing their compatibility and dispersion. Moreover, another approach for further improving flame resistance properties involves the synergistic effect between various nanomaterials and traditional FRs. Among the nanomaterials, 2D ones show more remarkable advantages in reducing heat release and toxic hazards because of their unique layered structure, indicating the promising application potential in FR-PU nanocomposites.

Prospective

Currently, numerous FRs based on nanomaterials have been reported as improving the FR properties of PU. Despite this major progress, there are still some problems and challenges. First, all nanomaterials without any modifications tend to aggregate, leading to poor dispersion in the polymer matrix. In the present chemical functionalization, nondegradable organic chemicals are commonly used, which is harmful to environment. In particular, some modifiers are very expensive. As a result, we may need to explore more environmentally friendly, cheaper agents to enhance the sustainability of the surface engineering strategy. Second, the current research trends are primarily focused on improving the FR properties rather than pursuing multifunction. With the increasing multifunctional requirements, the versatility of FR composites is in greater demand. The incorporated nanomaterials can endow PU with outstanding electroconductivity, hydrophilicity, and photocatalytic properties, which extend its application to fields such as electromagnetic shielding and oil-water separation. Therefore, further exploration of multifunctional PU nanocomposites can greatly expand its range of applications, as well as exploit their industrialization potential.

Acknowledgments

This chapter was co-financed by Anhui Provincial Natural Science Foundation for Distinguished Young Scholar (2008085J26) and the Anhui Provincial Key Technologies R&D Program (202004a05020044 and 1804a09020070).

References

1. Bayer, O. Das Di-lsocganat-Poluadditionsverfahren (Polyurethane). *Angew. Chem.* **1947**, *A59*, 257–288.
2. Akindoyo, J. O.; Beg, M. D. H.; Ghazali, S.; Islam, M. R.; Jeyaratnam, N.; Yuvaraj, A. R. Polyurethane Types, Synthesis and Applications – A Review. *RSC Adv.* **2016**, *6*, 114453–114482.
3. Krol, P. (2007) Synthesis Methods, Chemical Structures and Phase Structures of Linear Polyurethanes. Properties and Applications of Linear Polyurethanes in Polyurethane Elastomers, Copolymers and Ionomers. *Prog. Mater. Sci.* **2007**, *52*, 915–1015.
4. Chattopadhyay, D. K.; Raju, K. V. S. N. Structural Engineering of Polyurethane Coatings for High Performance Applications. *Prog. Polym. Sci.* **2007**, *32*, 352–418.

5. Chattopadhyay, D. K.; Webster, D. C. Thermal Stability and Flame Retardancy of Polyurethanes. *Prog. Polym. Sci.* **2009**, *34*, 1068–1133.
6. Levchik, S. V.; Weil, E. D. Thermal Decomposition, Combustion and Fire-Retardancy of Polyurethanes—A Review of the Recent Literature. *Polym. Int.* **2004**, *53*, 1585–1610.
7. Chen, X.; Huo, L.; Liu, J.; Jiao, C.; Li, S.; Qian, Y. Combustion Properties and Pyrolysis Kinetics of Flame-Retardant Polyurethane Elastomers. *J. Thermoplast. Compos. Mater.* **2016**, *30*, 255–272.
8. Xu, B.; Xu, W.; Liu, Y.; Chen, R.; Li, W.; Wu, Y.; Yang, Z. Surface Modification of α-zirconium Phosphate By Zeolitic Imidazolate Frameworks-8 and Its Effect on Improving the Fire Safety of Polyurethane Elastomer. *Polym. Adv. Technol.* **2018**, *29*, 2816–2826.
9. Verdejo, R.; Bernal, M. M.; Romasanta, L. J.; Lopez-Manchado, M. A. Graphene Filled Polymer Nanocomposites. *J. Mater. Chem.* **2011**, *21*, 3301–3310.
10. Zhang, Z.; Zhang, J.; Chen, P.; Zhang, B.; He, J.; Hu, G.-H. Enhanced Interactions Between Multi-Walled Carbon Nanotubes and Polystyrene Induced By Melt Mixing. *Carbon* **2006**, *44*, 692–698.
11. Yu, B.; Tawiah, B.; Wang, L.-Q.; Yin Yuen, A. C.; Zhang, Z.-C.; Shen, L.-L.; Lin, B.; Fei, B.; Yang, W.; Li, A.; Zhu, S.-E.; Hu, E.-Z.; Lu, H.-D.; Yeoh, G. D. Interface Decoration of Exfoliated MXene Ultra-Thin Nanosheets for Fire and Smoke Suppressions of Thermoplastic Polyurethane Elastomer. *J. Hazard. Mater.* **2019**, *374*, 110–119.
12. Yao, Y.; Jin, S.; Ma, X.; Yu, R.; Zou, H.; Wang, H.; Lv, X.; Shu, Q. (2020) Graphene-Containing Flexible Polyurethane Porous Composites with Improved Electromagnetic Shielding and Flame Retardancy. *Compos. Sci. Technol.* **2020**, *200*, 108457.
13. Wu, X.; Huang, S.; Zhang, Y.; Shi, L.; Luo, Y.; Deng, X.; Liu, Q.; Li, Z. (2020) Flame Retardant Polyurethane Sponge/MTMS Aerogel Composites with Improved Mechanical Properties Under Ambient Pressure Drying. *J. Nanopart. Res.* **2020**, *22*, 221.
14. Xie, H.; Yang, W.; Yuen, A. C. Y.; Xie, C.; Xie, J.; Lu, H.; Yeoh, G. H. Study on Flame Retarded Flexible Polyurethane Foam/Alumina Aerogel Composites with Improved Fire Safety. *Chem. Eng. J.* **2017**, *311*, 310–317.
15. Chen, H. B.; Shen, P.; Chen, M. J.; Zhao, H. B.; Schiraldi, D. A. Highly Efficient Flame Retardant Polyurethane Foam with Alginate/Clay Aerogel Coating. *ACS Appl. Mater. Interfaces* **2016**, *8*, 32557–32564.
16. Wi, S.; Berardi, U.; Loreto, S. D.; Kim, S. Microstructure and Thermal Characterization of Aerogel-Graphite Polyurethane Spray-Foam Composite for High Efficiency Thermal Energy Utilization. *J. Hazard. Mater.* **2020**, *397*, 122656.
17. Xu, W.; Chen, R.; Du, Y.; Wang, G. Design Water-soluble Phenolic/Zeolitic Imidazolate Framework-67 Flame Retardant Coating via Layer-By-Layer Assembly Technology: Enhanced Flame Retardancy and Smoke Suppression of Flexible Polyurethane Foam. *Polym. Degrad. Stab.* **2020**, *176*, 109152.
18. Xu, W.; Wang, G.; Xu, J.; Liu, Y.; Chen, R.; Yan, H. Modification of Diatomite with Melamine Coated Zeolitic Imidazolate Framework-8 as an Effective Flame Retardant to Enhance Flame Retardancy and Smoke Suppression of Rigid Polyurethane Foam. *J. Hazard. Mater.* **2019**, *379*, 120819.

19. Cheng, J.; Ma, D.; Li, S.; Qu, W.; Wang, D. Preparation of Zeolitic Imidazolate Frameworks and Their Application as Flame Retardant and Smoke Suppression Agent for Rigid Polyurethane Foams. *Polymers (Basel)* **2020**, *12*, 347.
20. Gao, X.; Guo, Y.; Tian, Y.; Li, S.; Zhou, S.; Wang, Z. Synthesis and Characterization of Polyurethane/Zinc Borate Nanocomposites. *Colloids Surf.* **2011**, *A 384*, 2–8.
21. Wang, H.; Jiao, C.; Zhao, L.; Chen, X. Preparation and Characterization of TiO2-Coated Hollow Glass Microsphere and Its Flame-Retardant Property in Thermoplastic Polyurethane. *J. Therm. Anal. Calorim.* **2017**, *131*, 2729–2740.
22. Chen, X.; Jiang, Y.; Liu, J.; Jiao, C.; Qian, Y.; Li, S. Smoke Suppression Properties of Fumed Silica on Flame-Retardant Thermoplastic Polyurethane Based on Ammonium Polyphosphate. *J. Therm. Anal. Calorim.* **2015**, *120*, 1493–1501.
23. Li, M. E.; Wang, S. X.; Han, L. X.; Yuan, W. J.; Cheng, J. B.; Zhang, A. N.; Zhao, H. B.; Wang, Y. Z. Hierarchically Porous SiO2/Polyurethane Foam Composites Towards Excellent Thermal Insulating, Flame-Retardant and Smoke-Suppressant Performances. *J. Hazard. Mater.* **2019**, *375*, 61–69.
24. Gao, W.; Qian, X.; Wang, S. Preparation of Hybrid Silicon Materials Microcapsulated Ammonium Polyphosphate and Its Application in Thermoplastic Polyurethane. *J. Appl. Polym. Sci.* **2018**, *135*, 45742.
25. Carretier, V.; Delcroix, J.; Pucci, M. F.; Rublon, P.; Lopez-Cuesta, J. M. Influence of Sepiolite and Lignin as Potential Synergists on Flame Retardant Systems in Polylactide (PLA) and Polyurethane Elastomer (PUE). *Materials (Basel)* **2020**, *13*, 2450.
26. Kausar, A. Waterborne Polyurethane-Coated Polyamide/Fullerene Composite Films: Mechanical, Thermal, and Flammability Properties. *Int. J. Polym. Anal. Charact.* **2016**, *21*, 275–285.
27. Brannum, D. J.; Price, E. J.; Villamil, D.; Kozawa, S.; Brannum, M.; Berry, C.; Semco, R.; Wnek, G. E. Flame-Retardant Polyurethane Foams: One-Pot, Bioinspired Silica Nanoparticle Coating. *ACS Appl. Polym. Mater.* **2019**, *1*, 2015–2022.
28. Ciesielczyk, F.; Szwarc-Rzepka, K.; Przybysz, M.; Czech-Polak, J.; Heneczkowski, M.; Oleksy, M.; Jesionowski, T. Comprehensive Characteristic and Potential Application of POSS-Coated MgO-SiO2 Binary Oxide System. *Colloids Surf. A* **2018**, *537*, 557–565.
29. Chen, X.; Wang, W.; Li, S.; Qian, Y.; Jiao, C. Synthesis of TPU/TiO2 Nanocomposites by Molten Blending Method. *J. Therm. Anal. Calorim.* **2018**, *132*, 793–803.
30. Dong, Q.; Chen, K.; Jin, X.; Sun, S.; Tian, Y.; Wang, F.; Liu, P.; Yang, M. Investigation of Flame Retardant Flexible Polyurethane Foams Containing DOPO Immobilized Titanium Dioxide Nanoparticles. *Polymers (Basel)* **2019**, *11*, 75.
31. Kanbur, Y.; Küçükyavuz, Z. Synthesis and Characterization of Surface Modified Fullerene. *Fullerenes, Nanotubes, Carbon Nanostruct.* **2012**, *20*, 119–126.
32. Kanbur, Y.; Tayfun, U. Development of Multifunctional Polyurethane Elastomer Composites Containing Fullerene: Mechanical, Damping, Thermal, and Flammability Behaviors. *J. Elastomers Plast.* **2018**, *51*, 262–279.
33. Flugel, E. A.; Ranft, A.; Haase, F.; Lotsch, B. V. Synthetic Routes Toward MOF Nanomorphologies. *J. Mater. Chem.* **2012**, *11*, 10119.

34. Xie, X.; Li, Y.; Liu, Z. Q.; Haruta, M.; Shen, W. Low-Temperature Oxidation of CO Catalysed by Co(3)O(4) Nanorods. *Nature* **2009**, *458*, 746–749.
35. Pan, H.; Pan, Y.; Song, L.; Hu, Y. Construction of β-FeOOH Nanorod-Filled Layer-By-Layer Coating with Effective Structure to Reduce Flammability of Flexible Polyurethane Foam. *Polym. Adv. Technol.* **2017**, *28*, 243–251.
36. Wang, W.; Pan, H.; Shi, Y.; Yu, B.; Pan, Y.; Liew, K. M.; Song, L.; Hu, Y. Sandwichlike Coating Consisting of Alternating Montmorillonite and β-FeOOH for Reducing the Fire Hazard of Flexible Polyurethane Foam. *ACS Sustainable Chem. Eng.* **2015**, *3*, 3214–3223.
37. Pan, H.; Lu, Y.; Song, L.; Zhang, X.; Hu, Y. Construction of Layer-By-Layer Coating Based on Graphene Oxide/β-FeOOH Nanorods and Its Synergistic Effect on Improving Flame Retardancy of Flexible Polyurethane Foam. *Compos. Sci. Technol.* **2016**, *129*, 116–122.
38. Iijima, S. Helical Microtubules of Graphitic Carbon. *Nature.* **1991**, *354*, 56–58.
39. Xia, H.; Song, M. Preparation and Characterization of Polyurethane-Carbon Nanotube Composites. *Soft Matter* **2005**, *1*, 386–394.
40. Sahoo, N. G.; Jung, Y. C.; Yoo, H. J.; Cho, J. W. Effect of Functionalized Carbon Nanotubes on Molecular Interaction and Properties of Polyurethane Composites. *Macromol. Chem. Phys.* **2006**, *207*, 1773–1780.
41. Holder, K. M.; Cain, A. A.; Plummer, M. G.; Stevens, B. E.; Odenborg, P. K.; Morgan, A. B.; Grunlan, J. C. Carbon Nanotube Multilayer Nanocoatings Prevent Flame Spread on Flexible Polyurethane Foam. *Macromol. Mater. Eng.* **2016**, *301*, 665–673.
42. Xie, H.; Ye, Q.; Si, J.; Yang, W.; Lu, H.; Zhang, Q. Synthesis of a Carbon Nanotubes/ZnAl-Layered Double Hydroxide Composite as a Novel Flame Retardant for Flexible Polyurethane Foams. *Polym. Adv. Technol.* **2016**, *27*, 651–656.
43. Pan, H.; Pan, Y.; Wang, W.; Song, L.; Hu, Y.; Liew, K. M. Synergistic Effect of Layer-by-Layer Assembled Thin Films Based on Clay and Carbon Nanotubes To Reduce the Flammability of Flexible Polyurethane Foam. *Ind. Eng. Chem. Res.* **2014**, *53*, 14315–14321.
44. Im, J. S.; Bai, B. C.; Bae, T.-S.; In, S. J.; Lee, Y.-S. Improved Anti-Oxidation Properties of Electrospun Polyurethane Nanofibers Achieved by Oxyfluorinated Multi-Walled Carbon Nanotubes and Aluminum Hydroxide. *Mater. Chem. Phys.* **2011**, *126*, 685–692.
45. Burgaz, E.; Kendirlioglu, C. Thermomechanical Behavior and Thermal Stability of Polyurethane Rigid Nanocomposite Foams Containing Binary Nanoparticle Mixtures. *Polym. Test.* **2019**, *77*, 105930.
46. Ji, X.; Chen, D.; Wang, Q.; Shen, J.; Guo, S. Synergistic Effect of Flame Retardants and Carbon Nanotubes on Flame Retarding and Electromagnetic Shielding Properties of Thermoplastic Polyurethane. *Compos. Sci. Technol.* **2018**, *163*, 49–55.
47. Ji, X.; Chen, D.; Shen, J.; Guo, S. Flexible and Flame-Retarding Thermoplastic Polyurethane-Based Electromagnetic Interference Shielding Composites. *Chem. Eng. J.* **2019**, *370*, 1341–1349.
48. Chen, C.; Zhao, X.; Shi, C.; Chen, J. Synergistic Effect Between Carbon Nanoparticle and Intumescent Flame Retardant on Flammability and Smoke Suppression of Copolymer Thermoplastic Polyurethane. *J. Mater. Sci.* **2018**, *53*, 6053–6064.

49. Zhao, X.-L.; Chen, C.-K.; Chen, X.-L. Effects of Carbon Fibers on the Flammability and Smoke Emission Characteristics of Halogen-Free Thermoplastic Polyurethane/Ammonium Polyphosphate. *J. Mater. Sci.* **2016**, *51*, 3762–3771.

50. Kim, Y. S.; Davis, R.; Cain, A. A.; Grunlan, J. C. Development of Layer-By-Layer Assembled Carbon Nanofiber-Filled Coatings To Reduce Polyurethane Foam Flammability. *Polymer* **2011**, *52*, 2847–2855.

51. Liu, W.; Zheng, Y.; Li, J.; Liu, L.; Huang, X.; Zhang, J.; Kang, X.; Tang, X. Novel Polyurethane Networks Based on Hybrid Inorganic/Organic Phosphazene-Containing Nanotubes with Surface Active Hydroxyl Groups. *Polym. Adv. Technol.* **2012**, *23*, 1–7.

52. Qiu, S.; Shi, Y.; Wang, B.; Zhou, X.; Wang, J.; Wang, C.; Gangireddy, C. S. R.; Yuen, R. K. K.; Hu, Y. Constructing 3D Polyphosphazene Nanotube@Mesoporous Silica@Bimetallic Phosphide Ternary Nanostructures via Layer-by-Layer Method: Synthesis and Applications. *ACS Appl. Mater. Interfaces* **2017**, *9*, 23027–23038.

53. Wu, F.; Pickett, K.; Panchal, A.; Liu, M.; Lvov, Y. Superhydrophobic Polyurethane Foam Coated with Polysiloxane-Modified Clay Nanotubes for Efficient and Recyclable Oil Absorption. *ACS Appl. Mater. Interfaces* **2019**, *11*, 25445–25456.

54. Wu, F.; Zheng, J.; Ou, X.; Liu, M. Two in One: Modified Polyurethane Foams by Dip-Coating of Halloysite Nanotubes with Acceptable Flame Retardancy and Absorbency. *Macromol. Mater. Eng.* **2019**, *304*, 1900213.

55. Smith, R. J.; Holder, K. M.; Ruiz, S.; Hahn, W.; Song, Y.; Lvov, Y. M.; Grunlan, J. C. Environmentally Benign Halloysite Nanotube Multilayer Assembly Significantly Reduces Polyurethane Flammability. *Adv. Funct. Mater.* **2018**, *28*, 1703289.

56. Qiu, S.; Zou, B.; Sheng, H.; Guo, W.; Wang, J.; Zhao, Y.; Wang, W.; Yuen, R. K. K.; Kan, Y.; Hu, Y. Electrochemically Exfoliated Functionalized Black Phosphorene and Its Polyurethane Acrylate Nanocomposites: Synthesis and Applications. *ACS Appl. Mater. Interfaces* **2019**, *11*, 13652–13664.

57. Zhang, X.; Li, S.; Wang, Z.; Sun, G.; Hu, P. Thermal Stability of Flexible Polyurethane Foams Containing Modified Layered Double Hydroxides and Zinc Borate. *Int. J. Polym. Anal. Charact.* **2020**, *25*, 499–516.

58. Quan, H.; Zhang, B.-Q.; Zhao, Q.; Yuen, R. K. K.; Li, R. K. Y. Facile Preparation and Thermal Degradation Studies of Graphite Nanoplatelets (GNPs) Filled Thermoplastic Polyurethane (TPU) Nanocomposites. *Composites Part A* **2009**, *40*, 1506–1513.

59. Qiu, S.; Zhou, Y.; Zhou, X.; Zhang, T.; Wang, C.; Yuen, R. K. K.; Hu, W.; Hu, Y. Air-Stable Polyphosphazene-Functionalized Few-Layer Black Phosphorene for Flame Retardancy of Epoxy Resins. *Small* **2019**, *15*, 1805175.

60. Feng, X.; Wang, X.; Cai, W.; Qiu, S.; Hu, Y.; Liew, K. M. Studies on Synthesis of Electrochemically Exfoliated Functionalized Graphene and Polylactic Acid/Ferric Phytate Functionalized Graphene Nanocomposites as New Fire Hazard Suppression Materials. *ACS Appl. Mater. Interfaces* **2016**, *8*, 25552–25562.

61. Shi, Y.; Wang, L.; Fu, L.; Liu, C.; Yu, B.; Yang, F.; Hu, Y. Sodium Alginate-Templated Synthesis of g-C3N4/Carbon Spheres/Cu Ternary Nanohybrids for Fire Safety Application. *J. Colloid. Interface Sci.* **2019**, *539*, 1–10.

62. Liu, X.; Guo, J.; Tang, W.; Li, H.; Gu, X.; Sun, J.; Zhang, S. Enhancing the Flame Retardancy of Thermoplastic Polyurethane by Introducing Montmorillonite Nanosheets Modified with Phosphorylated Chitosan. *Colloids Surf. A* **2019**, *119*, 291–298.
63. Berta, M.; Lindsay, C.; Pans, G.; Camino, G. Effect of Chemical Structure on Combustion and Thermal Behaviour of Polyurethane Elastomer Layered Silicate Nanocomposites. *Polym. Degrad. Stab.* **2006**, *91*, 1179–1191.
64. Guo, S.; Zhang, C.; Peng, H.; Wang, W.; Liu, T. Structural Characterization, Thermal and Mechanical Properties of Polyurethane/CoAl Layered Double Hydroxide Nanocomposites Prepared via In Situ Polymerization. *Compos. Sci. Technol.* **2011**, *71*, 791–796.
65. Lin, B.; Yuen, A. C. Y.; Li, A.; Zhang, Y.; Chen, T. B. Y.; Yu, B.; Lee, E. W. M.; Peng, S.; Yang, W.; Lu, H.-D.; Chan, Q. N.; Yeoh, G. H.; Wang, C. H. MXene/Chitosan Nanocoating for Flexible Polyurethane Foam Towards Remarkable Fire Hazards Reductions. *J. Hazard. Mater.* **2020**, *381*, 120952.
66. Cai, W.; Hong, N.; Feng, X.; Zeng, W.; Shi, Y.; Zhang, Y.; Wang, B.; Hu, Y. A Facile Strategy To Simultaneously Exfoliate and Functionalize Boron Nitride Nanosheets via Lewis Acid-Base Interaction. *Chem. Eng. J.* **2017**, *330*, 309–321.
67. Wang, J.; Zhang, D.; Zhang, Y.; Cai, W.; Yao, C.; Hu, Y.; Hu, W. Construction of Multifunctional Boron Nitride Nanosheet Towards Reducing Toxic Volatiles (CO and HCN) Generation and Fire Hazard of Thermoplastic Polyurethane. *J. Hazard. Mater.* **2019**, *362*, 482–494.
68. Sabet, M.; Soleimani, H.; Mohammadian, E.; Hosseini, S. The Effect of Graphene Oxide on the Mechanical, Thermal Characteristics and Flame Retardancy of Polyurethane. *Plast. Rubber Compos.* **2020**, *50*, 61–70.

Chapter 11

Industrial Flame Retardants for Polyurethanes

K. M. Faridul Hasan,[*,1] Péter György Horváth,[1] Seda Baş,[1] and Tibor Alpár[*,1]

[1]Simonyi Károly Faculty of Engineering, University of Sopron, Sopron 9400, Hungary
[*]Email: faridulwtu@outlook.com.
[*]Email: alpar.tibor@uni-sopron.hu.

Polyurethanes (PUs) are world-class versatile materials with great potential for industries, given characteristics like enhanced flame retardancy. Specifically, biological, physical, chemical, and mechanical properties of PUs have motivated researchers and manufacturers to tailor PUs to make them suitable and attractive materials. PUs are extensively used for automotive materials, carpeting, furniture, and so on for their lower density, remarkable thermal insulation, and substantial resistance properties toward harmful chemicals or toxicity. The effective fabrication of PU-based materials through tuning their production methods and raw materials is highly significant in making them usable. However, PUs possess some threats for the environment and health because of hazardous complexities, especially for living species. These could be minimized dramatically with functionalization through use of some chemical reagents, like incorporation of flame retardants (FRs). This can turn materials more environmentally friendly and sustainable. This study discusses PUs and their derivatives, synthesizing protocols, relevant chemistry, characteristics, incorporation with numerous FRs, and associated technologies. This work also investigates the merits and demerits of different FRs used for coating PUs industrially, along with their prominent marketing potential and applications through ensuring the demand toward sustainable products.

Introduction

The demand for polymeric material is increasing tremendously throughout the world. With time, inclusion of diversified functional properties—like improved flame retardancy, antibacterial performance, thermal conductivity, UV-resistance capability, mechanical performance, and so on—in polymeric materials/products is also getting scientists' attention (*1–7*). PUs are frequently used polymers with a wide variety of application potential. PU possesses superior resistance against abrasion, hardness, lower water absorption, and thermal conductivity. However, PUs also carry big drawbacks involving flammability problems, as they start to decompose quickly in the presence of flames. Fire retardancy of PUs improves through utilizing various flame retardants (FRs) (8–10). Example of some widely used FRs are aluminum polyphosphate, triphenyl phosphate, expandable graphite, layered silicate, melamine and its associated derivatives, boric components, and so on (11–14). The incorporation of FRs with PUs is mainly for reactive and additive types (15–17). Nevertheless, the usage of additive FRs also possess some detriments, like phase separation for loss of homogeneity, higher viscosity of constituent materials, and loads (18). PU foams (PUFs) are both rigid and flexible, which creates significant appeal for industrial applications, especially for seat cushioning, transportation materials, furniture, insulation, and packaging (19–25). However, the versatile applications of PUs need to be approved by different standards maintained by different national and international bodies or testing organizations (26, 27).

Generally, FRs are considered to be nitrogen-, silicon-, or phosphorous-containing components that are used for PU functionalization to cause fire resistance. They are also free from halogen-based compounds. Previously, halogenated FRs were also used in polymeric materials for improving the fire retardancy, but many of them are now banned around the world for their toxicity and corrosive gas emissions (28, 29). Some commonly used metallic hydroxide-based FRs are aluminum trihydroxide, magnesium dihydroxide, and so on. They are eco-friendly and cost-effective (26). Additionally, different inorganic nanoparticles—like graphene nanosheets, graphite carbon nitride, montmorillonite, carbon nanotubes, and polyhedral oligomeric silsesquioxane—also provide superior flame-retardancy performance in polymeric materials (30–35). These nanoparticles can create a strong barrier against heat and slow down the rate of heat release through suppressing the toxic gas and smoke during the combustion of polymeric materials. Phosphorous-based FRs are also considered just as eco-friendly. They are economical reagents containing PO and PO_2 chemical components in their polymeric structure. That is why they evaporate in gaseous phases and are capable of combustion reaction stoppage (36). Phosphorous- and nitrogen-based FRs are also known for intumescent properties. This enables further improved fire retardancy performance from PU polymers and associated composites (37–40).

PUs, meanwhile, are highly flammable and ignitable materials. The ignition of PUs generates vast amounts of combustion heat, poisonous gases, and huge smoke billows, which threatening lives as well as properties (22, 41, 42). Such problems can be minimized through using a coating of FRs. This coating methodology is also gaining popularity industrially, as it provides flame-retardancy functionalities without sacrificing mechanical properties. Various surface coating methods—like sol–gel processes, in situ processes, layer-by-layer deposition, and plasma treatments—are reported to improve flame retardancy (41, 43, 44) through deploying surface coating. Overall, FR reagents show a new route for utilizing PUs in diversified application areas through minimizing the combustion and toxic gas emissions.

PU Chemistry

The polymeric structure of PU facilitates them for reactions through chemical bonding with other polymeric materials. The components include epoxy, phenol, and unsaturated polyester (45, 46). PUs are synthesized through the reactions between the molecules of polyol and isocyanate, where ether functions as a catalyst or activated UV light (47). The isocyanate groups contain ($R'-(OH)_{n\geq2}$) and the polyol group contains ($R-(N=C=O)_{n\geq2}$) (45). However, the performance of PU depends on the polyol and isocyanate types from which it is produced (48). Accordingly, toughened rigid polymers require more cross-linking, whereas flexible longer segmented polymers can manufacture soft elastic polymeric components. Likewise, stretchy polymers need long chains, whereas hard polymers require short chains with more cross-linking. Conversely, the polymers with average amounts of cross-linking could be combined with long-chain polymers, thus producing adequate foam-making polymers. A PU could have an infinite molecular weight (M_W) because of the cross-linking phenomena through building three-dimensional networks. For this reason, PUs becomes a giant molecule even with a smaller fraction of polymer, which makes it feasible that it will not melt or turn softer. The addition of various additives aside from the polyols or isocyanate through modifying different processing parameters could carry out some potential characteristics for multifaceted applications (49). A reaction mechanism for PU formation is shown in Figure 1.

Polyols are comprised of two or more hydroxyl groups that are used for synthesizing PU. Derivatives of polyols, like polyether polyols, can be synthesized in the presence of compatible precursors through copolymerizations between ethylene and polypropylene oxides (50). The synthesis of polyester polyols follows similar protocols of preparing polyester polymers. Distinctly, poly(tetramethylene ether), glycol, and polyether polyols can be synthesized from tetrahydrofuran, which can be used efficiently for industrial applications (51). Polyols are also sometimes synthesized with a different M_W of the molecules but possessing similar characteristics. Molecules also possess different hydroxyl groups in their polymeric structures. However, because of the complex phenomena of polyols, industrial-scale productions are controlled carefully through ensuring homogenous properties from PUs consistently. Conversely, although the reactivity of isocyanate is slow at an environmental temperature, it is still highly reactive, hence it is incorporated with PUs through −OH group–possessing elements (52). For overcoming slow speed reaction problems, feasible catalysts and surfactants are needed to enhance reactivity.

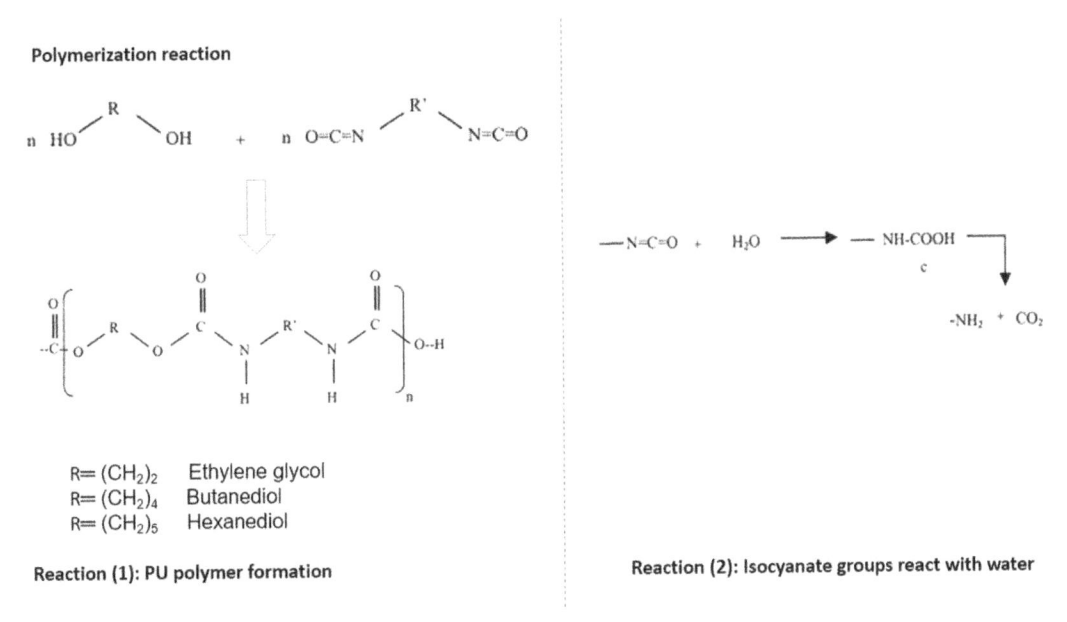

Figure 1. Formation of polymeric PU and reaction of isocyanate with water. Reproduced with permission from reference (53). Copyright 2004 Elsevier.

PU Types

A wide variety of PUs, either reticulated or linear, can be obtained through employing isocyanate and polyol in the reaction system. The various additives—like cross-linkers, catalysts, foaming agents, and chain extenders—are used to modify the ultimate properties, per requirements during the manufacturing of PUs. Generally, from an application perspective, PU can be categorized broadly in two ways: one is as a foam (flexible and rigid) type, and the other is as a special PU (coatings, elastomers, sealants, and adhesives) type.

Flexible PUFs

These are the PU-consuming polymers produced at the highest rate, representing around 45% of PU production, or 7.9 million tons annually (54, 55). However, flexible foams can further be categorized into another three types: viscoelastic, high-resilience, and conventional foams (56). Most flexible PUFs are produced from long-chain polyether polyols, depending on propylene oxide and ethylene. Conversely, conventional PUFs are manufactured from polyols with a M_w of 2800–4000 g/mol, whereas glycerin functions as an initiator. High-resilience foams are produced from polyether polyol with a M_w of 4000–6000 g/mol (long chain) in the presence of ethylene oxides (57). High-resilience foams have applications in automotive industry cushioning and viscoelastic foams in pillows and mattresses.

Rigid Foams

Rigid PUFs (RPUFs) are synthesized either from petroleum-originating polyols or from naturally originating polyols (lignin extracted from plants or vegetable oils). The characteristics of synthesized PUs depend on the available hydroxyl groups in polyols. The most significant difference between plant- and petroleum-based polyols is that a secondary hydroxy group is found for plant-based polyols, whereas a primary hydroxy group is found for petroleum-based polyols (45). However, a mixture of both types of polyols can reduce the costs of and dependence on petroleum-based products. Manufacturers are also trying continuously to move from petroleum-based PUs to naturally derived plant-based sustainable PUs. These are commonly used in insulation materials. RPUFs can minimize energy costs and be used for versatile commercial and residential appliances. Polymeric methylene diphenyl diisocyanate (MDI) is used for synthesizing PUFs in the case of isocyanates.

Special PUs: Coatings, Elastomers, Sealants, and Adhesives

Aside from the medium and long polymeric chains, some other extender chains, like 1,4-butadinol, are used for synthesizing these types of PUs. The hard segmented structure of these types is manufactured through a chain extender reaction with isocyanate or neat MDI. Meanwhile, soft segmented structures are produced from macrodiols like polytetrahydrofuran, with a M_w of 1000–2000 g/mol (54). Some examples of polyether polyols are polypropylene glycol capped diol (M_w: 400–4000 g/mol), polypropylene glycol diol (M_w: 400–4000 g/mol), and poly(trimethylene ether) glycol (M_w: 650–2400 g/mol). Meanwhile, polyester polyols are polycaprolactone polyol and polybutanediol adipate (54, 58). There is some potential seen, which could be an alternative for 1,4-butadinol (59–65). Other types of chain extenders show distinct potential, like 1,4-di(2-

hydroxyethylene) hydroquinone, diethylene glycol, N,N'-bis-(2-hydroxypropyl)aniline, and so on (54, 62). The polymeric structures of different chemicals are depicted in Figure 2.

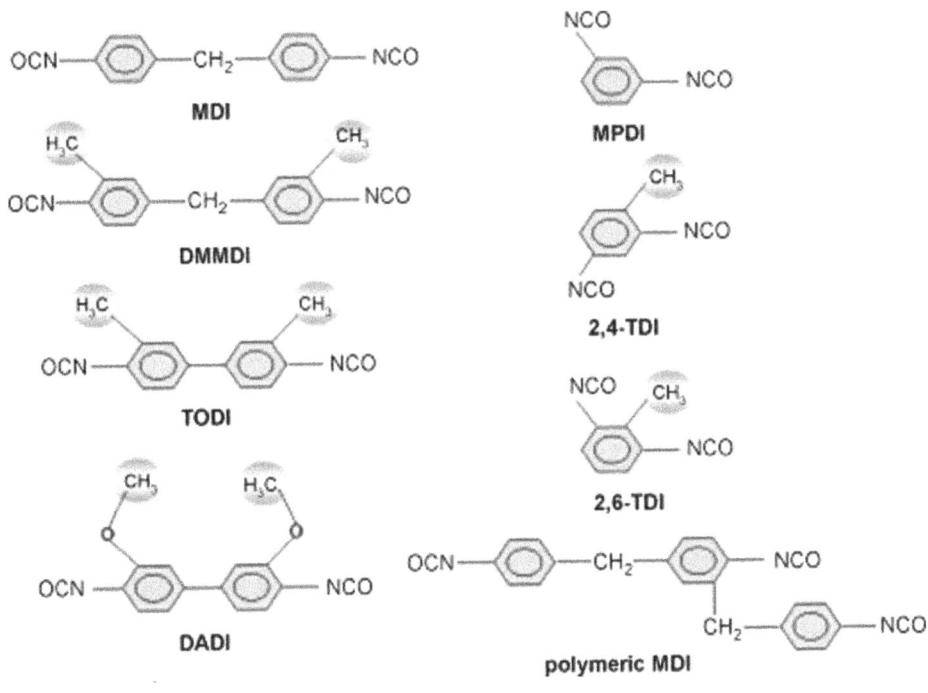

Figure 2. Polymeric structure of MDI, 1,3-phynelene diisocyanate, o-toluidine diisocyanate, aromatic diisocyanate (TDI), dianisidine diisocyanate, and dimethylol propanoic acid. Reproduced with permission from reference (63). Copyright 2009 Elsevier.

Properties of PUs

PU is a versatile polymeric material with extensive characteristics. That is why it attracts significant attention for numerous applications. Some of the exclusive properties of PUs are as follows:

- PU possesses a wide range of hardness, depending on the molecular structures of prepolymers.
- PU has a high loadbearing capacity against compression and tensions.
- PU can be made up of a very good selection where needed, with good flexural properties, as it possesses good elongation and recovery characteristics.
- PU can provide good resistance against abrasion and impact stress, even at lower temperatures.
- PU has high tensile strength and tear resistance.
- PU remains stable against oil, water, and greasy substances.
- PU has superior electrical insulation characteristics.
- PU possesses excellent bonding capability with other materials, such as wood, fiber, metals, and plastics.
- PU displays strong surviving capability against harsh environments, whereas many chemical-based materials show degradation.

- Most PU-based materials show strong resistance against mildew, mold, and fungus growths, which give them superior application potential, even at tropical environmental conditions.
- Colored pigments can be incorporated into PU during production processes. UV shielding can also be provided for potential outdoor applications.
- PU-based materials can be produced from prototypes or even large-scale volumes of production.
- PU needs a short manufacturing lead time.

PU Synthesis

There are numerous routes for synthesizing PUs. However, the most widely used route is to conduct a reaction between the two or more −OH groups present in polyols and diisocyanates (64). Different additives—like FRs, cross-linkers, pigments, blowing reagents, or even surfactants—are added during the PU synthesis as well. The density and hardness of the synthesized PU can be controlled and fabricated through incorporating the variations in isocyanates, polyols, and additives. The functions of different chemical reagents used during PU synthesis are outlined in Table 1.

Table 1. Functions of Different Components Used Throughout PU and FR-PU Manufacturing

Chemical Reagents	Functions/Purposes	Ref.
Polyols	Contributing to producing soft elastic materials	(62, 65)
Isocyanates	Ensuring the curing and reactivity of PU	(66, 67)
Non-isocyanates	Ensuring sustainable PU product with biobased non-isocyanates	(68, 69)
Fillers	Improving strength and stiffness of materials through using convenient fillers	(70)
Plasticizers	Minimizing materials' hardness	(71)
Pigments	Imparting color appearances to the PU materials	(72, 73)
Surfactants	Controlling bubble creation during PUF production	(74)
Chain extenders/cross-linkers	Modifying PU molecules structurally with improved mechanical properties	(75, 76)
FRs	Enhancing thermal stability against fire	(77, 78)

Isocyanates and Non-Isocyanates

Isocyanates are extremely important chemical elements. They are aliphatic/aromatic in nature and difunctional/heterofunctional categorically. The most common examples of isocyanate-based reagents are MDI, aliphatic diisocyanate, and aromatic diisocyanate (TDI). MDI and TDI are cheaper than cyanate-based reagents. Isocyanates can be modified for minimizing toxicity/volatility throughout reactions with polyol or other chemical additives. Some sustainable isocyanate manufacturing methods have also been attempted to minimize harmful impacts (79). There are also non-isocyanates (79–81) reported to overcome environmental and health risks. Ghasemlou et al. (82) documented crystallizable polyhydroxyurethanes by using non-isocyanate polymerizations of different diamines and ethylene carbonates. The same study revealed that the developed films provided tensile strengths between 1.7 and 3.2 MPa (82). A heat-resistant isocyanate formation reaction is displayed in Figure 3.

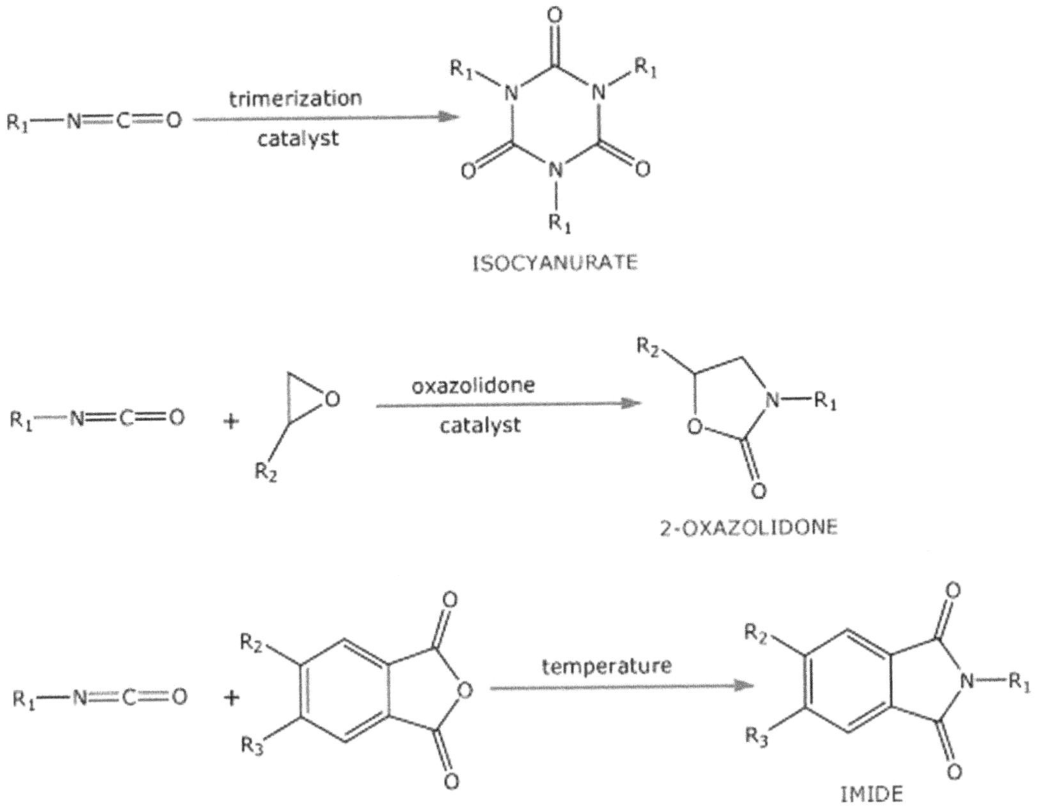

Figure 3. Heat-resistant isocyanate polymer formation. Reproduced with permission from reference (63). Copyright 2009 Elsevier.

Polyols

Polyol is an organic compound in either the group of polyesters or the group of polyether polyols. Polyether polyols are obtained through a reaction between an activated hydrogen-retaining element and an epoxide. Meanwhile, polyester polyols can be obtained through a polycondensation reaction of carboxylic acids (multifunctional) and hydroxyl components. The M_w of polyols can be 2000-10,000 g/mol, especially for flexible polyols, whereas rigid polyols also can have a low M_w (45, 64). There are also special types of polyols, which can provide potential environmental sustainability. Rao et al. (83) reported on a polyester polyol from diethanol amine and dimethyl methylphosphonate through implementing transesterifications to produce FR, flexible PUFs. They have further claimed that the developed FR, flexible PUFs could also pass vertical burning characterizations (83).

Catalysts, Cross-Linkers, and Surfactants

The most often incorporated catalysts for PU synthesis can be categorized mainly into two groups: one is amine compounds and another one is metallic complexes. Tertiary amine is possessed by amine complexes in general, such as dimethylciclohexylamine, trimethylenediamine, and so on.

The tertiary amine selection is dependent on the reaction capability between the isocyanates or polyols. Metallic substances such as zinc, lead, or bismuth can also be used for urethane catalysis (45). Wulf et al. (84) performed a study to investigate the effects of bifunctional metallic catalysts on non-isocyanate PUs and found that both of them displayed an increased M_w up to 19 kg/mol. Cross-linkers and chain extenders also play a vital role for PU synthesis (85). Gui et al. conducted a study on PU elastomers with six types of cross-linkers and side chains. They found that with an increase of side chains and cross-linkers, the vibration isolation and damping property is increased (86). These components are typically hydroxyl- and amine-terminated polymers with a lower M_w. The morphology of PU can be improved through using fibers, elastomers, adhesives, and so on. The hard and soft segments of the polymers facilitate elastomeric characteristics. Surfactant is another means of improving foam and nonfoam characteristics. In the case of PUFs, they function as emulsifiers for liquid materials to avoid void creation through controlling cell structures. On the other hand, for nonfoam-based applications, they function as wetting agents, or as antifoaming and air-releasing agents (45, 85–89). Some surface imperfections—such as pin holes, orange peels, and sink marks—can also be eliminated through utilizing surfactants. A reaction between diisocyanate and polyol is further shown in Scheme 1.

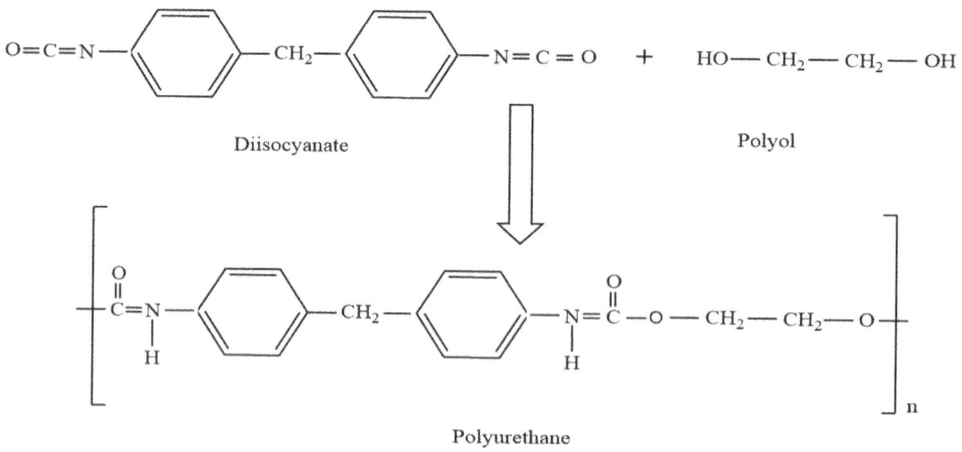

Scheme 1. A schematic of a PU formation reaction from diisocyanate and polyols.

Recycling of PU Waste

Increasing amounts of plastic-based waste are becoming a global threat. This explains the amount of plastic-based waste discharged industrially or after usage by consumers. Effective management is needed to protect the surrounding environment. Different types of recycling attempts—like feedstock and mechanical recycling, landfills, and incineration recycling—can be taken into account (Figure 4). It is expected that the scenario for discharged PU wastes after the end of their lifecycles could change drastically in the coming years. Researchers are also trying relentlessly to find out more feasible routes of effective solutions (90–97). Except for thermoplastic PU (TPU), PU materials are thermosetting polymers, generally having densities of 10–400 g/L (97), along with variable polymeric properties. The recycling of consumer PU waste is a more challenging task because of its variation of molecular structure, thermoset characteristics, and foamed nature. PU-based polymers are also used for producing composites (98–100) by using different polymers and artificial fibers. This could make recycling more difficult in terms of effective technology and cost. Thermomechanical, heat recovery, and incineration processes could be the most well-fitted technologies. Transforming plastic waste into gas or oil can be performed through implementing gasification, hydrogenation, and pyrolysis. The products could then be used by chemical manufacturing industries as raw materials.

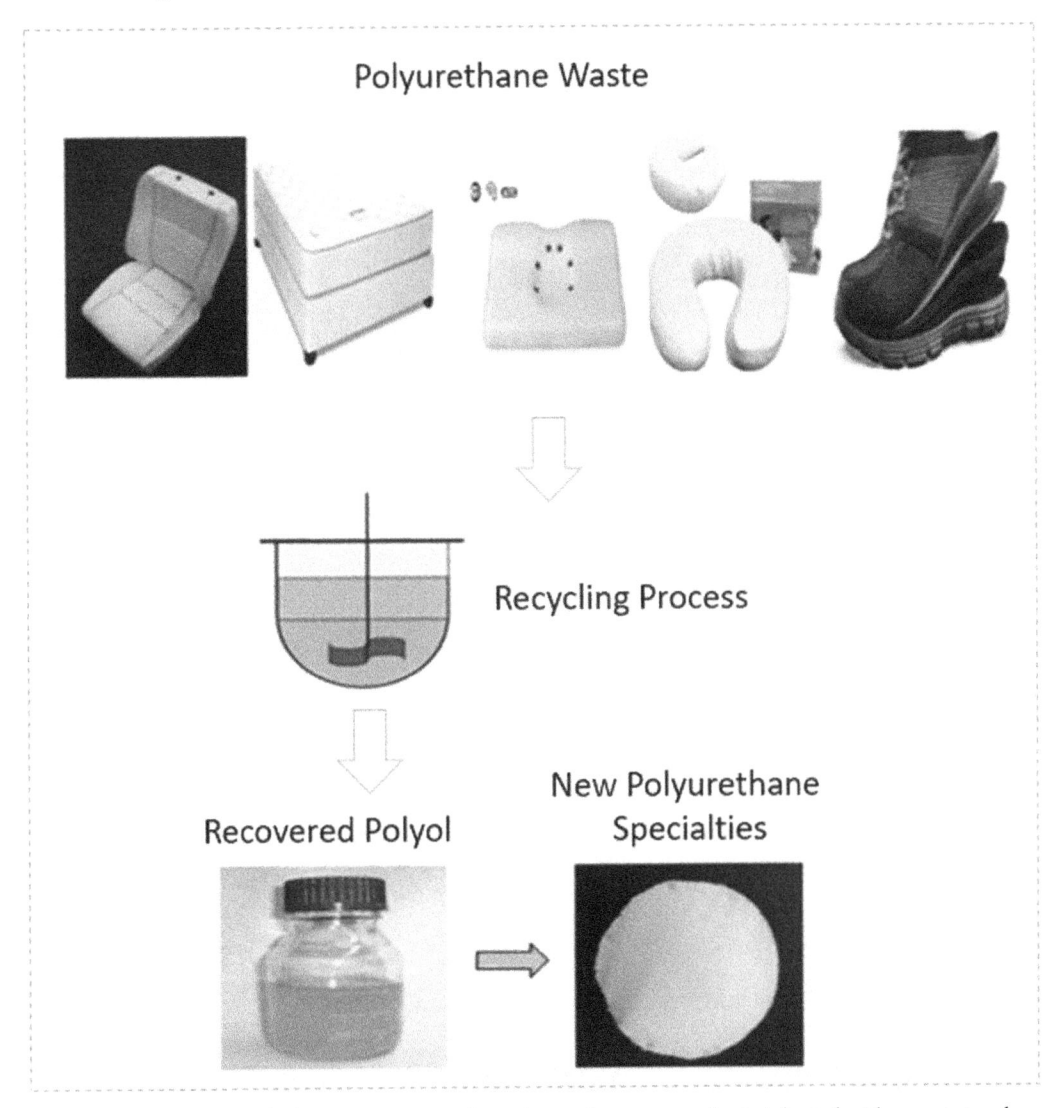

Figure 4. A schematic polyol recovery process through recycling waste polyol. Adapted with permission from reference (54). Copyright 2018 Elsevier.

The foam of PU can be recycled using chemical and mechanical methods from the final products. Yang et al. (*101*) executed a comparative study on disposed PUF recycling, in terms of physical and chemical approaches, and they summarized that a physical method is a simpler and more efficient technology. The recycled materials could be utilized directly for building blocks or as a precursor to them (*97*). In the case of chemical recycling methods, the urethane functional group is depolymerized gradually with the assistance of organic compounds with hydrogen atoms (active), which could invade the urethane bonds in the polymeric chains (*97*). The mechanical recycling process entails regrinding the materials to specific shapes and sizes, rebonding by wetting the particles through using isocyanate or its prepolymers, and adhesive pressing of particles (*97*). Mechanical recycling is also a cost-effective recycling method. Ragaert et al. (*102*) performed a study on FR plastic waste recycling in terms of the mechanical recycling process. They found that the effects are totally dependent on the types of polymers used with combinations of FRs. Chemical characteristics of FRs also govern the stability of fire retardancy. The temperature used during the mechanical recycling process (Figure 5) plays a crucial role. The selection of temperature is extremely important; it should be lower than the decomposition temperature of FRs (*103*). The PUF used in furniture and mattress companies can be recycled. Some initiatives have been taken or are going to be taken in some countries, such as France, the United Kingdom, Belgium, and the Netherlands (*97*).

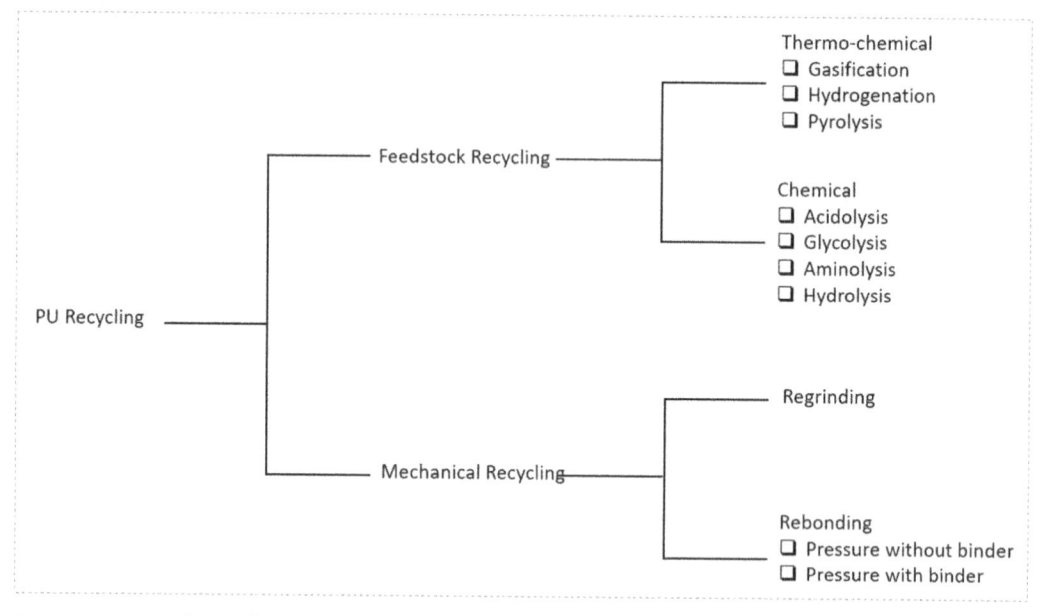

Figure 5. PU recycling techniques. Adapted with permission from reference (104). Copyright 2021 Elsevier.

Improving Flame Retardancy of PUs

Surface coating and addition of FR fillers are two promising methods for imparting FR to PU materials. A typical FR-PU manufacturing method is demonstrated in Figure 6.

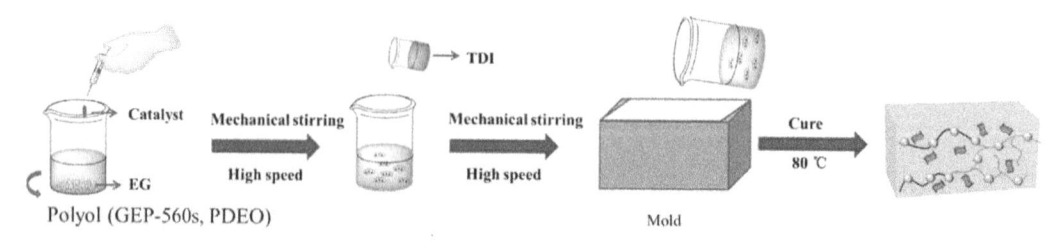

Figure 6. A typical FR-PU manufacturing method. Reproduced with permission from reference (105). Copyright 2018 Elsevier.

FR Filler Incorporation

The fire resistance of PU can be carried out both physically and chemically through incorporating FR additives into PU. FRs can be bonded with PU functional groups through covalent or noncovalent bonding. However, in the case of noncovalent bonding, higher amounts of FR agent are required for achieving better and more satisfactory performances against fire. This addition of FR could also negatively influence the mechanical and physical properties of the products. Conversely, if the FR is incompatible with the PU, leaching of the FR can occur. The chemical covalent bonding at the polymeric structures of PU backbone is created by the FRs.

Surface-Coating Approach

The surface-coating approach is considered efficient technology for imparting flame retardancy to materials (*106, 107*). This approach is also facile and economical. Mechanical properties are not sacrificed when using this technology. The coating typically becomes much thinner—even less than 100 nm. This method is widely studied for PUFs, textiles, woods, and synthetic polymers. There are different methods for surface coating of substrates such as sol–gel methods, layer-by-layer assembly, and plasma technology.

Sol–Gel Methods

The sol–gel process is a wet-based technology that is commonly used for providing FR coatings on substrates. This is performed through two-step hydrolysis reactions through forming organic–inorganic or inorganic coatings (*108, 109*). Consequently, there could be an efficient and strong FR property displayed by the materials because of the "synergic effects."

Layer-by-Layer Assembly

This is another simple, economically feasible, and eco-friendly approach for providing a thin coating, especially a nanomaterial-based coating. In this method, the materials are dipped/sprayed into the polyelectrolyte solutions with opposite charges (*110, 111*). Beyond this technology, there are also acceptor/donor, electrostatic, and hydrogen-bonding interactions used for incorporating FR coatings (*112, 113*).

Plasma Technology

Plasma coating is another type of eco-friendly coating approach for different substrates/polymeric materials. The surfaces of the materials can be modified easily through plasma treatment methods. This method is also feasible for nanocoating grafting/depositions on the surfaces. Some parameters—such as power types, atmospheric pressures, and polymerizable or nonpolymerizable gases—are needed to maintain these coatings (*43*).

FR Properties of PUs

PUs are combustible polymers with oxygen index values of 16–18 (*90, 114*). The porous lightweight foams tend to spread flames rapidly. PU contains enormous amounts of nitrogen in its structure, so it produces hydrogen cyanide during the combustion/pyrolysis stage. Conversely, the amount of hydrogen cyanide produced by other nitrogen-containing materials, like polyacrylonitrile and nylon, is lower than that produced by PUs. Researchers and industries are paying attention to imparting PU with significant flame retardancy properties. The incorporation of various FR additives—such as nitrogen, phosphorous, and halogen—could facilitate the flame retardancy property enhancements in Pus (*115*).

TPU

Convenient characteristics—like higher tensile and flexural fatigue strength; superior temperature flexibility; excellent abrasion and wear resistance; and eco-friendly features through resisting microbes, UV radiation, ozone, and humidity—have made TPU an ideal candidate for insulation and cable sheathing materials (*116*). As the FR of TPU is still poor, researchers are trying to improve it through fulfilling safety standards. Various types of FR materials—such as nitrogen-, phosphorous-, boron-, and silicon-containing materials—have been tried to improve fire resistance capabilities. Toldy et al. (*116*) conducted a study on TPU-based FRs and found that the FR property is enhanced with increased phosphorous content. Melt-dripping is also minimized significantly through use of phosphorous and boron (*115*). In the case of TPU, better electrical properties can be attained through use of zinc borate with magnesium hydroxide or alumina trihydrate conjunctly (*90*). An FR TPU was reported by Tabuani et al. (*117*) They used different quantities of melamine cyanurate (FR agent) and nanoclays and found synergism effects between them. This accelerated the flame retardancy and thermal stability that was investigated in terms of limiting oxygen index. The limiting oxygen index improved from 23 ± 0.2 (neat TPU) to 24.8 ± 0.2 from the incorporation of melamine cyanurate and organo-montmorillonite fillers (*117*).

FR on Rigid Foams

PUFs are widely used as insulation materials, such as in sheathing and roofing for building and construction sites. Accordingly, it becomes extremely important to make them safe from smoke and flammability through regulations. Rigid foams are also used for versatile applications such as doors, shelves, panels, housing equipment, and so on. This demands the property of flame retardancy. Generally, additives are used for applying flame retardancy to rigid foams (*118–120*). Rigid foam PU starts to degrade within 200–250 °C, so it is necessary to apply additives that can increase the stability against this low temperature degradation. The additive reagents used for rigid foams are tris(2-chloro-ethyl) phosphate, tris(1-chloro-ethyl) phosphate, tris(chloroisopropyl) phosphate, and so on. There are also applications found for triaphenyl phosphate, isopropylphenyl diisocyanate phosphate, and tricresyl phosphate, which are found as feasible FR additives on PU rigid foams. In addition, reactive FRs are used for improving flame retardancy of PU rigid foams. Wang et al. studied improving flame retardancy on RPUFs through a reactive FR triol, which was based on phosphate and triazine structures, using a chain extender (*121*). The developed materials had superior flame retardancy with improved compressive strength but a reduced thermal conductivity of 0.03 W/(m.K).

The scanning electron microscopy analysis of control RPUF and reactive FR triol (TDHTTP) is shown in Figure 7. It was reported by Wang et al. (*121*) The microstructural views show that the cell thicknesses and structures also increase with the increase of FRs (TDHTTP). After incorporation of TDHTTPs in the RPUFs, the surface becomes smoother, demonstrating a better compatibility between the FR and RPUFs.

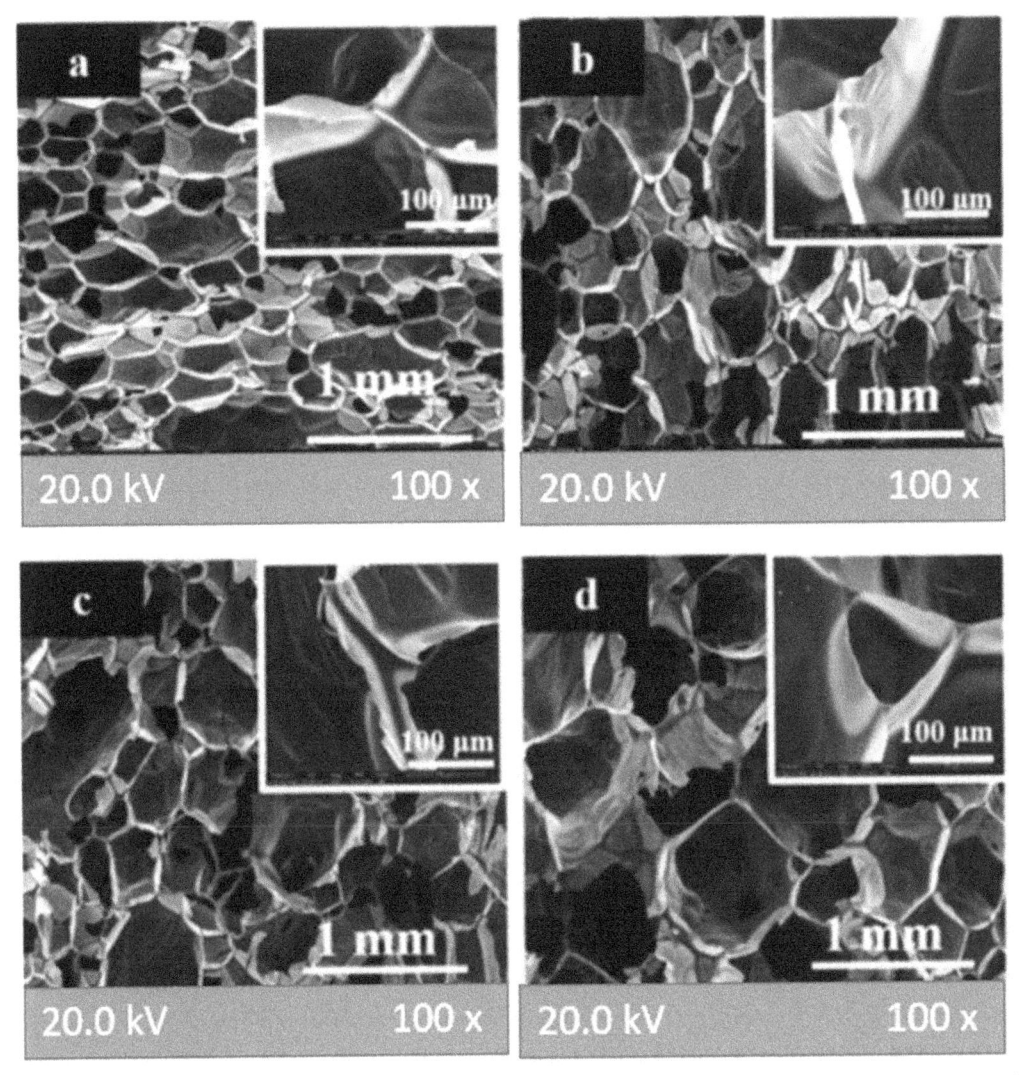

Figure 7. Morphological photographs of control RPUF: (a) RPUF-5, (b) RPUF-10, (c) RPUF-15, and (d) 15 diethyl phosphite (DHPP)/RPUF. All of the RPUF contains fixed contents of polyol (25 g), polyethylene glycol (20 g), deionized water (3 g), silicone oil (2 g), a catalyst (DMP-30, 1 g), and stannous octoate (0.3 g). RPUF-5, RPUF-10, RPUF-15, and 15 DHPP/RPUF contain 0, 10, 20, 30, and 0 g of TDHTTP, respectively; DHPP amounts of 0, 0, 0, 0, 30 g, respectively; and polymeric MDI amounts of 115, 120, 128, 132, and 132 g, respectively. Adapted with permission from reference (121). Copyright 2018 Elsevier.

The same study also investigated the thermal behavior further. A two-step degradation was found. The initial degradation was related to the hard segments of urethane bonds, and the second step of the degradation corresponded to soft segments (Figure 8). The thermogravimetric analysis showed that all the FR RPUFs displayed more stability when exposed to heat compared with the control RPUF.

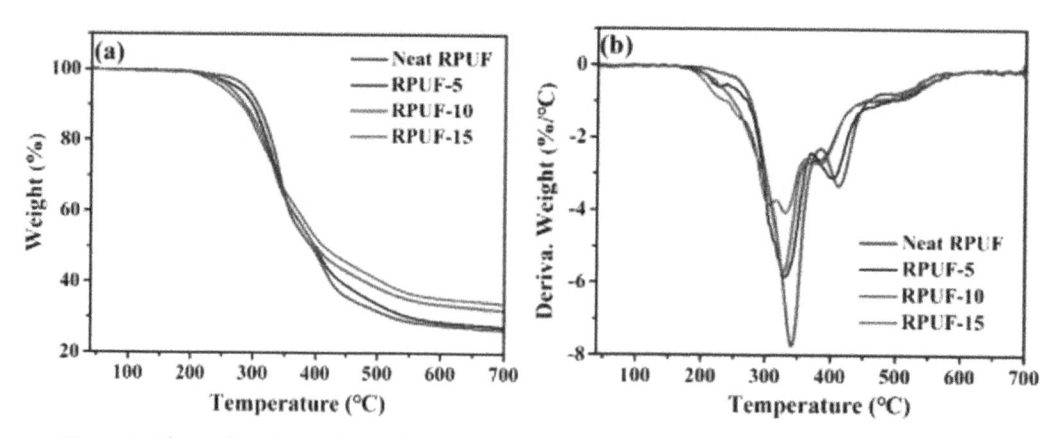

Figure 8. Thermal analysis of control RPUF and TDHTTP (a) thermogravimetric analysis and (b) derivative thermogravimetric analysis curves. Reproduced with permission from reference (121). Copyright 2018 Elsevier.

FR on Flexible Foams

The furniture industries use flexible foams that require stable flame retardancy characteristics. In this case, the ignition of temperature of the foams can be increased, which means the rate of flame spreading can be slowed down/reduced. According to Federal Motor Vehicle Safety Standard 302, for automotive materials, 16 parts per 100 of chloroalkyl phosphate in 1.0 lb/ft^3 and nearly 7.0 phr in lb/ft^3 are required for flexible PUFs. However, in the case of fabric-based laminated composites, it is more difficult to reach this standard if they are not treated by the FRs. Meanwhile, the United Kingdom and many other European countries follow BS 5852 standards. The presence of a small amount of melamine could influence the target flame retardancy negatively. The commonly used FRs for flexible PUFs are tris(2-chloroisopropyl) phosphate, tris(1,3-dichloro-2-propyl) phosphate, tris(2,3-dichloro-1-propyl) phosphate, and tris(2-chloroisopropyl) phosphate. Chen et al. (*115*) developed FR flexible PUFs from 2-carboxyetheyl(phenyl) phosphinic acid melamine (CMA) salt and found a very good flame retardancy effect, especially when they used 12% CMA.

Applications of PUs

In Europe, 7% of PU polymeric material is consumed. The plastic industries create employment for more than 1.5 million people throughout all of Europe, where nearly 60,000 companies are operating (impressively most of them are small to medium enterprises). The plastic industries in Europe showed a turnover of €340 billion or more in 2015. Among plastic materials, PU has an important position in Europe and throughout the world. Globally, PU is ranked as sixth among polymers' production annually, with a total production of 18 Mt in 2016 and $60.5 billion in consumptions in 2017. PU and FR-PU have extensive uses in different sectors, such as automotive materials, apparels, furniture, packaging, building and construction, and so on. There are six main fields of PU, which are used widely, as shown in Table 2.

Building materials must be lightweight, durable, and strong. These characteristics can be attained through PUs. Building materials also must have strong protection against fire. Therefore, FR-PU can play a vital role in this sector. The interest in FR-PU is also growing for its superior heat insulation performance and low energy consumption. Xie et al. reported that a eutectic salt (Na$_2$HPO$_4$·12H$_2$O−Na$_2$CO$_3$·10H$_2$O) composite with phase change materials with no leakage was produced initially (*122*). It was then combined further with porous diatomite and UV-cured PU acrylate to intensify the stability of the forms. This showed significant potential for construction sectors. Andersons et al. developed thermal breaking materials from rigid high-density PUFs where polyol was derived from renewable sources (*123*).

Table 2. Application of PU According to Different Types[a]

PU Types	Application Area	Production (%)
Rigid foams	Packaging, insulation panels, and household appliances	32
Flexible foams	Vehicle mattresses and seating items	36
Sealant and adhesives	Sealants, castings	6
Binders	Wood panel adhesives, elastomeric/rubber flooring surfaces	4
Elastomers	Medical facilities, glue, and so forth	8
Coatings	Side panels and bumpers of vehicles	14

[a] Reproduced with permission from reference (*104*). Copyright 2021 Elsevier.

PU is also frequently used in automotive sectors. It has applications for automotive cushioning, bumpers, ceilings, car bodies, windows, and so on. PU also enhances mileage by reducing weight, minimizing fuel consumption, creating higher corrosion resistance, providing better sound absorption, and creating overall better insulation properties. MDI-bonded flax/glass-woven fabric reinforced composites (*124*) can also be used for potential automotive applications.

PU-based epoxy resin provides diversified potential for marine applications, such as boat technology, through protecting them against water, corrosion, and weather (*125, 126*). PU-based polymeric materials also provide strong resistance against heat and sound. Better loadbearing capacity is provided as well, along with improved abrasion and tear resistance properties. Ship construction industries use PU-based materials for drive belts, wire coating, engine tubing, hydraulic seals, and hoses. PU has also provided significant potential for surface coating and painting materials for a long period of time (*127*). Polyester-based polyols can be used for marine antifouling.

The medical sector also uses PU-based materials for diversified applications in hospital bedding, surgical drapers, catheters, tubing, injection-molded instruments, and wound dressings (*128–130*). The better biocompatibility, physical properties, and mechanical properties enable them for these potential applications. The incorporation of PU in medical facilities provides longevity and toughness to the substrates, along with cost-effective features. PU also shows potential to produce yarns for textiles. These could be tuned to form the nylon fabric that is becoming a significant lightweight wearable item (*131–134*). PU is further used for thermoplastic elastomers and spandex fibers. With the advancement of technology, PU is also showing more advanced routes for artificial skins, bra cups, and leather-based products.

PU also displays application possibilities in packaging, flooring, and various appliances (*135–137*). Flooring carpet underlay and top coatings are some of the important PU-based applications that are durable with pleasant aesthetics. Meanwhile, PU plasticizers can be used for packaging applications. PU-based packaging materials provide better protection in transit for electronic items, mechanical parts, medical equipment, sensitive precious materials, and so on. The most significant applications of PU are found for natural fiber/wood polymer composite panels, providing superior thermomechanical properties. Likewise, cellulose nanocrystal is reinforced with

PU and has a glass transition temperature of 76 °C, whereas the perceived Young's modulus is 1.52 GPa (*138*).

Research Gap and Future Perspectives

Environmental concerns are significantly increasing throughout the world. Therefore, the demand for sustainable, hazard-free, and pollution-free materials is also increasing day by day. The replacement of petroleum-based FR-PU through biobased renewable source material could be a big achievement in this century. More research and industrial attention are required for this initiative to progress further. The addition of nanomaterials is also displaying prominent potential for developing FR-PU materials with superior mechanical and physical properties. However, there is still a long way to go in exploring new nanomaterials for imparting flame retardancy with minimized cost and higher production efficiency. Manufacturers also are focused on cost-effective raw materials and low-production, cost-based manufacturing. This also could put more attention on minimizing the cost of FR-PU. Aside from economic and environmental sustainability perspectives, the improvements in superior flame retardancy, better mechanical properties, gas barrier effects, electrical conductivity, and better hydrophobicity could also be potential directions for making FR-PU more attractive, usable, and commercially feasible. The digitalization of production processes on FR-PU systems could speed up the manufacturing processes industrially to a large extent. Digitalization could also help to predict the characteristics of FR-PU materials virtually before they go to production, in order to make necessary initiatives and advances more scientifically. The exploration of novel and potential application areas of FR-PU materials could also facilitate more scopes for industrial manufacturing units with diversified products.

Conclusion

PU is a robust and common potential research material. PUs are showing alternative routes for replacing metallic materials, rubber, and plastics in engineered structural materials. Industrially, they have potential for rigid insulation material, soft flexible foams, liquid coatings, paint, elastomers, and elastic fibers. This chapter reported on the summarized content of PU, its associated FRs, its chemistry, its structural characteristics and performance, and its relevant technology, which could help manufacturers and researchers to understand the industrial FRs of PU. Some of the developed FRs are not yet commercially suitable to minimize the risk of health hazards or cost issues through maintaining appropriate standards and quality parameters. The FR surface coating on PUs could significantly prevent smoke and toxic fumes without minimizing mechanical, electrical, or other prominent performance characteristics of polymeric materials needed for industrial-level production. More research and studies are still needed for minimizing environmental burdens and hazardous issues from PU-based FRs. Additional investigation should lead to discovery of more economically and commercially feasible FRs to functionalize PUs without sacrificing the required performance of developed materials in terms of thermal, physical, and mechanical properties.

References

1. Hai, Y.; Wang, C.; Jiang, S.; Liu, X. Layer-by-layer assembly of aerogel and alginate toward self-extinguishing flexible polyurethane foam. *Industrial Engineering Chemistry Research* **2019**, *59* (1), 475–483.

2. Zhou, S.; Zeng, H.; Qin, L.; Zhou, Y.; Hasan, K. F.; Wu, Y. Screening of enzyme-producing strains from traditional Guizhou condiment. *Biotechnology & Biotechnological Equipment* **2021**, *35* (1), 264–275.
3. Mahmud, S.; Hasan, K. F.; Jahid, M. A.; Mohiuddin, K.; Zhang, R.; Zhu, J. Comprehensive review on plant fiber-reinforced polymeric biocomposites. *J. Mater. Sci* **2021**, *56*, 7231–7264.
4. Hasan, K. F.; Horváth, P. G.; Horváth, A.; Alpár, T. Coloration of woven glass fabric using biosynthesized silver nanoparticles from Fraxinus excelsior tree flower. *Inorg Chem Commun* **2021**, *126*, 108477.
5. Hasan, K. M. F.; Péter, G. H.; Gábor, M.; Tibor, A. Thermo-mechanical characteristics of flax woven fabric reinforced PLA and PP biocomposites. *Green Mater* **2021**, 1–9.
6. Hasan, K. F.; Wang, H.; Mahmud, S.; Taher, M. A.; Genyang, C. Wool Functionalization Through AgNPs: Coloration, Antibacterial, and Wastewater Treatment. *Surf Innov* **2020**, *9* (1), 25–36.
7. Hasan, K. F.; Horváth, P. G.; Miklos, B.; Alpár, T. A state-of-the-art review on coir fiber-reinforced biocomposites. *RSC Advance* **2021**, *11*, 10548–10571.
8. Tirri, T.; Aubert, M.; Wilén, C.-E.; Pfaendner, R.; Hoppe, H. Novel tetrapotassium azo diphosphonate (INAZO) as flame retardant for polyurethane adhesives. *Polymer degradation and stability* **2012**, *97* (3), 375–382.
9. Song, L.; Hu, Y.; Tang, Y.; Zhang, R.; Chen, Z.; Fan, W. Study on the properties of flame retardant polyurethane/organoclay nanocomposite. *Polymer Degradation and Stability* **2005**, *87* (1), 111–116.
10. Zhang, L.; Zhang, M.; Hu, L.; Zhou, Y. Synthesis of rigid polyurethane foams with castor oil-based flame retardant polyols. *Industrial Crops and Products* **2014**, *52*, 380–388.
11. Cui, Y.; Liu, X.; Tian, Y.; Ding, N.; Wang, Z. Controllable synthesis of three kinds of zinc borates and flame retardant properties in polyurethane foam. *Colloids and Surfaces A: Physicochemical and Engineering Aspects* **2012**, *414*, 274–280.
12. Zheng, X.; Wang, G.; Xu, W. Roles of organically-modified montmorillonite and phosphorous flame retardant during the combustion of rigid polyurethane foam. *Polymer Degradation and Stability* **2014**, *101*, 32–39.
13. Zhang, L.; Zhang, M.; Zhou, Y.; Hu, L. The study of mechanical behavior and flame retardancy of castor oil phosphate-based rigid polyurethane foam composites containing expanded graphite and triethyl phosphate. *Polymer Degradation and Stability* **2013**, *98* (12), 2784–2794.
14. Modesti, M.; Lorenzetti, A.; Besco, S.; Hrelja, D.; Semenzato, S.; Bertani, R.; Michelin, R. Synergism between flame retardant and modified layered silicate on thermal stability and fire behaviour of polyurethane nanocomposite foams. *Polymer Degradation and Stability* **2008**, *93* (12), 2166–2171.
15. Yin, S.; Ren, X.; Lian, P.; Zhu, Y.; Mei, Y. Synergistic effects of black phosphorus/boron nitride nanosheets on enhancing the flame-retardant properties of waterborne polyurethane and its flame-retardant mechanism. *Polymers* **2020**, *12* (7), 1487.
16. Zhou, F.; Zhang, T.; Zou, B.; Hu, W.; Wang, B.; Zhan, J.; Ma, C.; Hu, Y. Synthesis of a novel liquid phosphorus-containing flame retardant for flexible polyurethane foam: Combustion behaviors and thermal properties. *Polymer Degradation and Stability* **2020**, *171*, 109029.

17. Xia, L.; Liu, J.; Li, Z.; Wang, X.; Wang, P.; Wang, D.; Hu, X. Synthesis and flame retardant properties of new boron-containing polyurethane. *Journal of Macromolecular Science, Part A* **2020**, *57* (8), 560–568.
18. Frigione, M.; Maffezzoli, A.; Finocchiaro, P.; Failla, S. Cure kinetics and properties of epoxy resins containing a phosphorous-based flame retardant. *Advances in Polymer Technology: Journal of the Polymer Processing Institute* **2003**, *22* (4), 329–342.
19. de Mello, D.; Pezzin, S. H.; Amico, S. C. The effect of post-consumer PET particles on the performance of flexible polyurethane foams. *Polymer Testing* **2009**, *28* (7), 702–708.
20. Kozlowski, R.; Malgorzata, M.; Bozena, M. Comfortable, flexible upholstery fire barriers on base of bast, wool and thermostable fibres. *Polymer degradation and stability* **2011**, *96* (3), 396–398.
21. Hillier, K.; King, D.; Henneuse, C. Study of odours coming out of polyurethane flexible foam mattresses. *Cellular polymers* **2009**, *28* (2), 113–144.
22. Singh, H.; Jain, A. Ignition, combustion, toxicity, and fire retardancy of polyurethane foams: a comprehensive review. *Journal of Applied Polymer Science* **2009**, *111* (2), 1115–1143.
23. Hirschler, M. Polyurethane foam and fire safety. *Polymers for Advanced Technologies* **2008**, *19* (6), 521–529.
24. Bashirzadeh, R.; Gharehbaghi, A. An investigation on reactivity, mechanical and fire properties of PU flexible foam. *Journal of Cellular Plastics* **2010**, *46* (2), 129–158.
25. Yang, F.; Nelson, G. L. Halogen-Free Flame Retardant Flexible Polyurethane Foams via a Combined Effect of Flame Retardants. In *Fire and Polymers VI: New Advances in Flame Retardant Chemistry and Science*; American Chemical Society: Washington, DC, USA, 2012; Vol. 1118, pp 139–149.
26. Visakh, P.; Semkin, A.; Rezaev, I.; Fateev, A. Review on soft polyurethane flame retardant. *Construction and Building Materials* **2019**, *227*, 116673.
27. Cooper, E. M.; Kroeger, G.; Davis, K.; Clark, C. R.; Ferguson, P. L.; Stapleton, H. M. Results from screening polyurethane foam based consumer products for flame retardant chemicals: assessing impacts on the change in the furniture flammability standards. *Environmental Science & Technology* **2016**, *50* (19), 10653–10660.
28. Chen, L.; Wang, Y. Z. A review on flame retardant technology in China. Part I: development of flame retardants. *Polymers for Advanced Technologies* **2010**, *21* (1), 1–26.
29. Covaci, A.; Harrad, S.; Abdallah, M. A.-E.; Ali, N.; Law, R. J.; Herzke, D.; de Wit, C. A. Novel brominated flame retardants: a review of their analysis, environmental fate and behaviour. *Environment International* **2011**, *37* (2), 532–556.
30. Zhang, W.; Camino, G.; Yang, R. Polymer/polyhedral oligomeric silsesquioxane (POSS) nanocomposites: An overview of fire retardance. *Progress in Polymer Science* **2017**, *67*, 77–125.
31. Song, L.; Xuan, S.; Wang, X.; Hu, Y. Flame retardancy and thermal degradation behaviors of phosphate in combination with POSS in polylactide composites. *Thermochimica Acta* **2012**, *527*, 1–7.
32. Song, P.; Zhao, L.; Cao, Z.; Fang, Z. Polypropylene nanocomposites based on C 60-decorated carbon nanotubes: thermal properties, flammability, and mechanical properties. *Journal of Materials Chemistry* **2011**, *21* (21), 7782–7788.

33. Xu, L.; Guo, Z.; Zhang, Y.; Fang, Z. Flame-retardant-wrapped carbon nanotubes for simultaneously improving the flame retardancy and mechanical properties of polypropylene. *Journal of Materials Chemistry* **2008**, *18* (42), 5083–5091.
34. Liu, S.; Fang, Z.; Yan, H.; Chevali, V. S.; Wang, H. Synergistic flame retardancy effect of graphene nanosheets and traditional retardants on epoxy resin. *Composites Part A: Applied Science and Manufacturing* **2016**, *89*, 26–32.
35. Hasan, K.; Horváth, P. G.; Alpár, T. Potential Natural Fiber Polymeric Nanobiocomposites: A Review. *Polymers* **2020**, *12* (5), 1072.
36. Nguyen, C.; Lee, M.; Kim, J. Relationship between structures of phosphorus compounds and flame retardancies of the mixtures with acrylonitrile–butadiene–styrene and ethylene–vinyl acetate copolymer. *Polymers for Advanced Technologies* **2011**, *22* (5), 512–519.
37. Zhao, Q.; Chen, C.; Fan, R.; Yuan, Y.; Xing, Y.; Ma, X. Halogen-free flame-retardant rigid polyurethane foam with a nitrogen–phosphorus flame retardant. *Journal of Fire Sciences* **2017**, *35* (2), 99–117.
38. Yang, R.; Wang, B.; Han, X.; Ma, B.; Li, J. Synthesis and characterization of flame retardant rigid polyurethane foam based on a reactive flame retardant containing phosphazene and cyclophosphonate. *Polymer Degradation and Stability* **2017**, *144*, 62–69.
39. Wu, L.; Guo, J.; Zhao, S. Flame-retardant and crosslinking modification of MDI-based waterborne polyurethane. *Polymer Bulletin* **2017**, *74* (6), 2099–2116.
40. Zhang, P.; He, Y.; Tian, S.; Fan, H.; Chen, Y.; Yan, J. Flame retardancy, mechanical, and thermal properties of waterborne polyurethane conjugated with a novel phosphorous-nitrogen intumescent flame retardant. *Polymer Composites* **2017**, *38* (3), 452–462.
41. Yang, H.; Yu, B.; Song, P.; Maluk, C.; Wang, H. Surface-coating engineering for flame retardant flexible polyurethane foams: A critical review. *Composites Part B: Engineering* **2019**, *176*, 107185.
42. Taheri, A.; Noroozifar, M.; Khorasani-Motlagh, M. Investigation of a new electrochemical cyanide sensor based on Ag nanoparticles embedded in a three-dimensional sol–gel. *Journal of Electroanalytical Chemistry* **2009**, *628* (1-2), 48–54.
43. Lin, D.; Zeng, X.; Li, H.; Lai, X.; Wu, T. One-pot fabrication of superhydrophobic and flame-retardant coatings on cotton fabrics via sol-gel reaction. *Journal of Colloid and Interface Science* **2019**, *533*, 198–206.
44. Zhou, L.; Yuan, L.; Guan, Q.; Gu, A.; Liang, G. Building unique surface structure on aramid fibers through a green layer-by-layer self-assembly technique to develop new high performance fibers with greatly improved surface activity, thermal resistance, mechanical properties and UV resistance. *Applied Surface Science* **2017**, *411*, 34–45.
45. Akindoyo, J. O.; Beg, M.; Ghazali, S.; Islam, M.; Jeyaratnam, N.; Yuvaraj, A. Polyurethane types, synthesis and applications–a review. *Rsc Advances* **2016**, *6* (115), 114453–114482.
46. Ulrich, H. *Chemistry and Technology of Isocyanates*; Wiley-Blackwell: Hoboken, NJ, USA, 1996; Vol. 36, p 514.
47. Soto, M.; Sebastián, R. M.; Marquet, J. Photochemical activation of extremely weak nucleophiles: Highly fluorinated urethanes and polyurethanes from polyfluoro alcohols. *The Journal of Organic Chemistry* **2014**, *79* (11), 5019–5027.

48. Charlon, M.; Heinrich, B.; Matter, Y.; Couzigné, E.; Donnio, B.; Avérous, L. Synthesis, structure and properties of fully biobased thermoplastic polyurethanes, obtained from a diisocyanate based on modified dimer fatty acids, and different renewable diols. *European Polymer Journal* **2014**, *61*, 197–205.
49. Pauzi, N. N. P. N.; Majid, R. A.; Dzulkifli, M. H.; Yahya, M. Y. Development of rigid bio-based polyurethane foam reinforced with nanoclay. *Composites Part B: Engineering* **2014**, *67*, 521–526.
50. Petrović, Z. S. Polyurethanes from vegetable oils. *Polymer Reviews* **2008**, *48* (1), 109–155.
51. Fox, R. B.; Edmund, B. *Mechanically Frothed Gel Elastomers and Methods of Making and Using Them*. U.S. Patent 14/730,867, Jan. 21, 2016.
52. Sonnenschein, M. F. *Polyurethanes: Science, Technology, Markets, and Trends*; John Wiley Sons: Hoboken, NJ, USA, 2021; p 481.
53. Bouchemal, K.; Briançon, S.; Perrier, E.; Fessi, H.; Bonnet, I.; Zydowicz, N. Synthesis and characterization of polyurethane and poly (ether urethane) nanocapsules using a new technique of interfacial polycondensation combined to spontaneous emulsification. *International Journal of Pharmaceutics* **2004**, *269* (1), 89–100.
54. Simón, D.; Borreguero, A.; De Lucas, A.; Rodríguez, J. Recycling of polyurethanes from laboratory to industry, a journey towards the sustainability. *Waste Management* **2018**, *76*, 147–171.
55. Zevenhoven, R. *Treatment and Disposal of Polyurethane Wastes: Options for Recovery and Recycling*; Technical Report TKK-ENY-19; Helsinki University of Technology Espoo: Finland, 2004; p 48, ISBN 9512271605.
56. Jiménez, E.; Cabañas, B.; Lefebvre, G. *Environment, Energy and Climate Change I: Environmental Chemistry of Pollutants and Wastes*; Springer: New York, USA, 2015; Vol. 32, p 429.
57. Behrendt, G.; Naber, B. W. The chemical recycling of polyurethanes. *Journal of the University of Chemical Technology and Metallurgy* **2009**, *44* (1), 3–23.
58. O'Connor, J. In *Polyurethane Sealants, Adhesives and Binders*; American Chemistry Council, Center for the Polyurethanes Industry: Washington, DC, USA, 2012.
59. Veras, S. T.; Rojas, P.; Florencio, L.; Kato, M. T.; Sanz, J. L. 1, 3-Propanediol production from glycerol in polyurethane foam containing anaerobic reactors: performance and biomass cultivation and retention. *Environmental Science and Pollution Research* **2020**, *27* (36), 45662–45674.
60. Włoch, M.; Datta, J. Synthesis, structure and properties of poly (ester-urethane-urea) s synthesized using biobased diamine. *Journal of Renewable Materials* **2016**, *4* (1), 72–77.
61. Casali, S.; Gungormusler, M.; Bertin, L.; Fava, F.; Azbar, N. Development of a biofilm technology for the production of 1, 3-propanediol (1, 3-PDO) from crude glycerol. *Biochemical Engineering Journal* **2012**, *64*, 84–90.
62. Cardoso, G. T.; Neto, S. C.; Vecchia, F. Rigid foam polyurethane (PU) derived from castor oil (Ricinus communis) for thermal insulation in roof systems. *Frontiers of Architectural Research* **2012**, *1* (4), 348–356.
63. Chattopadhyay, D. K.; Webster, D. C. Thermal stability and flame retardancy of polyurethanes. *Progress in Polymer Science* **2009**, *34* (10), 1068–1133.

64. Ionescu, M. *Chemistry and Technology of Polyols for Polyurethanes*; iSmithers Rapra Publishing: Shrewsbury, UK, 2005; p 586.

65. Datta, J.; Kosiorek, P.; Włoch, M. Calorimetry, Synthesis, structure and properties of poly (ether-urethane) s synthesized using a tri-functional oxypropylated glycerol as a polyol. *Journal of Thermal Analysis* **2017**, *128* (1), 155–167.

66. Fang, C.; Zhou, X.; Yu, Q.; Liu, S.; Guo, D.; Yu, R.; Hu, J. Synthesis and characterization of low crystalline waterborne polyurethane for potential application in water-based ink binder. *Progress in Organic Coatings* **2014**, *77* (1), 61–71.

67. Xi, X.; Wu, Z.; Pizzi, A.; Gerardin, C.; Lei, H.; Zhang, B.; Du, G. Non-isocyanate polyurethane adhesive from sucrose used for particleboard. *Wood Science Technology* **2019**, *53* (2), 393–405.

68. Ghasemlou, M.; Daver, F.; Ivanova, E. P.; Adhikari, B. Bio-based routes to synthesize cyclic carbonates and polyamines precursors of non-isocyanate polyurethanes: A review. *European Polymer Journal* **2019**, *118*, 668–684.

69. Maisonneuve, L. *Vegetable Oils As a Platform for the Design of Sustainable and Non-Isocyanate Thermoplastic Polyurethanes*; Université Sciences et Technologies-Bordeaux I: France, 2013. https://tel.archives-ouvertes.fr/tel-01249386/file/MAISONNEUVE_LISE_2013.pdf (accessed June 25, 2021).

70. Taheri, N.; Sayyahi, S. Effect of clay loading on the structural and mechanical properties of organoclay/HDI-based thermoplastic polyurethane nanocomposites. *e-Polymers* **2016**, *16* (1), 65–73.

71. Claeys, B.; Vervaeck, A.; Hillewaere, X. K.; Possemiers, S.; Hansen, L.; De Beer, T.; Remon, J. P.; Vervaet, C. Thermoplastic polyurethanes for the manufacturing of highly dosed oral sustained release matrices via hot melt extrusion and injection molding. *European Journal of Pharmaceutics Biopharmaceutics* **2015**, *90*, 44–52.

72. Randall, D.; Lee, S. *The Polyurethanes Book*; Wiley-Blackwell: Washington, DC, USA, 2002; p 494.

73. Mahmoodi, A.; Ebrahimi, M.; Khosravi, A.; Mohammadloo, H. E. A hybrid dye-clay nano-pigment: Synthesis, characterization and application in organic coatings. *Dyes Pigments* **2017**, *147*, 234–240.

74. Savelyev, Y.; Veselov, V.; Markovskaya, L.; Savelyeva, O.; Akhranovich, E.; Galatenko, N.; Robota, L.; Travinskaya, T. Preparation and characterization of new biologically active polyurethane foams. *Materials Science Engineering: C* **2014**, *45*, 127–135.

75. Blackwell, J.; Nagarajan, M.; Hoitink, T. *The Structure of the Hard Segments in MDI/Diol/PTMA Polyurethane Elastomers*; ACS Publications: Washington, DC, USA, 1981; Vol. 172, pp 179–196.

76. Sheikhy, H.; Shahidzadeh, M.; Ramezanzadeh, B.; Noroozi, F. Studying the effects of chain extenders chemical structures on the adhesion and mechanical properties of a polyurethane adhesive. *Journal of Industrial Engineering Chemistry* **2013**, *19* (6), 1949–1955.

77. Patel, R.; Shah, M.; Patel, H. Synthesis and characterization of structurally modified polyurethanes based on castor oil and phosphorus-containing polyol for flame-retardant coatings. *International Journal of Polymer Analysis Characterization* **2011**, *16* (2), 107–117.

78. Huang, Y.; Jiang, S.; Liang, R.; Liao, Z.; You, G. A green highly-effective surface flame-retardant strategy for rigid polyurethane foam: Transforming UV-cured coating into intumescent self-extinguishing layer. *Composites Part A: Applied Science Manufacturing* **2019**, *125*, 105534.

79. Błażek, K.; Datta, J. Renewable natural resources as green alternative substrates to obtain bio-based non-isocyanate polyurethanes-review. *Critical Reviews in Environmental Science Technology* **2019**, *49* (3), 173–211.

80. Ghasemlou, M.; Daver, F.; Ivanova, E. P.; Murdoch, B. J.; Adhikari, B. Use of Synergistic Interactions to Fabricate Transparent and Mechanically Robust Nanohybrids Based on Starch, Non-Isocyanate Polyurethanes, and Cellulose Nanocrystals. *ACS Applied Materials Interfaces* **2020**, *12* (42), 47865–47878.

81. Zareanshahraki, F.; Asemani, H.; Skuza, J.; Mannari, V. Synthesis of non-isocyanate polyurethanes and their application in radiation-curable aerospace coatings. *Progress in Organic Coatings* **2020**, *138*, 105394.

82. Ghasemlou, M.; Daver, F.; Ivanova, E. P.; Adhikari, B. Synthesis of green hybrid materials using starch and non-isocyanate polyurethanes. *Carbohydrate polymers* **2020**, *229*, 115535.

83. Rao, W.-H.; Xu, H.-X.; Xu, Y.-J.; Qi, M.; Liao, W.; Xu, S.; Wang, Y.-Z. Persistently flame-retardant flexible polyurethane foams by a novel phosphorus-containing polyol. *Chemical Engineering Journal* **2018**, *343*, 198–206.

84. Wulf, C.; Reckers, M.; Perechodjuk, A.; Werner, T. Catalytic Systems for the Synthesis of Biscarbonates and Their Impact on the Sequential Preparation of Non-Isocyanate Polyurethanes. *ACS Sustainable Chemistry Engineering* **2019**, *8* (3), 1651–1658.

85. Abdel-Wakil, W. S.; Kamoun, E. A.; Fahmy, A.; Hassan, W.; Abdelhai, F.; Salama, T. M. Assessment of vinyl acetate polyurethane-based graft terpolymers for emulsion coatings: Synthesis and characterization. *Journal of Macromolecular Science, Part A* **2020**, *57* (4), 229–243.

86. Gui, T.; Xia, T.; Wei, H.; Zhang, Z.; Ouyang, X. Investigation on Effects of Chain Extenders and Cross-linking Agents of Polyurethane Elastomers Using Independent Building Vibration Isolation Sensor. *Sensors and Materials* **2019**, *31*, 4069–4078.

87. Bao, L.; Fan, H.; Chen, Y.; Yan, J.; Yang, T.; Guo, Y. Effect of surface free energy and wettability on the adhesion property of waterborne polyurethane adhesive. *RSC Advances* **2016**, *6* (101), 99346–99352.

88. Esmaeilpour, M.; Niroumand, B.; Monshi, A.; Ramezanzadeh, B.; Salahi, E. The role of surface energy reducing agent in the formation of self-induced nanoscale surface features and wetting behavior of polyurethane coatings. *Progress in Organic Coatings* **2016**, *90*, 317–323.

89. Jia-Hu, G.; Yu-Cun, L.; Tao, C.; Su-Ming, J.; Hui, M.; Ning, Q.; Hua, Z.; Tao, Y.; Wei-Ming, H. Synthesis and properties of a nano-silica modified environmentally friendly polyurethane adhesive. *RSC Advances* **2015**, *5* (56), 44990–44997.

90. Levchik, S. V.; Weil, E. D. Thermal decomposition, combustion and fire-retardancy of polyurethanes—a review of the recent literature. *Polymer International* **2004**, *53* (11), 1585–1610.

91. Cregut, M.; Bedas, M.; Durand, M.-J.; Thouand, G. New insights into polyurethane biodegradation and realistic prospects for the development of a sustainable waste recycling process. *Biotechnology Advances* **2013**, *31* (8), 1634–1647.

92. Calvo-Correas, T.; Ugarte, L.; Trzebiatowska, P. J.; Sanzberro, R.; Datta, J.; Corcuera, M. Á.; Eceiza, A. Thermoplastic polyurethanes with glycolysate intermediates from polyurethane waste recycling. *Polymer Degradation and Stability* **2017**, *144*, 411–419.

93. Kemona, A.; Piotrowska, M. Polyurethane Recycling and Disposal: Methods and Prospects. *Polymers* **2020**, *12* (8), 1752.

94. Dannecker, P.-K.; Meier, M. A. Facile and sustainable synthesis of erythritol bis (carbonate, a Valuable Monomer for Non-Isocyanate polyurethanes (NIpUs). *Scientific Reports* **2019**, *9* (1), 1–6.

95. Zhao, L.; Semetey, V. Recycling Polyurethanes through Transcarbamoylation. *ACS Omega* **2021**, *6* (6), 4175–4183.

96. Marson, A.; Masiero, M.; Modesti, M.; Scipioni, A.; Manzardo, A. Life Cycle Assessment of Polyurethane Foams from Polyols Obtained through Chemical Recycling. *ACS Omega* **2021**, *6* (2), 1718–1724.

97. Eling, B.; Tomović, Ž.; Schädler, V. Current and future trends in polyurethanes: An industrial perspective. *Macromolecular Chemistry and Physics* **2020**, *221* (14), 2000114.

98. Jiao, C.; Wang, H.; Li, S.; Chen, X. Fire hazard reduction of hollow glass microspheres in thermoplastic polyurethane composites. *Journal of Hazardous Materials* **2017**, *332*, 176–184.

99. Atiqah, A.; Jawaid, M.; Sapuan, S.; Ishak, M.; Ansari, M.; Ilyas, R. Physical and thermal properties of treated sugar palm/glass fibre reinforced thermoplastic polyurethane hybrid composites. *Journal of Materials Research and Technology* **2019**, *8* (5), 3726–3732.

100. Hasan, K. F.; Horváth, P. G.; Kóczán, Z.; Alpár, T. Thermo-mechanical properties of pretreated coir fiber and fibrous chips reinforced multilayered composites. *Sci. Rep.* **2021**, *11* (1), 1–13.

101. Yang, W.; Dong, Q.; Liu, S.; Xie, H.; Liu, L.; Li, J. Recycling and disposal methods for polyurethane foam wastes. *Procedia Environmental Sciences* **2012**, *16*, 167–175.

102. Ragaert, K.; Delva, L.; Geem, K. Mechanical and chemical recycling of solid plastic waste. *Waste Management* **2017**, *69*, 24–58.

103. Afzaluddin, A.; Jawaid, M.; Salit, M. S.; Ishak, M. R. Physical and mechanical properties of sugar palm/glass fiber reinforced thermoplastic polyurethane hybrid composites. *Journal of Materials Research and Technology* **2019**, *8* (1), 950–959.

104. Deng, Y.; Dewil, R.; Appels, L.; Ansart, R.; Baeyens, J.; Kang, Q. Reviewing the thermo-chemical recycling of waste polyurethane foam. *Journal of Environmental Management* **2021**, *278*, 111527.

105. Rao, W.-H.; Liao, W.; Wang, H.; Zhao, H.-B.; Wang, Y.-Z. Flame-retardant and smoke-suppressant flexible polyurethane foams based on reactive phosphorus-containing polyol and expandable graphite. *Journal of Hazardous Materials* **2018**, *360*, 651–660.

106. Abd El-Wahab, H.; Abd El-Fattah, M.; Ahmed, A. H.; Elhenawy, A. A.; Alian, N. Synthesis and characterization of some arylhydrazone ligand and its metal complexes and their potential application as flame retardant and antimicrobial additives in polyurethane for surface coating. *Journal of Organometallic Chemistry* **2015**, *791*, 99–106.

107. Abd El-Fattah, M.; El Saeed, A. M.; Dardir, M.; El-Sockary, M. A. Studying the effect of organo-modified nanoclay loading on the thermal stability, flame retardant, anti-corrosive and

mechanical properties of polyurethane nanocomposite for surface coating. *Progress in Organic Coatings* **2015**, *89*, 212–219.

108. Malucelli, G. Surface-engineered fire protective coatings for fabrics through sol-gel and layer-by-layer methods: An overview. *Coatings* **2016**, *6* (3), 33.

109. Altıntaş, Z.; Çakmakçı, E.; Kahraman, M.; Apohan, N.; Güngör, A. Preparation of photocurable silica–titania hybrid coatings by an anhydrous sol–gel process. *Journal of Sol-Gel Science and Technology* **2011**, *58* (3), 612–618.

110. Cai, J.; Heng, H.-M.; Hu, X.-P.; Xu, Q.-K.; Miao, F. A facile method for the preparation of novel fire-retardant layered double hydroxide and its application as nanofiller in UP. *Polymer Degradation and Stability* **2016**, *126*, 47–57.

111. Wang, X.; Pan, Y.-T.; Wan, J.-T.; Wang, D.-Y. An eco-friendly way to fire retardant flexible polyurethane foam: layer-by-layer assembly of fully bio-based substances. *Rsc Advances* **2014**, *4* (86), 46164–46169.

112. Pan, H.; Wang, W.; Pan, Y.; Song, L.; Hu, Y.; Liew, K. M. Formation of layer-by-layer assembled titanate nanotubes filled coating on flexible polyurethane foam with improved flame retardant and smoke suppression properties. *ACS Applied Materials & Interfaces* **2015**, *7* (1), 101–111.

113. Pan, Y.; Liu, L.; Cai, W.; Hu, Y.; Jiang, S.; Zhao, H. Effect of layer-by-layer self-assembled sepiolite-based nanocoating on flame retardant and smoke suppressant properties of flexible polyurethane foam. *Applied Clay Science* **2019**, *168*, 230–236.

114. Cullis, C. F.; Hirschler, M. M. *The Combustion of Organic Polymers*; Oxford University Press: London, UK, 1981; Vol. 5, p 419.

115. Chen, M.-J.; Shao, Z.-B.; Wang, X.-L.; Chen, L.; Wang, Y.-Z. Halogen-free flame-retardant flexible polyurethane foam with a novel nitrogen–phosphorus flame retardant. *Industrial Engineering Chemistry Research* **2012**, *51* (29), 9769–9776.

116. Toldy, A.; Harakály, G.; Szolnoki, B.; Zimonyi, E.; Marosi, G. Flame retardancy of thermoplastics polyurethanes. *Polymer Degradation and Stability* **2012**, *97* (12), 2524–2530.

117. Tabuani, D.; Bellucci, F.; Terenzi, A.; Camino, G. Flame retarded Thermoplastic Polyurethane (TPU) for cable jacketing application. *Polymer Degradation Stability* **2012**, *97* (12), 2594–2601.

118. Xu, W.; Wang, G.; Zheng, X. Research on highly flame-retardant rigid PU foams by combination of nanostructured additives and phosphorus flame retardants. *Polymer Degradation Stability* **2015**, *111*, 142–150.

119. Akdogan, E.; Erdem, M.; Ureyen, M. E.; Kaya, M. Rigid polyurethane foams with halogen-free flame retardants: Thermal insulation, mechanical, and flame retardant properties. *Journal of Applied Polymer Science* **2020**, *137* (1), 47611.

120. Wang, C.; Wu, Y.; Li, Y.; Shao, Q.; Yan, X.; Han, C.; Wang, Z.; Liu, Z.; Guo, Z. Flame-retardant rigid polyurethane foam with a phosphorus-nitrogen single intumescent flame retardant. *Polymers for Advanced Technologies* **2018**, *29* (1), 668–676.

121. Wang, S.-X.; Zhao, H.-B.; Rao, W.-H.; Huang, S.-C.; Wang, T.; Liao, W.; Wang, Y.-Z. Inherently flame-retardant rigid polyurethane foams with excellent thermal insulation and mechanical properties. *Polymer* **2018**, *153*, 616–625.

122. Xie, N.; Niu, J.; Zhong, Y.; Gao, X.; Zhang, Z.; Fang, Y. Development of polyurethane acrylate coated salt hydrate/diatomite form-stable phase change material with enhanced thermal stability for building energy storage. *Construction and Building Materials* **2020**, *259*, 119714.

123. Andersons, J.; Kirpluks, M.; Cabulis, P.; Kalnins, K.; Cabulis, U. Bio-based rigid high-density polyurethane foams as a structural thermal break material. *Construction Building Materials* **2020**, *260*, 120471.

124. Hasan, K. M. F.; Péter György, H.; Tibor, A. Thermomechanical Behavior of Methylene Diphenyl Diisocyanate-Bonded Flax/Glass Woven Fabric Reinforced Laminated Composites. *ACS Omega* **2020**, *6* (9), 6124–6133.

125. Davies, P.; Evrard, G. Accelerated ageing of polyurethanes for marine applications. *Polymer Degradation and Stability* **2007**, *92* (8), 1455–1464.

126. Ekin, A.; Webster, D. C.; Daniels, J. W.; Stafslien, S. J.; Cassé, F.; Callow, J. A.; Callow, J. A.; Callow, M. E. Synthesis, formulation, and characterization of siloxane–polyurethane coatings for underwater marine applications using combinatorial high-throughput experimentation. *Journal of Coatings Technology and Research* **2007**, *4* (4), 435–451.

127. Erdinler, E. S.; Koc, K. H.; Dilik, T.; Hazir, E. Layer thickness performances of coatings on MDF: Polyurethane and cellulosic paints. *Maderas. Ciencia y tecnología* **2019**, *21* (3), 317–326.

128. Zhang, Y.; Li, T.-T.; Shiu, B.-C.; Sun, F.; Ren, H.-T.; Zhang, X.; Lou, C.-W.; Lin, J.-H. Eco-friendly versatile protective polyurethane/triclosan coated polylactic acid nonwovens for medical covers application. *Journal of Cleaner Production* **2021**, *282*, 124455.

129. Kashyap, D.; Kumar, P. K.; Kanagaraj, S. 4D printed porous radiopaque shape memory polyurethane for endovascular embolization. *Additive Manufacturing* **2018**, *24*, 687–695.

130. Shin, E. J.; Choi, S. M. Advances in Waterborne Polyurethane-Based Biomaterials for Biomedical Applications. In *Novel Biomaterials for Regenerative Medicine*; Springer: The Gateway East, Singapore, 2018; Vol. 1077, pp 251–283.

131. Sáenz-Pérez, M.; Bashir, T.; Laza, J. M.; García-Barrasa, J.; Vilas, J. L.; Skrifvars, M.; León, L. M. Novel shape-memory polyurethane fibers for textile applications. *Textile Research Journal* **2019**, *89* (6), 1027–1037.

132. Olcay, H.; Kocak, E. D.; Yıldız, Z. Sustainability in Polyurethane Synthesis and Bio-based Polyurethanes. In *Sustainability in the Textile and Apparel Industries*; Springer: Gewerbestrasse, Cham, Switzerland, 2020; pp 139–156.

133. Sultana, M. Z.; Mahmud, S.; Pervez, M. N.; Hasan, K. F.; Heng, Q. Green synthesis of glycerol monostearate-modified cationic waterborne polyurethane. *Emerg. Mater. Res* **2019**, *8* (2), 137–147.

134. Hasan, K. M. F.; Wang, H.; Mahmud, S.; Jahid, M. A.; Islam, M.; Jin, W.; Genyang, C. Colorful and antibacterial nylon fabric via in-situ biosynthesis of chitosan mediated nanosilver. *J. Mater. Res. Technol.* **2020**, *9* (6), 16135–16145.

135. Somarathna, H.; Raman, S.; Mohotti, D.; Mutalib, A.; Badri, K. The use of polyurethane for structural and infrastructural engineering applications: A state-of-the-art review. *Construction and Building Materials* **2018**, *190*, 995–1014.

136. Golling, F. E.; Pires, R.; Hecking, A.; Weikard, J.; Richter, F.; Danielmeier, K.; Dijkstra, D. Polyurethanes for coatings and adhesives–chemistry and applications. *Polymer International* **2019**, *68* (5), 848–855.

137. Das, M.; Mandal, B.; Katiyar, V. Environment-friendly synthesis of sustainable chitosan-based nonisocyanate polyurethane: A biobased polymeric film. *Journal of Applied Polymer Science* **2020**, *137* (36), 49050.

138. Hasan, R. *Ways of Saying: Ways of meaning: Selected Papers of Ruqaiya Hasan*; Bloomsbury Publishing: London, UK, 2015; p 191–242.

Chapter 12

Recycling of Polyurethanes Containing Flame-Retardants and Polymer Waste Transformed into Flame-Retarded Polyurethanes

Marcin Włoch[*,1]

[1]Department of Polymers Technology, Faculty of Chemistry, Gdańsk University of Technology, G. Narutowicza 11/12 Str., 80-233 Gdańsk, Poland
[*]Email: marcin.wloch@pg.edu.pl

The growing number of polyurethanes (PUs) produced every year has developed methods for their mechanical and chemical recycling which yield valuable products like substitutes for commercial polyols or flame-retardants. PUs can be produced in different shapes and forms (i.e., elastomers, flexible or rigid foams, coatings, etc.) using several different components (i.e., di- or polyisocyanates, ester- or ether-based polyols, low-molecular weight chain extenders, fillers, and other modifiers). Therefore, different recycling methods should be considered for a wide range of materials, including postproduction of postconsumer wastes) depending on their chemical structure and properties. This chapter presents a review of selected mechanical (e.g., regrinding and using as a filler, rebonding, and compression molding) and chemical (e.g., glycolysis, glycerolysis, acidolysis and phosphorolysis) recycling methods applicable for PUs. This chapter also presents examples of flame-retardants and flame-retarded PUs obtained by PU recycling, poly(ethylene terephthalate), and melamine formaldehyde foam.

Introduction

The recycling of polyurethanes (PUs) involves four processes: (1) Mechanical recycling, which is related to the processing of polymer into secondary raw material or products without significantly changing the chemical structure of the material; (2) Chemical recycling, which relates to changing of polymer chemical structure (due to presence of urethane and ester moieties in PU macromolecules) by using alcohols, amines, ammonia, carboxylic acids, phosphoric acid esters, or water at elevated temperatures, sometimes in the presence of catalysts; (3) Thermo-chemical recycling (including pyrolysis, gasification, or hydrogenation) which permit to obtain chemicals, fuels, energy, and others; (4) Recovery of energy (including combustion and incineration) connected with thermal decomposition of wastes (*1–4*).

© 2021 American Chemical Society

Recycling of PUs containing flame-retardants (FRs) is not well described in the literature. May works present the possibilities of PUs waste recycling, but the composition (formulation) of waste is not clearly indicated, which means the effect of type and amount of FRs in PU wastes on their recycling cannot be defined. Some works are connected with recycling of PU waste (e.g., from car seats (5), refrigerators (6–8), or insulation panels (9)), which are commonly modified by the addition of FRs. It is supposed that currently developed recycling methods will be suitable for the recycling of PUs containing FRs.

Mechanical Recycling of PUs

Mechanical recycling of PUs involves cutting, crushing, grinding, and powdering of waste to obtain suitable feedstock (scraps, flakes, or powder) for further processing. The prepared materials can be rebounded, compression molded, injection molded, extruded, or used as a filler (Table 1). The most important advantages of mechanical recycling are the process's simple procedure and the connected low cost. On the other hand, products obtained can be characterized by poor mechanical properties (e.g., due to aging of virgin PUs during their usage) (10).

Table 1. Mechanical Recycling Methods of Polyurethanes

Method	Description
Regrinding	- group of methods involving cutting, shredding, milling - lowering of waste dimensions is the processing technique for materials subjected to further mechanical and chemical recycling
Rebonding	- method that involves bonding of flakes or scraps using adhesives such as, MDI[a] and PMDI[b] and pressing
Compression molding	- method involves softening of waste, self-bonding under heat (without adhesive), and pressing
Injection molding	- method involves mixing, heating, and pressing to transform PU waste into a thermoplastic material (as a result of macromolecules chain shortening)
Extrusion	- continuous process that involves the mixing, heating, and pressing of PU waste

[a] MDI: 4,4′-methylenephenyl diisocyanate.   [b] PMDI; polymeric MDI.

The type of method applied results from the type of processed PU waste. Thermoplastics and some flexible foams can be reprocessed (due to thermoplastic nature of soft segments) under the suitable temperature and pressure (e.g., by injection or compression molding). After powdering, rigid PU foams (PUFs) should be processed using adhesives (e.g., MDI and PMDI) or used as filler for concrete (11–14), thermosets (e.g., polyester resin (15)), or thermoplastics (e.g., polyethylene (16) or polypropylene (17)). Figure 1 presents powdered PUF waste used for preparation of lightweight plaster materials, while Figure 2 presents grounded PUF waste applied as lightweight aggregates in concrete. Polymer composites with powdered filler can be processed by injection molding or extrusion (10). Examples of mechanical recycling of PUs and applications of obtained products are presented in Table 2.

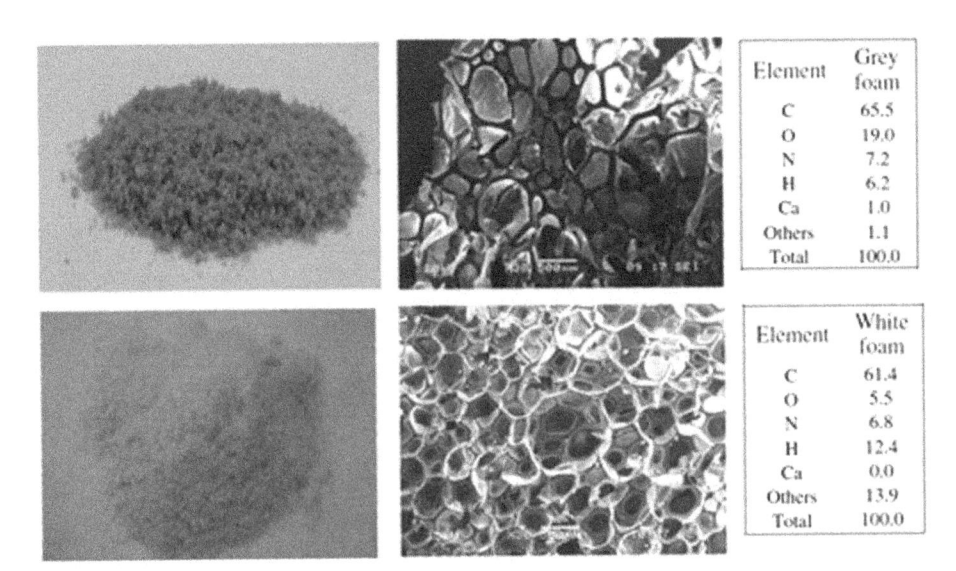

Figure 1. The form, microstructure, and elemental composition of PUF wastes from different sources used to prepare lightweight plaster materials. Gray foam is a byproduct of the automobile industry, while the white foam is from the manufacturing of PU for thermal insulation in the construction industry. Reproduced with permission from reference (9). Copyright 2012 Elsevier.

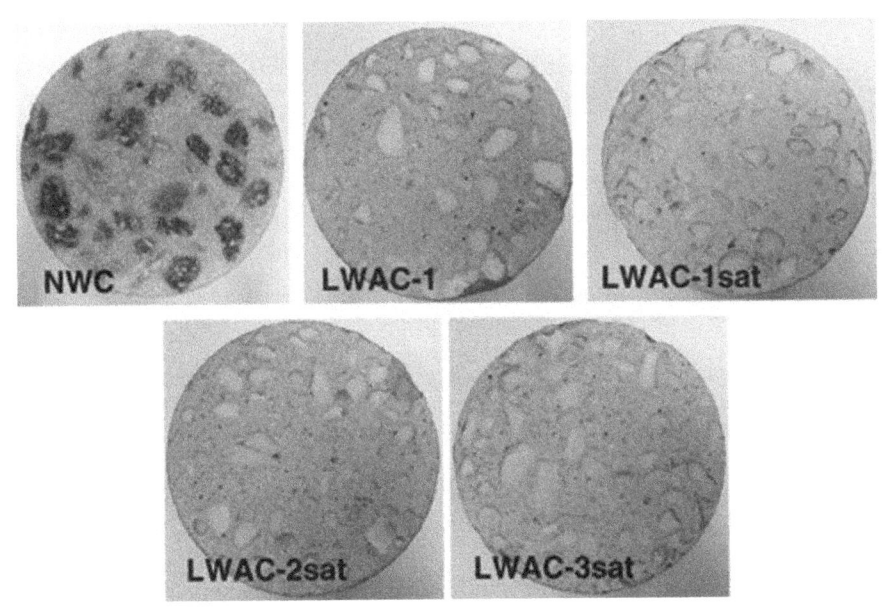

Figure 2. Normal weight concrete (NWC) and four lightweight aggregate concretes (LWAC-1, LWAC-1sat, LWAC-2sat, and LWAC-3sat) containing PUF waste aggregates. Reproduced with permission from reference (11). Copyright 2010 Elsevier.

Table 2. Examples of PUs Mechanical Recycling Described in the Literature

PU type	Applied recycling method	Ref.
PUF wastes obtained by recycling of insulation panels used in building industry	- incorporation of PUF wastes into cementitious mixtures (volume fraction of PU was from 13.1% to 33.7%) to produce lightweight concrete - PUF was ground to particle size ≤0.090 mm before being used	(14)
PUF waste obtained from the recycling of panels used in construction industry	- production of lightweight white cement pastes by partial replacement with different ratios of PUF waste based on the weight of cement (10%, 20%, 30%, or 40%) - PUF was used to obtain (particles with a diameter lower than 2 mm were predominant fraction)	(18)
PUF waste (byproduct from automobile industry and manufacturing of PU for thermal insulation in the construction industry)	- preparation of plaster mixtures (the main constituent of plaster was calcium sulfate hemihydrate) using differing volumes of PUF waste from different sources (PUF/plaster volume ratio was equal 0.5, 1, 2, 3, or 4) - PUFs were ground to different granulometric sizes	(9)
Polyisocyanurate foam was obtained as production waste of an insulation	- production of lightweight concrete block from PU waste and ordinary Portland cement (blends contain from 6 to 20 wt% of PUF) - foam waste was crushed by a granulator with staggered rotor blades to obtain suitable particle size	(13)
RPUF[a] waste obtained from recycling of panels used in the automotive industry	- RPUF wastes were mixed with cement-based mixtures to produce lightweight mortars - PU waste was ground to particle sizes of less than 4 mm before the use as an aggregate substitute	(12)
Reaction injection molded flexible PUF waste	- recycling method of reaction injection molding PUF waste by controlled degradation during twin-screw extrusion - coextrusion of PU waste with poly(vinyl chloride) (amount of poly(vinyl chloride: 510 wt. %)	(19)
Reaction injection molded PU (scrapped vehicle bumpers)	- PU was granulated and introduced as filler (29 wt%) in bulk molding compounds based on polyester resin (filled also with chopped glass fibers and talc) - PU recycled as filler improved the flexibility and toughness of polyester resin (in comparison to conventional resin)	(15)
PU waste collected from railway waste disposal	- polyethylene-graft-maleic anhydride-grafted PU/high density polyethylene composites were prepared by melt-blending at various concentrations (010 phr) of PE-g-MA-grafted PU	(16)
RPUF waste	- preparation of polypropylene blends with RPUF waste (030 wt%) - maleic anhydride grafted polypropylene was used as harmonizing agent	(17)

Table 2. (Continued). Examples of PUs Mechanical Recycling Described in the Literature

PU type	Applied recycling method	Ref.
RPUF waste from the footwear industry	- preparation of composites of recycled poly(vinyl butyral) and RPUF (amount of foam waste was equal 20, 35, or 50 wt%) by extrusion and injection molding - flasks of poly(vinyl butyral) were obtained from the recycling of automotive glass sandwich - PU waste was milled in a plastic granulator with knives and after milling particles with the size between 0.25 and 1.68 mm were obtained	(20)

[a] RPUF: rigid PUF.

Chemical Recycling of PUs

Chemical recycling of PUs is well described in the literature and can be realized by several different techniques (i.e., hydrolysis, glycolysis, glycerolysis, ammonolysis, aminolysis, phosphorolysis, and phenolysis), which are presented in Figure 3 (1, 2). Description of the methods is presented in Table 3.

Table 3. Description of Chemical Recycling Methods of PU Materials

Method	Description (reagents and conditions)
Glycolysis	- depolymerizing agents: low-molecular weight diols (glycols), for example: 1,2-ethanediol, 1,3-propanediol, 1,4-butanediol, 1,6-hexandiol, and others - catalysts are required: acetates (e.g., sodium, potassium, or lithium acetate), hydroxides (e.g., sodium, potassium, or lithium hydroxide), metal-organic compounds (e.g., dibutyltin dilaurate, stannous octoate), or amines (e.g., diethanolamine, diazabiscyclooctane, triethylamine) - process can be realized with the mass excess of PU or glycol ("split-phase" glycolysis) - temperature of the process is depending on depolymerizing agents and is in the range from 180 to 240°C - reaction is mostly realized under atmospheric pressure (sometimes in the presence of inert gas, like nitrogen)
Glycerolysis	- depolymerizing agent: glycerol (which can be derived from bio-based resources) - process is realized at temperature around 220–240°C under atmospheric pressure - catalysts are required and can be the same as in glycolysis - process can be realized with the mass excess of PU or glycerol ("split-phase" glycerolysis)
Acidolysis	- depolymerizing agents: dicarboxylic acids (e.g., succinic acid, phthalic acid, and adipic acid) - process is realized at temperature around 190–200°C for 4–5h - obtained products can be used as polyols in the synthesis of PUFs or adhesives
Aminolysis	- depolymerizing agents: diamines or polyamines (e.g., diethylenetriamine, triethylenetetramine, tetraethylenepentamine) - process can be realized without using a catalyst, but can be supported by the addition of (e.g., sodium hydroxide) - obtained products can be used as curing agents for epoxy resins

Table 3. (Continued). Description of Chemical Recycling Methods of PU Materials

Method	Description (reagents and conditions)
Ammonolysis	- depolymerizing agent is ammonia (NH_3) used in the mass excess with respect to PU material - process is realized at elevated temperatures under high pressures (up to 140 atm.) - the main products of the reaction are diamines, low- and high-molecular weight diols (or polyols) and unsubstituted urea
Hydrolysis	- depolymerizing agent: water (H_2O) - catalyst are required: acids (e.g., hydrochloric acid) or bases (e.g., sodium hydroxide) - process is realized at elevated temperatures (200–400°C) under the pressure from atmospheric to 15–50 atm. - main products of the reaction are diamines, low- and high-molecular weight diols (or polyols) and carbon dioxide
Phosphorolysis	- depolymerizing agents: phosphonate esters, for example: dimethyl phosphonate, diethyl phosphonate, triethylphosphonate, and others - process is realized with the mass excess of phosphonate ester with respect to PU materials, and unreacted depolymerizing agent is removed under the vacuum - temperature is ranged from 140 to 190°C and catalysts are not required - phosphorous containing oligomers are the main products
Phenolysis	- depolymerizing agent: phenol - process is realized in the temperature above melting point of phenol in the presence of catalysts (bases or acids)

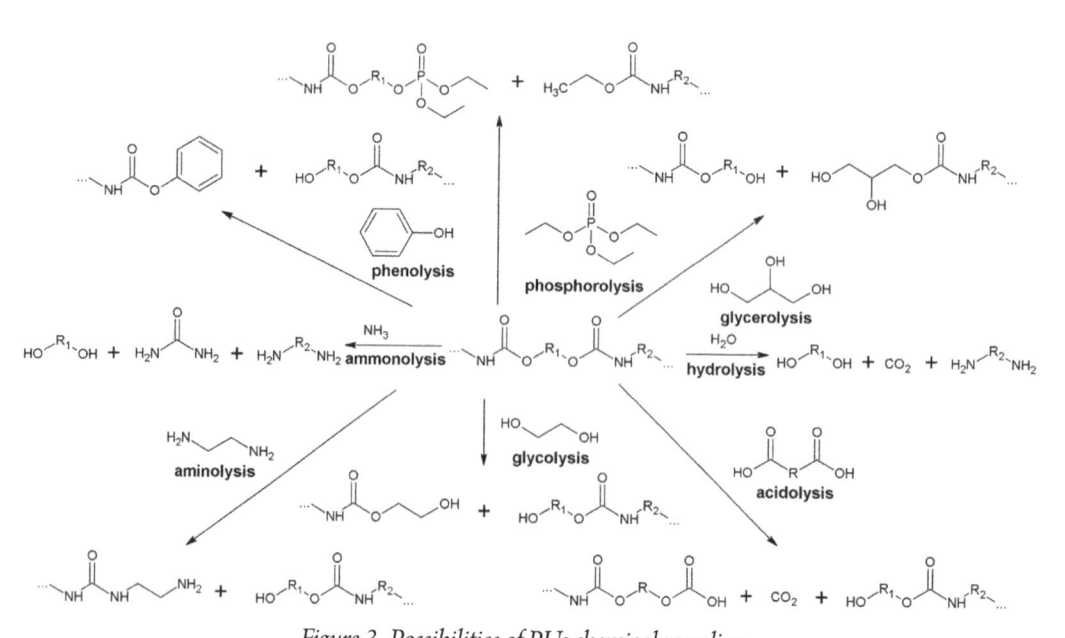

Figure 3. Possibilities of PUs chemical recycling.

Today, glycolysis is one the most important method of PUs chemical recycling, which is successfully realized in the industry. Glycolysis can be performed using: (1) mass excess of PU waste with respect to diol products of the reaction can be homogenous or two-phased and are generally oligomeric intermediates containing urethane moieties in the structure and terminated mostly by hydroxyl groups; any purification or removing of unreacted glycol is not required (*21*, *22*); (2) mass excess of low-molecular weight diol with respect to PU material (commonly known as "split-phase" glycolysis technique) permit recovery of polyol and filler from the PU waste; products are generally two-phased: the upper phase contain recovered polyol, while the bottom phase contains glycol, carbamates, and amines; after the process recovered polyol should be purified and unreacted glycol should be removed by vacuum distillation (after purification glycol can be used in another glycolysis process) (*23*, *24*).

During the glycolysis process two major reactions take place: (1) transesterification of urethane and urea bonds (with the formation of oligomers ended by hydroxyl and amine moieties); (2) breakup of the crosslinking allophanate and biuret moieties (*1*). In some applications the presence of the amine groups is not allowed, so transformation of amine-terminated oligomers to hydroxy-terminated ones can be realized by their reaction with cyclic carbonates (e.g., ethylene carbonate), which is presented in Figure 4 (*21*).

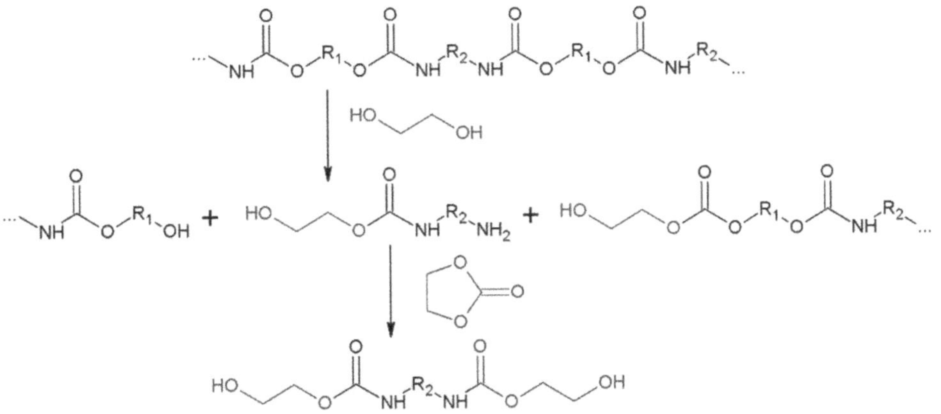

Figure 4. Glycolysis of PU material (using ethylene glycol) and transformation of amine-terminated oligomers (using ethylene carbonate) into oligomeric diols.

The process of glycolysis and properties of obtained products are affected by: the type of PU waste (elastomer, rigid, or flexible foam), the chemical structure and properties of depolymerized polyurethane, the size of depolymerized materials, the type of used low-molecular weight diol, the type and amount of a catalyst, the molar/mass ratio of PU material to depolymerizing agent; the reaction temperature, time, and atmosphere; the stirring speed and feeding rate (*1*).

Glycerolysis is a modification of the glycolysis process, where glycerol is used as a depolymerizing agent. Glycerol can be derived from plants (also as byproduct of biodiesel production). Studies confirm that glycerol characterized by different purity (from 62 to 99.5%) can be successfully applied in the chemical recycling of PUs (*23*, *25–29*). In comparison to glycolysis, glycerolysis performed with the mass excess of PU materials resulted in the formation of oligomeric products with hydroxyl functionality higher than two. Therefore, the obtained product are rather dedicated for the preparation of PUFs.

Acidolysis is a relatively new chemical method of PUs recycling that relates to the use of saturated carboxylic acids (e.g., succinic or adipic acid). Obtained products can be directly used in the synthesis of PUFs or adhesives (30, 31).

Glycolysis, glycerolysis, and acidolysis products are mainly characterized by chemical structure (using Fourier transform infrared spectroscopy or nuclear magnetic resonance spectroscopy) and physicochemical properties (mainly hydroxyl value, acid value, viscosity, average molecular weight, and dispersity index by gel permeation chromatography).

Examples related to chemical recycling of PUs are presented in Table 4

Table 4. Examples of PUs Chemical Recycling (by Glycolysis, Glycerolysis, Acidolysis, and Aminolysis) Described in the Literature

PU type	Method	Reagents and conditions	Ref.
Industrial scraps of flexible PU foam (from car seats) prepared by Reactive Injection Molding	Glycolysis	- depolymerizing agent: diethylene glycol or dipropylene glycol - catalyst: diethanolamine - temperature: 220 °C - reaction time: 2.5h - polyurethane/diol mass ratio: 1:1	(5)
Flexible PUF	Glycolysis	- depolymerizing agent: 1,6-hexanediol - catalyst: potassium acetate - temperature: 230–245 °C - reaction time: 9–70 min (after dosing of last portion of polyurethane) - polyurethane/diol mass ratio: 1:1, 2:1, 4:1, 6:1,8:1, or 10:1	(22)
Flexible PUF	Glycolysis	- depolymerizing agent: 1,3-propanediol or 1,6-hexanediol - catalyst: potassium acetate - temperature: 190–245 °C - reaction time: 30–82 min (after dosing of last portion of polyurethane) - polyurethane/diol mass ratio: 6:1,8:1, or 10:1	(21)
Flexible PUF synthesized using ether-based polyol and toluene diisocyanate (TDI)	Glycolysis	- depolymerizing agent: diethylene glycol - catalyst: stannous octoate - temperature: 179–189 °C - reaction time: 180 min - polyurethane/diol mass ratio: 1:0.9, 1:1.25, or 1:1.5	(33)
RPUF from refrigerators	Glycolysis	- depolymerizing agent: diethylene glycol - catalyst: potassium acetate - temperature: 220–226 °C - reaction time: max. 6h - polyurethane/diol mass ratio: 1:1, 1:2, or 1:3 - stirring speed: 1000–1100 rpm - feeding rate: 1.5–3 g min^{-1}	(6)

Table 4. (Continued). Examples of PUs Chemical Recycling (by Glycolysis, Glycerolysis, Acidolysis, and Aminolysis) Described in the Literature

PU type	Method	Reagents and conditions	Ref.
Flexible PUF synthesized using ether-based polyol and TDI	Glycolysis	- depolymerizing agent: diethylene glycol - catalyst: potassium octoate - temperature: 175–195 °C - reaction time: max. 350 min (depending on temperature and catalyst concentration) - polyurethane/diol mass ratio: 1:1.5	(34)
Flexible PUF synthesized using ether-based polyol and TDI	Glycolysis	- depolymerizing agent: diethylene glycol - catalyst: diethanolamine, titanium(IV) butoxide, potassium octoate, and calcium octoate - temperature: 189 °C - reaction time: 5h 10 min - polyurethane/diol mass ratio: 1:1.5 - feeding rate: 5 g min^{-1}	(24)
Flexible PUF from recycled cars	Glycolysis	- depolymerizing agent: diethylene glycol - catalyst: potassium acetate - temperature: 215–225 °C - reaction time: 400 min - polyurethane/diol mass ratio: 1:0.5, 1:1, 1:1.5, or 1:2.5 - feeding rate: 17–34 g min^{-1} - stirring speed: 1000–1100 rpm - system pressure: 1 atm	(35)
RPUF from refrigerators	Glycolysis	- depolymerizing agent: ethylene glycol or diethylene glycol - catalyst: sodium acetate, sodium hydroxide or triethanolamine - temperature: 198 °C for EG and 245 °C for diethylene glycol - reaction time: 400 min - polyurethane/diol mass ratio: 1:0.8, 1:1, 1:1.5, 1:2, or 1:3 - feeding rate: 1.5–3 g min^{-1}	(7)
RPUF with nanosilica as a filler	Glycolysis	- depolymerizing agent: diethylene glycol - catalyst: tin octoate - reaction temperature: 190 °C - reaction time: 6h - inert atmosphere - vigorous stirring (300 rpm) - PUF/diol mass ratio: 1:2 - process was realized as split-phase glycolysis and two-phase product were obtained - recovered polyol and filler was used for the synthesis of new PUF	23

Table 4. (Continued). Examples of PUs Chemical Recycling (by Glycolysis, Glycerolysis, Acidolysis, and Aminolysis) Described in the Literature

PU type	Method	Reagents and conditions	Ref.
RPUF with nanosilica as a filler	Glycerolysis	- depolymerizing agent: crude glycerol - catalyst: tin octoate - reaction temperature: 190 °C - reaction time: 6h - inert atmosphere - vigorous stirring (300 rpm) - PUF/glycerol mass ratio: 1:1.5 - PUF/diol mass ratio: 1:1.5 or 1:2 - process was realized as split-phase glycolysis and two-phase product were obtained - recovered polyol and filler was used for the synthesis of new PUF	(23)
Rigid polyisocyanurate foam	Glycolysis	- depolymerizing agent: dipropylene glycolyl - catalyst: potassium acetate - temperature: 200 °C - reaction time: 3h - PUF/diol mass ratio: 1:1.5 - recovered polyol can be applied in the synthesis of PUFs	(36)
Flexible PUF synthesized using ether-based polyol and PMDI	Glycerolysis	- depolymerizing agent: glycerol (purity: 62 or 84%) - catalysts: potassium acetate, sodium hydroxide, 1,4-diazabiscyclo[2.2.2]octane, tin(II) ethylhexanoate, dibutyltin dilaurate or triethylamine - temperature: 220 °C - reaction time: 13h - PUF/diol mass ratio: 3:1 - feeding rate: 5 g min^{-1} - two-phase glycerolysates were obtained and used for the synthesis of cast PUs	(25)
PU elastomer synthesized using ether-based polyol and MDI	Glycerolysis	- depolymerizing agent: glycerol (purity: 80 or 99.5%) - catalyst: potassium acetate - temperature: 225-230 °C - reaction time: 95–145 min - PU /glycerol mass ratio: 2:1, 4:1, 6:1, 8:1, or 10:1	(26)

Table 4. (Continued). Examples of PUs Chemical Recycling (by Glycolysis, Glycerolysis, Acidolysis, and Aminolysis) Described in the Literature

PU type	Method	Reagents and conditions	Ref.
Flexible PUF synthesized using ether-based polyol and PMDI	Glycerolysis	- depolymerizing agent: glycerol (purity: 40, 62, 84, or 99.5%) - catalyst: sodium hydroxide - temperature: 150, 210, 220, and 220°C (depending on glycerol purity) - reaction was carried out 30 min after dosing the last portion of polyurethane - polyurethane/glycerol mass ratio: 3:1 - feeding rate: 5 g min^{-1} - obtained glycerolysates were single- or two-phase products and were used to synthesize of cast PUs	(28)
Flexible PUF synthesized using ether-based polyol and PMDI	Glycerolysis	- depolymerizing agent: glycerol (purity: 84%) - catalysts: potassium acetate, sodium hydroxide, 1,4-diazabiscyclo[2.2.2]octane, tin(II) ethylhexanoate, dibutyltin dilaurate, or triethylamine - temperature: 225–230 °C - reaction was carried out 30 min after dosing the last portion of polyurethane - polyurethane/glycerol mass ratio: 3:1 - obtained glycerolysates were single- or two-phase products and were used to synthesize of cast PUs	(27, 29)
PUF	Acidolysis	- depolymerizing agents: succinic acid, phthalic acid, or mixture of acids - temperature: 190 °C - reaction time: 5h - polyurethane/diacid mass ratio: 3.0:1 (for phthalic acid), 4.2:1 (for mixture), or 4.6:1 (for succinic acid) - recycled polyols were used as substitute of petroleum-based polyol in the production of PU adhesives	(32)
Different types of flexible PUs foams	Acidolysis	- depolymerizing agent: succinic acid - temperature: 195 °C - reaction time: 5h - polyurethane/diacid mass ratio: 4.5:1 - recycled polyols was used as partial substitute of conventional polyol (up to 20 wt%) in the production of RPUFs	(31)

Table 4. (Continued). Examples of PUs Chemical Recycling (by Glycolysis, Glycerolysis, Acidolysis, and Aminolysis) Described in the Literature

PU type	Method	Reagents and conditions	Ref.
PUF synthesis using MDI, PMDI or TDI and ether-based polyol / polyisocyanurate foam synthesized using PMDI and ester-based polyol	Aminolysis	- depolymerizing agent: diethylenetriamine, triethylenetetramine, tetraethylenepentamine - temperature: 150, 160, 170, or 180°C - reaction time: 40, 70, 100, 130, or 150 min - polyurethane/amine mass ratio: 2:1 - obtained aminolysates were used as curing agents for epoxy resins	(37)

Preparation of FR PUs by Recycling of Polymers

Products of chemical recycling of polymers can be successfully used as substrates for the preparation FR PUs, for example: (1) phosphorous containing oligomers (prepared by phosphorolysis of PUF) can be applied as FRs in the preparation of RPUFs (38); (2) glycolysis product of rigid polyurethane-polyisocyanurate foams (prepared using boron-containing polyol) with reduced flammability can be applied as polyol in new rigid polyurethane-polyisocyanurate foams with improved flame-retardancy (39); (3) glycolysis product of postconsumer poly(ethylene terephthalate) bottles (40) and textiles can be used as a polyol in preparation of PUFs with improved thermal stability and flame-retardancy (41); (4) melamine formaldehyde foam waste can be powdered and added into PUF as a FR filler (42).

Phosphorolysis of PUs was developed by Troev et al. and is realized using phosphonate esters (PPE) to decompose PU material (Figure 5 and Table 5) (43–47). The mass ratio of PU to PPE is equal 1:3 and before the phosphorolysis PU materials are cut into small pieces (3–5 mm in size) and placed with phosphonate ester in a flask, then the reaction mixture is heated to required temperature (140–180 °C). After the reaction is completed, the temperature is lowered, and the unreacted phosphonate ester is removed under vacuum. Phosphorus containing oligomers can be applied as FRs in preparation of PUFs.

Chung, Kim, and Kim performed phosphorolysis of PUF using three different PPE (i.e., triethyl phosphate, trimethyl phosphonate, or tris(1-methyl-2-chloroethyl) phosphanate) and, after removing of unreacted depolymerizing agent products, were applied as FRs in the preparation of RPUFs (38). The PUFs containing products of phosphorolysis (5, 10, 15, and 20 wt % to polyol) were characterized by reduced flammability, improved thermal stability, and tensile strength in comparison to PUFs without recycling products. It was found that recycling products obtained using tris(1-methyl-2-chloroethyl) phosphanate were the most effective FRs for PUFs.

Grancharov et al. describe a new approach for the recycling of microporous PU elastomers using tris(1-methyl-2-chloroethyl) phosphate, after removing unreacted phosphonate ester, followed by reaction of obtained product with propylene oxide to obtain a reactive phosphorus containing oligomer (Figure 6) suitable for the preparation of RPUFs. Modification of phosphorolysis products enable significant reduction in acidity (48). The obtained foams (containing 5, 10, or 15 wt % of recycling product) have higher densities, better mechanical properties (bending and compression strength), and comparable thermal stability when compared to the standard RPUF.

Table 5. Exemplary Works Related to the Phosphorolysis of PU Materials

PU type	Reagents and conditions	Ref.
Flexible PUF (synthesized using toluene diisocyanate and ester-based polyol)	- PPE: dimethyl phosphonate - PU:PPE mass ratio: 1:3 - temperature: 160 °C - time: 1, 2, or 3h	(43)
	- PPE: diethyl phosphonate - PU:PPE mass ratio: 1:3 - temperature: 160 °C - time: 1h	
Flexible PUF (synthesized using TDI and ether-based polyol)	- PPE: dimethyl phosphonate - PU:PPE mass ratio: 1:3 - temperature: 160 °C - time: 3h	(45)
Microporous PU elastomer (synthesized using MDI and ester-based polyol)	- PPE: dimethyl phosphonate - PU:PPE mass ratio: 1:2 - temperature: 142 °C or 150 °C - time: 30, 45, 60, 90, or 120 min	(44)
Microporous PU elastomer (synthesized using MDI and ester-based polyol)	- PPE: triethyl phosphonate - PU:PPE mass ratio: 1:3 - temperature: 180 °C - time: 5, 6, 7, or 8h	(46)
	- PPE: tris(2-chloroethyl) phosphate - PU:PPE mass ratio: 1:3 - temperature: 180 °C - time: 5h	
Microporous PU elastomer (synthesized using MDI and ester-based polyol)	- PPE: diethyl phosphonate - PU:PPE mass ratio: 1:3 - temperature: 170 °C - time: 4, 5, 6, or 8h	(47)
	- PPE: tris(1-methyl-2-chloroethyl) phosphanate - PU:PPE mass ratio: 1:3 - temperature: 180 °C - time: 4, 6, or 8h	
PUF (synthesized using MDI and ester-based polyol)	- PPEs: triethyl phosphate, trimethyl phosphonate or tris(1-methyl-2-chloroethyl) phosphanate - PU:PPE mass ratio: 1:3 - temperature: 190 °C - time: 6h	(38)
Microporous PU elastomer (synthesized using MDI and ester-based polyol)	- PPE: tris(1-methyl-2-chloroethyl) phosphanate - PU:PPE mass ratio: 1:3 - temperature: 180 °C - time: 8h - product (after removing of unreacted PPE and further reaction with propylene oxide) was used for the preparation of RPUF	(48)

Figure 5. Phosphorolysis of PUs: general scheme of the reaction and used PPE.

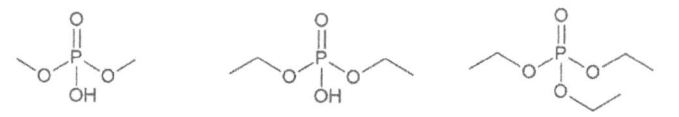

Figure 6. Reaction of phosphorus containing oligomers with propylene oxide resulted in the formation of reactive intermediates for the synthesis of PUFs.

Paciorek-Sadowska, Czupryński, and Liszkowska described glycolysis of rigid polyurethane-polyisocyanurate foams with reduced flammability, which were obtained using tri[(N,N'-dimethyleneoxy-3-hydroxypropyl)urea]borate as a polyol. The synthesis of boron-containing polyol is presented in Figure 7 (39).

Glycolysis was performed with the mass excess of RPUF with respect to a glycolyzing medium (polyurethane/glycolyzing medium mass ratio was equal 2:1). Glycolyzing medium consisted of diethylene glycol, ethanolamine, and zinc stearate (mass ratio of mentioned compounds was equal 2:1:0.75). The glycolyzing medium was heated to 210 °C, then milled PUF was dosed (60 min) in small portions (approximately 3g) and, when the foam was fully introduced into the flask, the reaction mixture was maintained at 180°C for 30 min. The glycolysis product was a tea-colored liquid characterized by followed properties: density: 1.529 g cm^{-3}; viscosity: 17.04 Pa s; hydroxyl value: 450.9 mg KOH/g; acid value: 7.1 mgKOH/g; and water content: 0.4%.

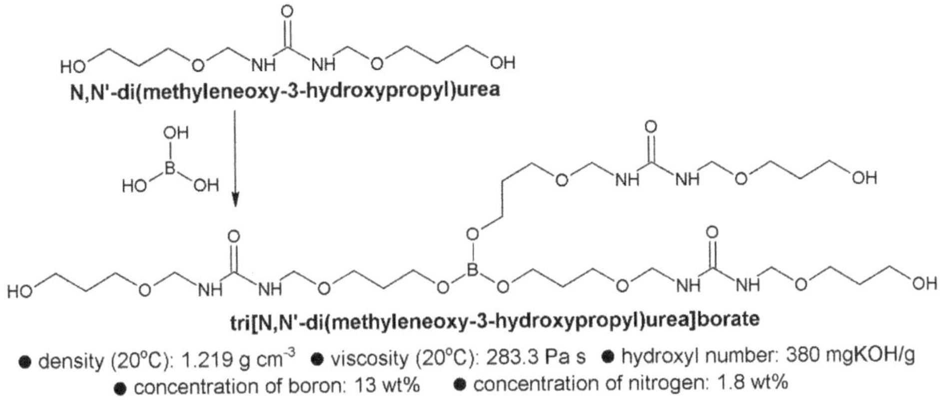

Figure 7. Synthesis of boron-containing polyol realized by Paciorek-Sadowska and physicochemical properties of obtained product.

Moreover, authors prepared new rigid polyurethane–polyisocyanurate foams containing glycolysis product that was used as a polyol raw material. Investigations confirmed that the glycosylate obtained was the FR in the new foams.

Pham et al. described highly thermally stable PUF with excellent FR performance obtained using the product of chemical recycling of postconsumer poly(ethylene terephthalate) (PET) bottles (40). The glycolysis of PET flakes using diethylene glycol was performed in presence of zinc sulfate as a catalyst inside a microwave oven at a constant power of 250 W for 80 min. The molar ratio of diethylene glycol to repeating unit of PET was equal 2.5:1. Compared to conventional PUF prepared from commercial polyol obtained, PUF containing recycling product PUF is characterized by higher thermal stability and higher effectiveness of flame-retardancy, even without adding of additional FR (e.g., diammonium phosphate). The presence of aromatic rings in the prepared oligodiols structure improved mechanical properties (e.g., compressive strength) and thermal stability of the prepared materials. Notably, prepared recycling product PUF/ diammonium phosphatematerials meet the fire safety requirements of polymer applications proposed by the authors procedure permit to reduce PET waste and improve waste management due to application of relatively a high amount of postconsumer PET in the synthesis of flame-retarded PUFs.

Li et al. proposed application of bis(2-hydroxyethyl) terephthalate, synthesized by glycolysis of PET textile waste as polyol in the synthesis of flame-retarded RPUFs (41). Glycolysis of PET were performed with the mass excess of 1,2-ethanediol (three times higher weight of PET) with using zinc chloride as a catalyst (Figure 8). The process was conducted at 196–200 °C for 4h. The presence of aromatic moieties in polyol used to obtain PUFs improves their flame-retardancy (connected with increasing of limiting oxygen index), which was also exhibited by addition of dimethyl methylphosphonate.

Wang et al. proposed application of powdered melamine formaldehyde foam waste as a FR filler (added in 20, 30, or 40 wt % with respect to polyol mixture) in PUFs (42). The effectiveness of a recycled FR can be enhanced by adding guanidine phosphate.

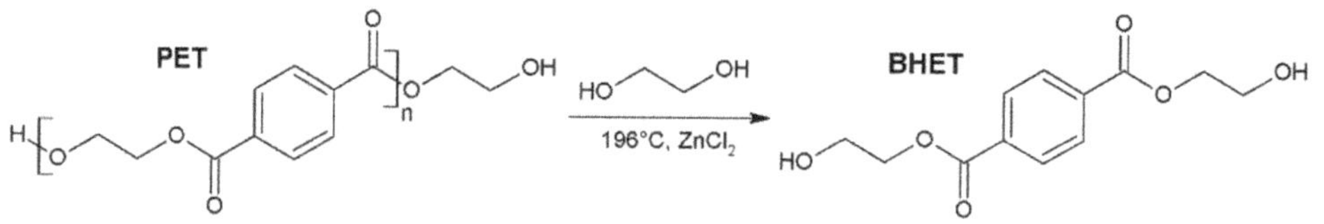

Figure 8. Glycolysis of poly(ethylene terephthalate) with mass excess of ethylene glycol resulted in formation of bis(2-hydroxyethyl) terephthalate.

Summary

PU waste can be transformed into valuable semi-products or products by their mechanical or chemical recycling. Mechanical recycling involves the change of size and form of PU waste, which can be further processed by methods suitable for thermoplastics or used as fillers for ceramic or polymeric materials. Chemical recycling methods produce valuable intermediates (e.g., polyols or FRs) for the synthesis of new PUs. The most important methods are glycolysis, glycerolysis, acidolysis, and phosphorolysis. Glycolysis and glycerolysis can be performed with the mass excess of PU or depolymerizing agent, which permit to obtain oligomeric products containing urethane moieties and terminated mostly by hydroxyl groups (suitable for the reaction with di- or polyisocyanates) or to recovery of polyol, respectively. Acidolysis also enables obtaining polyol suitable for the preparation of PUs (in the form of foams or adhesives). Phosphorolysis leads to flame-retardants or reactive intermediates (when obtained product is subjected to additional chemical treatment).

It should also be noted that some polymers, like poly(ethylene terephthalate), can be successfully transformed (including glycolysis of postconsumer bottles or textiles) into FR PUs.

References

1. Datta, J.; Włoch, M. Recycling of Polyurethanes. *In Polyurethane Polymers: Blends and Interpenetrating Polymer Networks*; Elsevier, Cambridge, MA, 2017, pp 323–358 https://doi.org/10.1016/B978-0-12-804039-3.00014-2.
2. Simón, D.; Borreguero, A. M.; de Lucas, A.; Rodríguez, J. F. Recycling of Polyurethanes from Laboratory to Industry, a Journey towards the Sustainability. *Waste Manag.* **2018**, *76*, 147–171. https://doi.org/10.1016/j.wasman.2018.03.041.
3. Deng, Y.; Dewil, R.; Appels, L.; Ansart, R.; Baeyens, J.; Kang, Q. Reviewing the Thermo-Chemical Recycling of Waste Polyurethane Foam. *J. Environ. Manage.* **2021**, *278*, 111527. https://doi.org/10.1016/j.jenvman.2020.111527.
4. Zia, K. M.; Bhatti, H. N.; Ahmad Bhatti, I. Methods for Polyurethane and Polyurethane Composites, Recycling and Recovery: A Review. *React. Funct. Polym.* **2007**, *67* (8), 675–692. https://doi.org/10.1016/j.reactfunctpolym.2007.05.004.
5. Beneš, H.; Rösner, J.; Holler, P.; Synková, H.; Kotek, J.; Horák, Z. Glycolysis of Flexible Polyurethane Foam in Recycling of Car Seats. *Polym. Adv. Technol.* **2007**, *18* (2), 149–156. https://doi.org/10.1002/pat.810.
6. Wu, C.-H.; Chang, C.-Y.; Li, J.-K. Glycolysis of Rigid Polyurethane from Waste Refrigerators. *Polym. Degrad. Stab.* **2002**, *75* (3), 413–421. https://doi.org/10.1016/S0141-3910(01)00237-3.

7. Zhu, P.; Cao, Z. B.; Chen, Y.; Zhang, X. J.; Qian, G. R.; Chu, Y. L.; Zhou, M. Glycolysis Recycling of Rigid Waste Polyurethane Foam from Refrigerators. *Environ. Technol.* **2014**, *35* (21), 2676–2684. https://doi.org/10.1080/09593330.2014.918180.
8. Shin, S.-R.; Kim, H.-N.; Liang, J.-Y.; Lee, S.-H.; Lee, D.-S. Sustainable Rigid Polyurethane Foams Based on Recycled Polyols from Chemical Recycling of Waste Polyurethane Foams. *J. Appl. Polym. Sci.* **2019**, *136* (35), 47916. https://doi.org/10.1002/app.47916.
9. Gutiérrez-González, S.; Gadea, J.; Rodríguez, A.; Junco, C.; Calderón, V. Lightweight Plaster Materials with Enhanced Thermal Properties Made with Polyurethane Foam Wastes. *Constr. Build. Mater.* **2012**, *28* (1), 653–658. https://doi.org/10.1016/j.conbuildmat.2011.10.055.
10. Datta, J.; Głowińska, E.; Włoch, M. Mechanical Recycling via Regrinding, Rebonding, Adhesive Pressing, and Molding. In *Recycling of Polyurethane Foams*; William Andrew Publishing, Cambridge, MA, 2018, pp 57–65. https://doi.org/10.1016/b978-0-323-51133-9.00005-x
11. Ben Fraj, A.; Kismi, M.; Mounanga, P. Valorization of Coarse Rigid Polyurethane Foam Waste in Lightweight Aggregate Concrete. *Constr. Build. Mater.* **2010**, *24* (6), 1069–1077. https://doi.org/10.1016/j.conbuildmat.2009.11.010.
12. Gadea, J.; Rodríguez, A.; Campos, P. L.; Garabito, J.; Calderón, V. Lightweight Mortar Made with Recycled Polyurethane Foam. *Cem. Concr. Compos.* **2010**, *32* (9), 672–677. https://doi.org/10.1016/j.cemconcomp.2010.07.017.
13. Kanchanapiya, P.; Methacanon, P.; Tantisattayakul, T. Techno-Economic Analysis of Light Weight Concrete Block Development from Polyisocyanurate Foam Waste. *Resour. Conserv. Recycl.* **2018**, *138*, 313–325. https://doi.org/10.1016/j.resconrec.2018.07.027.
14. Mounanga, P.; Gbongbon, W.; Poullain, P.; Turcry, P. Proportioning and Characterization of Lightweight Concrete Mixtures Made with Rigid Polyurethane Foam Wastes. *Cem. Concr. Compos.* **2008**, *30* (9), 806–814. https://doi.org/10.1016/j.cemconcomp.2008.06.007.
15. Hulme, A. J.; Goodhead, T. C. Cost Effective Reprocessing of Polyurethane by Hot Compression Moulding. *J. Mater. Process. Technol.* **2003**, *139* (1), 322–326. https://doi.org/10.1016/S0924-0136(03)00548-X.
16. Park, J.-S.; Lim, Y.-M.; Nho, Y.-C. Preparation of High Density Polyethylene/Waste Polyurethane Blends Compatibilized with Polyethylene-Graft-Maleic Anhydride by Radiation. *Materials* **2015**, *8*, 1626–1635. https://doi.org/10.3390/ma8041626.
17. Becker, D.; Roeder, J.; Oliveira, R. V. B.; Soldi, V.; Pires, A. T. N. Blend of Thermosetting Polyurethane Waste with Polypropylene: Influence of Compatibilizing Agent on Interface Domains and Mechanical Properties. *Polym. Test.* **2003**, *22* (2), 225–230. https://doi.org/10.1016/S0142-9418(02)00086-7.
18. Abdel-Rahman, H. A.; Younes, M. M.; Khattab, M. M. Recycling of Polyurethane Foam Waste in the Production of Lightweight Cement Pastes and Its Irradiated Polymer Impregnated Composites. *J. Vinyl Addit. Technol.* **2019**, *25* (4), 328–338. https://doi.org/10.1002/vnl.21698.
19. Heneczkowski, M.; Galina, H. Material Recycling of RIM Flexible Polyurethane Foams Wastes. *Polimery* **2002**, *47* (7–8), 523–527. https://doi.org/10.14314/polimery.2002.523.
20. Sônego, M.; Costa, L. C.; Ambrósio, J. D. Flexible Thermoplastic Composite of Polyvinyl Butyral (PVB) and Waste of Rigid Polyurethane Foam. *Polímeros* **2015**, *25* (2), 175–180. https://doi.org/10.1590/0104-1428.1944

21. Datta, J.; Haponiuk, J. T. Influence of Glycols on the Glycolysis Process and the Structure and Properties of Polyurethane Elastomers. *J. Elastomers Plast.* **2011**, *43* (6), 529–541. https://doi.org/10.1177/0095244311413447.
22. Datta, J.; Rohn, M. Thermalproperties of Polyurethanes Synthesized Using Waste Polyurethane Foam Glycolysates. *J. Therm. Anal. Calorim.* **2007**, *88* (2), 437–440. https://doi.org/10.1007/s10973-006-8041-0.
23. del Amo, J.; Borreguero, A. M.; Ramos, M. J.; Rodríguez, J. F. Glycolysis of Polyurethanes Composites Containing Nanosilica. *Polymers* **2021**, *13* (9), 1418. https://doi.org/10.3390/polym13091418.
24. Molero, C.; de Lucas, A.; Rodríguez, J. F. Recovery of Polyols from Flexible Polyurethane Foam by "Split-Phase" Glycolysis with New Catalysts. *Polym. Degrad. Stab.* **2006**, *91* (4), 894–901. https://doi.org/10.1016/j.polymdegradstab.2005.06.023.
25. Jutrzenka Trzebiatowska, P.; Beneš, H.; Datta, J. Evaluation of the Glycerolysis Process and Valorisation of Recovered Polyol in Polyurethane Synthesis. *React. Funct. Polym.* **2019**, *139*, 25–33. https://doi.org/10.1016/j.reactfunctpolym.2019.03.012.
26. Datta, J.; Kopczyńska, P.; Simón, D.; Rodríguez, J. F. Thermo-Chemical Decomposition Study of Polyurethane Elastomer Through Glycerolysis Route with Using Crude and Refined Glycerine as a Transesterification Agent. *J. Polym. Environ.* **2018**, *26* (1), 166–174. https://doi.org/10.1007/s10924-016-0932-y.
27. Kopczyńska, P.; Datta, J. Rheological Characteristics of Oligomeric Semiproducts Gained via Chemical Degradation of Polyurethane Foam Using Crude Glycerin in the Presence of Different Catalysts. *Polym. Eng. Sci.* **2017**, *57* (8), 891–900. https://doi.org/10.1002/pen.24466.
28. Jutrzenka Trzebiatowska, P.; Dzierbicka, A.; Kamińska, N.; Datta, J. The Influence of Different Glycerine Purities on Chemical Recycling Process of Polyurethane Waste and Resulting Semi-Products. *Polym. Int.* **2018**, *67* (10), 1368–1377. https://doi.org/https://doi.org/10.1002/pi.5638.
29. Jutrzenka Trzebiatowska, P.; Santamaria Echart, A.; Calvo Correas, T.; Eceiza, A.; Datta, J. The Changes of Crosslink Density of Polyurethanes Synthesised with Using Recycled Component. Chemical Structure and Mechanical Properties Investigations. *Prog. Org. Coatings* **2018**, *115*, 41–48. https://doi.org/10.1016/j.porgcoat.2017.11.008.
30. Sołtysiński, M.; Piszczek, K.; Romecki, D.; Narożniak, S.; Tomaszewska, J.; Skórczewska, K. Conversion of Polyurethane Technological Foam Waste and Post-Consumer Polyurethane Mattresses into Polyols – Industrial Applications. *Polimery* **2021**, *63*, 234–238. https://doi.org/10.14314/polimery.2018.3.8.
31. Gama, N.; Godinho, B.; Marques, G.; Silva, R.; Barros-Timmons, A.; Ferreira, A. Recycling of Polyurethane Scraps via Acidolysis. *Chem. Eng. J.* **2020**, *395*, 125102. https://doi.org/10.1016/j.cej.2020.125102.
32. Godinho, B.; Gama, N.; Barros-Timmons, A.; Ferreira, A. Recycling of Polyurethane Wastes Using Different Carboxylic Acids via Acidolysis to Produce Wood Adhesives. *J. Polym. Sci.* **2021**, *59* (8), 697–705. https://doi.org/10.1002/pol.20210066.
33. Simón, D.; García, M. T.; de Lucas, A.; Borreguero, A. M.; Rodríguez, J. F. Glycolysis of Flexible Polyurethane Wastes Using Stannous Octoate as the Catalyst: Study on the Influence of

Reaction Parameters. *Polym. Degrad. Stab.* **2013**, *98* (1), 144–149. https://doi.org/10.1016/j.polymdegradstab.2012.10.017.

34. Molero, C.; de Lucas, A.; Rodríguez, J. F. Recovery of Polyols from Flexible Polyurethane Foam by "Split-Phase" Glycolysis: Study on the Influence of Reaction Parameters. *Polym. Degrad. Stab.* **2008**, *93* (2), 353–361. https://doi.org/10.1016/j.polymdegradstab.2007.11.026.

35. Wu, C.-H.; Chang, C.-Y.; Cheng, C.-M.; Huang, H.-C. Glycolysis of Waste Flexible Polyurethane Foam. *Polym. Degrad. Stab.* **2003**, *80* (1), 103–111. https://doi.org/https://doi.org/10.1016/S0141-3910(02)00390-7.

36. Modesti, M.; Costantini, F.; dal Lago, E.; Piovesan, F.; Roso, M.; Boaretti, C.; Lorenzetti, A. Valuable Secondary Raw Material by Chemical Recycling of Polyisocyanurate Foams. *Polym. Degrad. Stab.* **2018**, *156*, 151–160. https://doi.org/10.1016/j.polymdegradstab.2018.08.011.

37. Xue, S.; Omoto, M.; Hidai, T.; Imai, Y. Preparation of Epoxy Hardeners from Waste Rigid Polyurethane Foam and Their Application. *J. Appl. Polym. Sci.* **1995**, *56* (2), 127–134. https://doi.org/10.1002/app.1995.070560202.

38. Chung, Y.; Kim, Y.; Kim, S. Flame Retardant Properties of Polyurethane Produced by the Addition of Phosphorous Containing Polyurethane Oligomers (II). *J. Ind. Eng. Chem.* **2009**, *15* (6), 888–893. https://doi.org/10.1016/j.jiec.2009.09.018.

39. Paciorek-Sadowska, J.; Czupryński, B.; Liszkowska, J. Glycolysis of Rigid Polyurethane–Polyisocyanurate Foams with Reduced Flammability. *J. Elastomers Plast.* **2015**, *48* (4), 340–353. https://doi.org/10.1177/0095244315576244.

40. Pham, C. T.; Nguyen, B. T.; Nguyen, H. T. T.; Kang, S.-J.; Kim, J.; Lee, P.-C.; Hoang, D. Comprehensive Investigation of the Behavior of Polyurethane Foams Based on Conventional Polyol and Oligo-Ester-Ether-Diol from Waste Poly(Ethylene Terephthalate): Fireproof Performances, Thermal Stabilities, and Physicomechanical Properties. *ACS Omega* **2020**, *5* (51), 33053–33063. https://doi.org/10.1021/acsomega.0c04555.

41. Li, M.; Luo, J.; Huang, Y.; Li, X.; Yu, T.; Ge, M. Recycling of Waste Poly(Ethylene Terephthalate) into FR Rigid Polyurethane Foams. *J. Appl. Polym. Sci.* **2014**, *131* (19) https://doi.org/10.1002/app.40857.

42. Wang, X.; Shi, Y.; Liu, Y.; Wang, Q. Recycling of Waste Melamine Formaldehyde Foam as FR Filler for Polyurethane Foam. *J. Polym. Res.* **2019**, *26* (3), 57. https://doi.org/10.1007/s10965-019-1717-5.

43. Troev, K.; Tsekova, A.; Tsevi, R. Chemical Degradation of Polyurethanes: Degradation of Flexible Polyester Polyurethane Foam by Phosphonic Acid Dialkyl Esters. *J. Appl. Polym. Sci.* **2000**, 78 (14), 2565–2573. https://doi.org/10.1002/1097-4628(20001227)78:14<2565::AID-APP180>3.0.CO;2-H.

44. Troev, K.; Atanasov, V. I.; Tsevi, R.; Grancharov, G.; Tsekova, A. Chemical Degradation of Polyurethanes. Degradation of Microporous Polyurethane Elastomer by Dimethyl Phosphonate. *Polym. Degrad. Stab.* **2000**, *67* (1), 159–165. https://doi.org/10.1016/S0141-3910(99)00105-6.

45. Troev, K.; Tsekova, A.; Tsevi, R. Chemical Degradation of Polyurethanes2. Degradation of Flexible Polyether Foam by Dimethyl Phosphonate. *Polym. Degrad. Stab.* **2000**, *67* (3), 397–405. https://doi.org/10.1016/S0141-3910(99)00106-8.

46. Troev, K.; Atanassov, V.; Tzevi, R. Chemical Degradation of Polyurethanes. II. Degradation of Microporous Polyurethane Elastomer by Phosphoric Acid Esters. *J. Appl. Polym. Sci.* **2000**, *76* (6), 886–893. https://doi.org/10.1002/(SICI)1097-4628(20000509)76:6<886::AID-APP15>3.0.CO;2-O.
47. Troev, K.; Grancharov, G.; Tsevi, R. Chemical Degradation of Polyurethanes 3. Degradation of Microporous Polyurethane Elastomer by Diethyl Phosphonate and Tris(1-Methyl-2-Chloroethyl) Phosphate. *Polym. Degrad. Stab.* **2000**, *70* (1), 43–48. https://doi.org/10.1016/S0141-3910(00)00086-0.
48. Grancharov, G.; Mitova, V.; Shenkov, S.; Topliyska, A.; Gitsov, I.; Troev, K. Smart Polymer Recycling: Synthesis of Novel Rigid Polyurethanes Using Phosphorus-Containing Oligomers Formed by Controlled Degradation of Microporous Polyurethane Elastomer. *J. Appl. Polym. Sci.* **2007**, *105* (2), 302–308. https://doi.org/10.1002/app.25676.

Editor's Biography

Ram K. Gupta

Dr. Ram K. Gupta is an associate professor at Pittsburg State University. Dr. Gupta's research focuses on biobased polymers; flame-retardant polyurethanes; conducting polymers and composites; and green energy production and storage using two-dimensional materials, optoelectronics and photovoltaics devices, organic–inorganic heterojunctions for sensors, biocompatible nanofibers for tissue regeneration, scaffold and antibacterial applications, and biodegradable metallic implants. Dr. Gupta has published more than 230 peer-reviewed articles; has made more than 280 national, international, and regional presentations; has chaired many sessions at national and international meetings; has edited many books; and has written several book chapters. He has received more than $2.5 million for research and educational activities from many funding agencies. He is serving as editor-in-chief, associate editor, and editorial board member of numerous journals.

Indexes

Author Index

Abu Elella, M., 189
Alpár, T., 239
Bajwa, D., 173
Barbosa, R., 141
Bartoli, M., 59
Baş, S., 239
Chanda, S., 173
Chen, K., 37
Gamal, H., 189
Genisoglu, M., 125
Ghosh, T., 83
Goda, E., 189
Gupta, R., x, 1
Hasan, K., 239
Hong, S., 189
Horváth, P., 239
Jeong, H., 103
Jung, Y., 103
Karak, N., 83
Kim, Y., 103
Li, Z., 37
Lu, H., 221
Malucelli, G., 59
Mulinari, D., 141
Rosa, D., 141
Ryu, S., 103
Silva, N., 141
Sofuoglu, S., 125
Sofuoglu, A., 125
Sulaiman, M., 1
Tagliaferro, A., 59
Wang, M., 37
Wei, C., 221
Włoch, M., 265
Yang, W., 221
Yang, W., 221
Yoon, K., 189
Yuen, R., 221
Zanini, N., 141
de Souza, F., 1
de Souza, A., 141

Subject Index

B

Boron-based compounds, mechanistic study
  boron-based FRs, 175
    boron, incorporation, 176f
    carbazole, reaction, 176f
    char residues, digital pictures, 182f
    CPBN, graphical preparation scheme, 182f
    h-BN nanosheets for TPU, scheme of proposed FR mechanism, 184f
    hydroxyalkylation to produce boron-induced oligoetherol, 177f
    MEL and H_3BO_3 reaction, schematic illustration, 179f
    PEI/h-BN–coated PUF, transmission electron microscopy cross-sectional micrograph of one bilayer, 184f
    preparation of h-BN nanosheets, schematic illustration, 183f
    PUA coatings, LOI analysis, 180f
    synthesis of boron-containing PU, schematic diagram, 181f
    synthesized boron-containing oligoetherol, 178
  boron-based FRs, significance, 175
  conclusion, 185
  introduction, 173

F

Flame-retardant additives for polymeric matrix, overview on classification
  conclusion, 72
  inorganic FR additives, 60
    ABS copolymer, HRR versus time curves, 62f
    combined ammonium polyphosphate, 67
    HRR of poly(methyl methacrylate) (PMMA), effect of boehmite (AlOOH) content, 63f
    pristine (AAP) and modified (IMAPP) ammonium polyphosphate, 66f
    production of silica-nanohybrid-based PP composites, synthetic route, 65f
    ultrafine magnesium hydroxide particles, scanning electron microscope capture, 61f
    zinc borate containing PP, HRR, 64f
  introduction, 59
    occurring during burning of a polymeric matrix, main mechanisms, 60
  organic FR additives, 68
    biomacromolecules and biosourced products, 71
    THR, effect of graphite oxide, 70f
Flame retardant polyurethane nanocomposites, 221
  conclusion, 233
  FR-PU nanocomposites, preparation, 224
    ZIF-67, WSPR, and FPUF, assembly mechanism, 226f
  introduction, 222
    PU, typical combustion process, 223f
    PU nanocomposites based on different nanomaterials, 223f
    PUs, synthesis route, 222f
  various nanomaterial-based FR-PU nanocomposites, 226
    1D nanomaterial-based PU nanocomposites, 228
    2D nanomaterial-based PU nanocomposites, 230
    selected PU nanocomposites, FR properties, 232t
    silica-based PU nanocomposites, general synthesis process, 227f
    Theother tubular structural 1D nanomaterials, 229

H

Highly flame-retardant polyurethane
  analyzing FR properties, methods, 105
    cone calorimeter (CC) configuration, 106f
    limiting oxygen index (LOI) test method, 105f
    UL-94 method, 107f
  conclusion, 121
    FR-PU, research, 122
  highly FR-PU, 107

BH/EG, flame-retarding mechanism, 118*f*
foams after exposure to torch flame, images, 110*f*
function of time during CC testing, HRR, 109*f*
GCO- and COFPL-filled PUF, yotal heat evolved (THE), 108*f*
GCO-filled PUF, COFPL-filled PUF, HRR curves, 108*f*
neat RPUF and FR–RPUF, HRR curves, 117*f*
neat TPU, TPU/SPE1, and TPU/SPE5, combustion process, 120*f*
possible FR mechanism for MoS_2-DOPO/FPUF composite, schematic illustration, 116*f*
proposed FR mechanism, schematic illustration, 113*f*
pure and FR FPUFs, smoke density curves, 115*f*
pure and FR FPUFs, smoke production rate, 114*f*
RPUF and FR–RPUFs, HRR curves, 119*f*
RPUF and FR–RPUFs, thermogravimetric curves, 119*f*
TPU and its composites, HRR and THR *versus* time curves, 113*f*
TPU and TPU/SPE systems, CC curves, 121*f*
TPU composite samples, HRR and THR curves, 111*f*
TPU–SCS composite, proposed pyrolysis mechanism, 112*f*
introduction, 103
PU, molecular structure, 104

P

Polyurethane foam indoors, role
conclusion, 136
fate of SVOCs, fugacity-based modelling, 132
flame retardants, 126
some BFRs, molecular structure, 127*f*
FRs, PUF-affected indoor levels, 136
indoor SVOCs, fate, 133
indoor environment, SVOC fate process between air and PUF, 134*f*
introduction, 125
organic compounds on PUFs, pp-LFER-based sorption model, 134
PU ether and ester foams, sorbent descriptors, 135*t*
polychlorinated biphenyls, 128
PCBs, chemical structure, 129*f*
polycyclic aromatic hydrocarbons, 126
six priority PAH compounds, molecular structure, 126*f*
PUFs, 129
between PUFs and SVOCs, equilibrium partitioning, 130
organic compounds, MW values, 131*t*
SVOCs between air and PUFs, diffusive transport, 132
Polyurethanes, halogen-based flame retardants, 141
BFRs, 148
Font et al. model, illustration, 153*f*
HBCD decomposition, proposed mechanism, 154*f*
main BFRs and their chemical structures, 149*f*
TBBA, thermal degradation mechanism, 152*f*
TBBA and HBCDs, 150
TBBA pyrolysis process, decomposition steps involved, 151*f*
CFRs, 154
dehydrochlorination of CP, illustrative process, 156*f*
long-chain highly CP, thermal degradation, 157*f*
main CFRs and their chemical structures, 155*f*
major ClOPFRs, chemical structure, 158*f*
TCPP thermal degradation pathway, 159*f*
TDCPP thermal degradation pathway, 159*f*
conclusion, 164
future trends, 162
summary of publications per year, 163*f*
HFRs, 147
introduction, 142
main FR classes, schematic representation, 144*f*
typical PU combustion process, schematic representation, 143*f*
main halogen compounds and suppliers, 160

NBFR compounds and their varieties, 161t
PBDE compounds and their varieties, 160t
PU flame retardancy, chlorinated
  compounds, 161t
PU thermal and fire resistance, 145
  PIR–PUR foams to flame retardance, main
    reactions, 146f
Polyurethanes, industrial flame retardants
  catalysts, 245
    PU formation reaction, schematic, 246
  conclusion, 254
  FR filler incorporation, 249
  FR on flexible foams, 252
  FR on rigid foams, 250
    control RPUF, morphological photographs, 251f
    control RPUF and TDHTTP, thermal analysis, 252f
  introduction, 239
  isocyanates and non-isocyanates, 244
    heat-resistant isocyanate polymer formation, 245f
  polyols, 245
  PU chemistry, 240
    polymeric PU, formation, 241f
  PUs, applications, 252
    PU, application, 253t
  PUs, FR properties, 249
  PUs, improving flame retardancy, 248
    typical FR-PU manufacturing method, 248f
  PUs, properties, 243
  PU synthesis, 244
    different components used throughout PU and FR-PU manufacturing, functions, 244t
  PU types, 242
    MDI, 1,3-phynelene diisocyanate, polymeric structure, 243f
  PU waste, recycling, 246
    PU recycling techniques, 248f
    schematic polyol recovery process, 247f
  research gap, 254
  surface-coating approach, 249
  TPU, 250
Polyurethanes, materials and chemistry
  conclusion, 30
    PU chemistry, 30
  introduction, 1
    chain extenders or cross-linkers, 2
    between the isocyanate and hydroxyl groups, mechanism for the catalyzed reaction, 5f
    polyol and diisocyanate reaction, general addition polymerization reaction, 1f
    soft and hard domains, schematic representation, 3f
  PUs, applications, 15
    antimicrobial cationic WPU, chemical structure, 28f
    click reaction to form a cross-linked WPU, schematics, 30f
    CNC arrangement, scheme, 17f
    diol monomers, synthetic process, 29f
    flexible PU composite, micrographs and pictures, 19f
    general design of a laminated hybrid composite, schematics, 17f
    polydioxolane PUI, synthesis, 24f
    PTMEG block polymeric TPU, synthetic procedure, 22f
    PTMEG content, increase, 21
    PU composite, schematic, 26f
    PU matrix, chemical structure, 25f
    stable WPU suspension, 27
    strength-to-weight ratio, 18
    thermoplastic PU (TPU), 20
    TPU containing an allyl–ether side group, synthetic procedure, 23f
  PUs, materials and chemistry, 4
    acrylic polyol derived from soybeans, synthetic route, 7f
    aromatic isocyanate, resonance effect, 13f
    caprolactone, ring-opening reaction, 6f
    cardanol oil using performic acid, epoxidation and ring-opening reaction performed, 11f
    commonly known terpenes, chemical structures, 9f
    corn and soybean oil, 8
    lignin sources, phenolic derivatives, 9f
    most common aromatic and aliphatic isocyanates, chemical structures, 14f
    polyester polyols, synthesis, 6f
    polyol, general structure, 11f
    synthesis of conventional polyether polyols, schematics, 5f
    tetrahydrofuran, ring-opening reaction, 5f
    thiol-ene, 10

thiol-ene coupling reaction, example, 12f
three-step process of epoxidation, isocyanate-free synthetic route, 14f
triglyceride, hydroformylation process, 12f
Polyurethanes containing flame-retardants, recycling
  introduction, 265
    chemical recycling methods of PU materials, description, 269t
    polyurethanes, mechanical recycling methods, 266t
    PUF waste aggregates, normal weight concrete, 267f
    PUF wastes, form, microstructure, and elemental composition, 267f
    PU material, glycolysis, 271f
    PUs chemical recycling, examples, 272t
    PUs chemical recycling, possibilities, 270f
    PUs mechanical recycling, examples, 268t
  recycling of polymers, preparation of FR PUs, 276
    boron-containing polyol, synthesis, 279f
    phosphorolysis of PU materials, exemplary works related, 277t
    phosphorus containing oligomers, reaction, 278f
    poly(ethylene terephthalate), glycolysis, 280f
    PUs, phosphorolysis, 278f
  summary, 280

S

Self-extinguishing polyurethanes
  covalently modified FR-PUs, 92
    char and scanning electron micrographs, digital images, 93f
    PU and different compositions of phosphorus-containing polyol-based PUs, digital images, 94f
  flame-retardant (FR) behavior, basics, 85
  FR-PUs, applications, 97
  FR-PUs using suitable additives, generation, 86
    additives through the formation of char layer, FR behavior, 89
    char residue, scanning electron micrographs, 90f
    flammability parameters, variations, 87t
    PUF before and after horizontal burning tests, digital images, 91f
  future directions, 98
  introduction, 83
    synthetic polymers in millions of tons, production, 84f
  PU surface coated with FR materials, 95
    LBL approach, schematic representation, 96f
    LBL approach used in Pan et al., schematic representation, 97f
  summary, 98
Smart flame retardants for polyurethane, two-dimensional nanomaterials, 189
  2D FRs, advances, 194
    various 2D FRs used for PUs, schematic diagram, 195f
  introduction, 190
    formation of PU, main reaction, 190
    FR actions, 193
    gelling and blowing reactions for forming PU foams, schematic diagram, 191f
    polymers, combustion cycle, 192f
  PU nanocomposites, layered carbon, 195
    black phosphorous, 206
    BN as a PU nanocomposite, 207
    EG, structure, 199f
    FGNS PU nanocomposites, thermogravimetric analysis data, 198t
    layer-by-layer method, 197
    MMT as a PU nanocomposite, 204
    phosphorylated CS-modified MMT sheets, schematic preparation, 205f
    preparation of the Si/EG-coated PU sample, schematic illustration, 200f
    preparing GO/β-FeOOH-coated PU foam, layer-by-layer assembly technique, 198
    PU nanocomposites, carbides/carbonitrides, 201
    PU nanocomposites, flammability properties, 208t
    synthesis of functionalized GO, graphical representation, 196f
    synthesizing functionalized Ti_3C_2, graphical representation, 203f
    synthesizing MoS_2–DOPO hybrid, schematic representation, 202
  summary, 209

Synthesis of polyurethanes, green materials
  biobased chain extender, 50
  castor oil, 48
    itaconic acid-based polyesteramide polyol, synthesis, 49f
  conclusion, 55
    effective functional groups, 55
  cottonseed oil, 47
    epoxidized cottonseed oil, preparation of polyols, 48f
    unsaturation in terms of fatty acid and fatty acid composition, degree, 47t
  introduction, 37
    NIPU, synthetic scheme, 39f
    PU foaming system, primary reaction, 38f
    PU foaming system, secondary reaction, 38f
  lignin, 45
    lignin and its structure, 46f
  market conditions, 42
  non- and green isocyanate materials, 51
    synthetized NIPU, main chemical routes, 52f
  other, 52
    PU–CNSL, synthesis route, 53f
    synthesis of fatty amide, reaction scheme, 54f
  soybean oil, 49
    synthesis of epoxidized soybean oil, polyester polyol, and PU, reaction scheme, 50f
  typical green polyol materials, 42
    dimer fatty acid-based PU acrylate resin oligomer, synthesis, 45f
    RO-based polyol, idealized synthesis scheme, 43f
    vegetable oil-based polyols, comparison of different methods, 44t
  vegetable oils, 39
    chemical structure of thermoplastic PU, schematic illustration, 40f
    most common vegetable oils, properties and fatty acid compositions, 41t
    vegetable oils, triglyceride structure, 41f

Printed in the USA/Agawam, MA
October 5, 2022

799370.059